Wideband Microwave Materials Characterization

For a listing of recent titles in the
Artech House Microwave Library,
turn to the back of this book.

Wideband Microwave Materials Characterization

John W. Schultz

ARTECH HOUSE

BOSTON | LONDON
artechhouse.com

Library of Congress Cataloging-in-Publication Data
A catalog record for this book is available from the U.S. Library of Congress

British Library Cataloguing in Publication Data
A catalog record for this book is available from the British Library.

ISBN-13: 978-1-63081-946-0

Cover design by Mark Bergeron

685 Canton St.
Norwood, MA

10 9 8 7 6 5 4 3 2 1

*To Becky, who patiently
taught me how to throw stars*

Contents

Preface

It has been 10 years since the predecessor to this book, *Focused Beam Methods*, was published. Shortly after publishing that book, I switched from academic researcher to chief scientist of a small engineering company. The academic environment provides many opportunities for working on difficult engineering problems and is a rigorous and challenging setting. However, my transition to the engineering business world led to a discovery that the task of turning research into useful products is significantly more demanding than the world of academic research. In the business environment it is not enough to publish research results in journals or to present them at conferences. Instead, research that we conduct in the business environment isn't complete until it has become a product that can be used by someone else. Success is measured not by a peer reviewer or two, but by customers who see the merit in a product, and then commit their own money to purchase that product.

Engineers and scientists in the business world need resources to help them do their job not only in conducting fundamental research, but in transitioning that research into a widget that someone else will want. This book is intended to be such a resource. It can certainly be used in an academic setting either for learning or for guiding fundamental research. However, it is also intended to go beyond that by providing practical information for conducting wideband material measurements, whether in support of new product development or manufacturing quality assurance.

Determining intrinsic radio frequency properties or extrinsic performance of materials is important for a variety of applications such as wireless propagation, antenna and microwave circuit design, remote sensing, electromagnetic interference mitigation, material state awareness, and defect detection. Measuring electromagnetic material properties has traditionally happened in the laboratory. However, modern technology and manufacturing applications are driving an increased need to adapt these measurement methods for in-line quality assurance, in-situ process control, or even field inspection of materials and components. Therefore, this book is intended to be a practical guide to electromagnetic material measurements for both laboratory and manufacturing/field environments. Its target audience includes scientists or engineers with an undergraduate understanding of calculus and basic electrical engineering principles.

A number of methods exist for characterizing materials at RF and microwave frequencies, including both resonant and wide-bandwidth techniques. These different techniques are like tools in a toolbox, and each has its advantages and disadvantages. However, this book focuses on the wideband, nonresonant methods as they are applicable to the widest range of materials and are often more practical to use in nonlaboratory environments. The most versatile of the wideband material measurement methods are the free-space techniques. Chapter 2 describes not only the various configurations for free-space measurements, but also provides guidance on calibration methods and signal processing. Chapter 2 also covers the different methods for extracting dielectric and magnetic properties, including the necessary equations for implementing these methods. Next, Chapter 3 explains the use of microwave nondestructive evaluation (NDE) methods including probe design. Chapter 3 also gives an in-depth look into applications such as thickness determination or defect detection.

The interaction of electromagnetic waves in real-world applications often includes concepts around scatter. Chapter 6 is devoted to free-space methods for characterizing scatter whether from inhomogeneous materials or structures. Related to electromagnetic scatter is the concept of surface-traveling waves, which is a phenomenon related to the propagation of energy around a body or component. Understanding surface-traveling waves is necessary in the field of radar detection and cross-section reduction. The theory of traveling wave phenomena along with methods and techniques for evaluating traveling wave effects on materials are also discussed in Chapter 6.

Chapter 5 covers wideband guided-wave methods such as rectangular waveguide, coaxial airline, and stripline transmission line fixtures. The calibration and inversion methods are described for these techniques, and common experimental issues and uncertainty sources such as airgaps are detailed. Going beyond conventional waveguide methods, Chapter 7 discusses a newer method for material property determination called computational electromagnetic (CEM) inversion. The modern evolution of electromagnetic material measurements has involved CEM tools. The introduction of CEM to material measurements not only improves fixture design but has enabled a new paradigm for inverting material properties, not possible with traditional methods. Chapter 7 details this emerging idea of CEM-based material property inversion and provides concrete examples of how to implement the method.

Finally, Chapter 8 describes impedance analysis methods such as dielectric spectroscopy and magnetic permeameter devices. Impedance analysis, a traditional method that has been primarily limited to lower frequencies, is a powerful technique for understanding material behavior such as phase transitions, or for monitoring material changes such as cure or drying. Chapter 8 also discusses the modern adaptation of impedance analysis to CEM inversion methods and shows how this powerful new technique can be used to significantly improve conventional measurement methods.

In summary, this book will acquaint engineers and scientists with the theory and practice of wideband electromagnetic characterization of materials. It also provides the necessary equations for implementing these methods and gives hints and techniques for their practical use. It is hoped that this foundation will support the continued advancement of electromagnetic material measurements techniques and their use in both fundamental research and technology development.

1

Introduction to Electromagnetic Materials Properties

1.1 Dielectric Properties

While this book is primarily concerned with determining electromagnetic properties of materials, measuring these material properties works best with some insight about the physical mechanisms behind them. Moreover, measuring electromagnetic properties sometimes has unexpected results that might look like measurement errors. In fact, these unexpected results may stem from idiosyncrasies of the materials under test rather than a limitation of the measurement apparatus. For this reason, intuition about the underlying material mechanisms is an important tool for understanding measurement results. This chapter reviews some of the fundamental physical aspects of materials, starting in this section with the origin of dielectric properties.

Simple materials usually fall into one of three classifications: polymer, ceramic, or metal. For any of these material types, an applied electric field induces electric polarization within the material. The usual convention is to express the electric field as **E** where the bold type designates this as a vector, meaning that it will have both amplitude and direction. As we will see later, this idea of directionality is important since properties of a material may be different for different directions. The electric field, which is typically derived in terms of the change in electric potential at a given location, has units such as volts per meter, although it can also be derived in terms of forces exerted on electric charges.

There is a second expression that relates to the electric field called the electric flux density or **D**. Flux is an effect that appears to go through an area, and the electric flux density includes not only the electric field, but also the effects of the medium through which the electric field is passing. For example, charges within a medium can react to the electric field by rearranging themselves so that the medium or material becomes polarized. The magnitude of this reaction is usually linear with the applied electric field, and the proportionality constant is called the permittivity and is designated by the symbol ε. The electric flux density is then related to the electric field by, $\mathbf{D} = \varepsilon\mathbf{E}$.

A fundamental constant of nature is the permittivity of vacuum, $\varepsilon_0 = 8.854 \times 10^{-12}$F/m. Usually, the permittivity is expressed as *relative* permittivity, which is the ratio of the absolute permittivity to the permittivity of a vacuum, $\varepsilon_r = \varepsilon/\varepsilon_0$. This can be a source of confusion since it is common to drop the subscript r from the symbol for relative permittivity. It is usually up to the reader to infer whether ε means permittivity or relative permittivity based on its context. To add to the confusion, the relative permittivity is sometimes also called simply permittivity or the dielectric constant. The convention of this book is to leave off the subscript r, and the dielectric permittivities (and magnetic permeabilities) are assumed to be relative unless otherwise designated.

In a time-varying or oscillating electric field, the permittivity is best represented by a complex number, $\varepsilon = \varepsilon' - i\varepsilon''$. In this notation, ε' is the real part of the permittivity and is often called permittivity for short. ε'' is the imaginary part of the permittivity and is also called the dielectric loss factor. The loss factor is usually associated with energy absorption by the material. With the above definition for complex permittivity, ε'' should always be a positive number since energy conservation dictates that a passive material cannot exhibit gain. In some cases, the complex permittivity is defined with a + instead of a – (i.e., $\varepsilon = \varepsilon' + i\varepsilon''$), in which case the ε'' will be a negative number. This book uses the "– convention" for complex permittivity so the loss factor should be positive.

Another quantity associated with energy absorption by a material is the loss tangent, defined by $\tan\delta = \varepsilon''/\varepsilon'$. This loss tangent is another way to express how a material absorbs energy and is simply the tangent of the angle defined by the real and imaginary permittivity in the complex plane. Because it effectively normalizes the loss factor by the real part of the permittivity, loss tangent can be a convenient way to compare the dielectric loss of materials that have differing real permittivities. Yet another definition that is useful where conduction processes occur, is an "apparent" conductivity. This quantity, σ, is usually calculated from the dielectric loss factor by, $\sigma = \omega\varepsilon'' = \omega\varepsilon_0\varepsilon''_r$, where ω is the angular frequency, $\omega = 2\pi f$. Conductivity is normally thought of as a steady-state quantity, and the idea of apparent conductivity is a way to extend the concept to oscillating currents.

With these basic definitions, we can look at how they manifest in a material. In particular, the response of a simple dielectric material tends to be driven by two dominant physical phenomena: dipole reorientation and conduction. In simple terms, dipoles are created by charge separation within crystals, molecules, or molecular fragments. In an element, the charge separation is between the nuclei and orbiting electrons. In compounds of two or more atomic species, charge separation also exists between the species because they have different affinities for electric charge.

Notional dipoles are illustrated in Figure 1.1, which shows a hypothetical polymer fragment on the left and a crystalline array of ionically bonded atoms (e.g., a ceramic) on the right. Dipole moments exist between atoms with differing charge, and these dipoles are vectors that describe the charge distribution in units of charge times displacement. Polymer chains are usually made up of thousands to millions of bonds that rotate or stretch in response to external stimuli. Thus, an applied electric field induces the dipole fragments to realign themselves to partially cancel the effects of the applied electric field. In a ceramic material, the charge centers displace from their equilibrium position when an electric field is present. In essence, an applied electric field causes the electron clouds bound to each atom to shift relative to the nuclei and for different nuclei to shift relative to each other, thus changing the spatial charge balance.

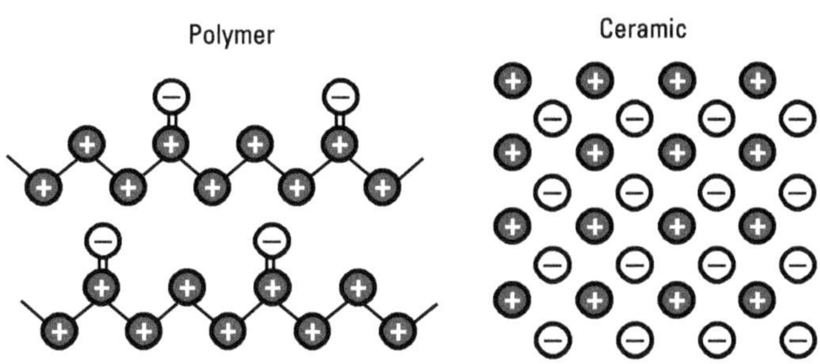

Figure 1.1 Schematic representations of charge distribution, which leads to dipoles within different types of materials.

The time it takes for a dipole to realign to an applied electric field varies according to the properties of the material and external conditions such as temperature and pressure. Thus, a material's dielectric response can also be characterized by the relaxation time (or more precisely by a distribution of relaxation times) of the intrinsic dipoles. When the applied electric field is periodic, it will have a certain frequency or frequencies associated with it. Whether or not a material responds to an incident electric field also depends on if the characteristic relaxation time of the dipoles aligns with the incident E-field frequency. Shorter relaxation times correspond to higher frequencies while longer relaxation times correspond to lower frequency behaviors. Put another way, the period of the oscillating field is the inverse of the frequency. When that period is similar to or longer than the dipole relaxation time, the dipoles respond. When the period is shorter than the dipole relaxation time, then the intrinsic dipoles don't have time to respond.

The second way a material responds to an electric field is through conduction, which involves the physical translation of charged species. Charged species can be either ions or electrons. In semiconductor materials there is also the concept of holes, which represent the lack of an electron in a crystalline lattice where one should normally exist. When an electric field is applied, opposites attract, so a positively charged species is attracted to the negative potential, while the negatively charged species is attracted to the positive potential. More precisely, an applied electric field perturbs the Brownian motion of charged species within a material so that they tend to drift toward oppositely charged electrodes depending on their charge.

Like dipole reorientation, conduction is also affected by various chemical and environmental variables. As charged species travel toward a positive potential, they are slowed by their surroundings. The slower they travel, the more resistant the material. The faster they travel, the more conductive the material. The parameter that quantifies how well electrons and ions can travel is called conductivity. Electron or hole conduction happens when there are electrons not strongly bound to nuclei. These unbound electrons are prevalent in graphitic materials, semiconductors, and metals. Electron or hole conduction can be affected by imperfections in the crystal lattice, temperature, or pressure. Figure 1.2 shows a notional representation of how an electron may travel through a crystalline lattice and how it is slowed down by interactions with that lattice. Ionic conduction can happen in materials with sufficient free volume for the larger ions to travel. An example of this would be a liquid or gel, such as inside a battery. Ionic conduction is similarly affected by its environment, including temperature and pressure.

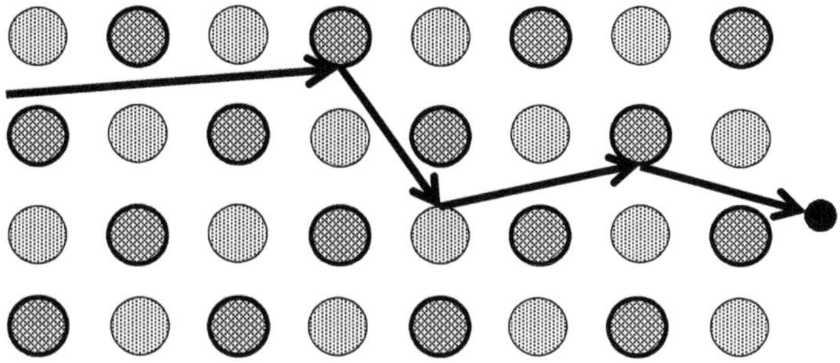

Figure 1.2 Schematic representation of an electron traveling through a lattice.

1.2 Magnetic Properties

Magnetic properties in a material are a response to an externally applied magnetic field, and electromagnetic radiation includes both electric and magnetic fields. Magnetic permeability is denoted by the symbol, μ, which is the proportionality factor that relates the magnetic flux density to the magnetic field, $\mathbf{B} = \mu\mathbf{H}$, where **B** is the flux density and **H** is the field vector. Magnetic permeability depends on intrinsic phenomena such as magnetic moment and domain magnetization. A fundamental constant is the permeability of vacuum, $\mu_0 = 4\pi \times 10^{-7}$ H/m. Like permittivity, the magnetic permeability is usually expressed as a relative permeability, which is the ratio of the absolute permeability to the permeability of a vacuum, $\mu_r = \mu/\mu_0$. This can also be a source of confusion, since it is common to drop the subscript *r* from the symbol for relative permeability. It is usually up to the reader to infer whether μ means absolute or relative permeability based on its context.

In a time-varying or oscillating electric field, the permeability is best represented by a complex number, $\mu = \mu' - i\mu''$, where μ' is the real part of the permeability and μ'' is the imaginary part. Analogous to permittivity, μ'' is associated with energy absorption by the material interacting with the magnetic field and is called the magnetic loss factor. Also, in this book, the sign convention used for permeability is the same as that for permittivity, and μ'' is always a positive number. A magnetic loss tangent can also be defined as an alternate way to compare the loss associated with different magnetic materials, $\tan\delta_m = \mu''/\mu'$.

While electric properties are related to charge, nonnegligible magnetic properties require another quantum mechanical effect called spin. Electrons associated with a nucleus exist in orbitals, which are relationships that describe the probable location or wave function of the electron. Orbitals are organized into shells and subshells and have rules defined by a set of three quantum numbers within the context of the Schrodinger equation. A fourth quantum number, termed spin, is also necessary to fit material behavior to the quantum mechanical model [1]. The electron spin is discrete, taking the values of either +1/2 or −1/2. Electrons surrounding a nucleus occupy unique states within the available orbitals, and no two electrons can occupy the same quantum state. This rule is known as the Pauli exclusion principle, and the periodic table can be constructed by the interplay of this principle with the available quantum numbers. Electrons fill orbitals within the shells and subshells so that they are in a minimum energy configuration, which is also known as Hund's rule [2]. As orbitals fill, the Pauli exclusion principal dictates that electrons pair up so that they have opposite spins within each orbital. Paired spins have a minimum net magnetic moment. However, there are elements in the transition metal and rare Earth series where the ground energy state is such that electrons remain unpaired, resulting in a nontrivial magnetic moment. The most common element exhibiting this behavior is iron, and magnetic absorbers designed for the microwave frequency range most often employ iron or an iron alloy.

Materials with atoms that have no net spin have a negligible magnetic response, and their macroscopic magnetic permeability is close to that of free space. Electrons do have orbital motion that creates the effect of microscopic currents contributing to a magnetic response. This effect is called diamagnetism; however, these effects are too small to be of consequence in RF applications. On the other hand, paramagnetic materials consist of atoms that do have unpaired electron spins, but where those spins do not have any strong coupling to each other. In this case, the material responds more strongly to an applied magnetic field but still does not retain any long-range order of the magnetic moments after the applied field is removed. Paramagnetism, though stronger than diamagnetism, is still relatively weak and has limited utility for RF applications.

The most important magnetic effect is ferromagnetism, where atoms have unpaired spins, and there is a coupling between the spins of neighboring atoms called exchange interaction. Counterintuitively, the exchange interaction is related to electrostatic energy. Specifically, when outer electron orbitals from neighboring atoms overlap, the Pauli exclusion principle dictates that they have opposite spins. However, overlapping electron orbitals have a strong electrostatic repulsion. The occurrence of overlapping electron charge, and therefore electrostatic energy, is minimized when those electrons' spins are aligned so that they cannot be near each other. In other words, parallel electron spins of unpaired electrons in neighboring atoms is favored under these conditions since it leads to minimized electrostatic energy.

Iron, cobalt, and nickel are examples of elements with unfilled outer shell electrons that exhibit this ferromagnetic behavior. These materials are also metallic, and the outer electrons have attributes of both free electrons and bound electrons that contribute to the ferromagnetic behavior. However, the mobility of electrons in these materials also leads to significant conductivity that shields the material from interacting with external electromagnetic energy. For this reason, magneto-dielectric materials that include ferromagnetic metals often are formed from ferromagnetic particles within a nonconducting matrix such as a polymer. This enables electromagnetic waves to penetrate rather than only to be reflected.

Variations of ferromagnetism also exist in nonmetallic compounds such as oxides. For example, at room temperature, pure iron metal exists in a crystalline lattice with a body-centered-cubic (BCC) arrangement of the atoms. Magnetite on the other hand is a compound of iron (Fe) and oxygen (O) that forms a more complex crystalline structure. The chemical formula for magnetite can be written as Fe_3O_4; however, it is also more descriptively written as $FeO \cdot Fe_2O_3$, which indicates that it has two sublattices. This more complex structure results in what is called ferrimagnetic behavior, where the magnetic moments of the sublattices are opposite. Because the sublattices are different, the opposite magnetic moments are also different and only partially cancel. Ferrimagnetic materials are generally ceramics such as oxides or garnets, and they have a net magnetic moment that is not quite as strong as the ferromagnetic transition metals. However, they are not highly conductive and do not suffer from the problem of being too reflective for RF applications. In some cases, a material is antiferromagnetic, where the exchange interaction results in equal but opposite magnetic moments of the sublattices.

As evident by the variety of magnetic behaviors in the above-described materials, there are a variety of models for describing exchange interactions, which depend on the specific electronic environment within the material [3]. These exchange interactions provide an understanding of the source of magnetic permeability within a material, which is the magnetic moment. Magnetic moment is analogous to the electrical dipole moment and therefore drives the real part of the magnetic permeability. There are also mechanisms for magnetic loss, which is the conversion of incident magnetic field into heat or motion within the lattice of the magnetic material. For example, an applied external magnetic field induces precession of the electron, where the axis of the spin rotates around the applied field. This idea is illustrated in Figure 1.3, which shows a single electron and its spin vector in response to an applied H-field. This loss mechanism is a source for the ferromagnetic resonance (FMR). Below the FMR, RF magnetic materials can have a high magnetic

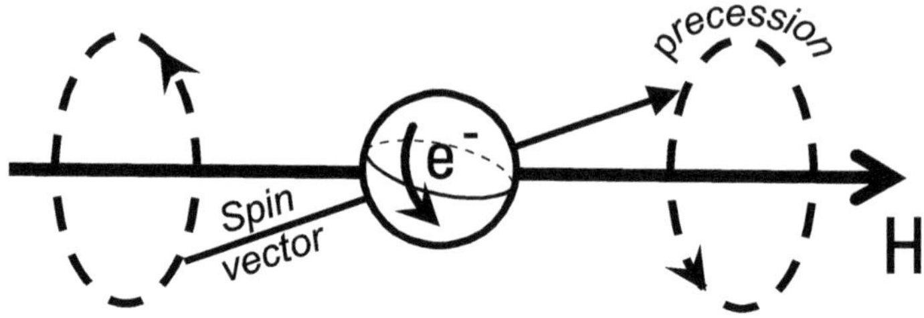

Figure 1.3 The spin of a single electron precessing around an applied H-field.

permeability and are useful as substrates for reducing antenna size. Near the FMR these materials are efficient at absorbing microwave energy and can be used as radar absorbers or for reducing electromagnetic interference between components or antennas.

The description of loss in a magnetic material is more complicated than just the precession of individual electrons and includes a concept called "domains." As noted, the interplay between electrostatic forces and the Pauli exclusion principle organizes electron spins to have a common orientation. The region in which this order is maintained is called a magnetic domain. Under certain circumstances a material may consist of just one domain. However, more commonly, a macroscopic material will consist of numerous magnetic domains, and these domains will have different orientations, because randomization of the domain's magnetic moment is energetically favorable. This idea of electron spin domains is sketched in Figure 1.4. An external magnetic field can magnetize a material by aligning or consolidating the domains. In some cases, removal of that magnetizing field will rerandomize the domains, such as in a soft magnetic material. In "hard" magnets, the alignment of the domains will remain after the removal of the magnetizing field, unless some other external influence such as mechanical or thermal energy causes them to disorganize. Magnetic materials in RF applications such as absorption are typically soft magnetics, and loss mechanisms include not just precession and reorientation of domains, but also movement of domain walls and interaction of the domains and domain walls with the crystalline grain boundaries of a material.

As an aside, domains are also possible in nonmagnetic materials such as ferroelectrics. These materials are so named because of their analogous behavior to ferromagnetics with formation of dielectric domains. The mechanisms are somewhat different, and domains are created when a slight distortion of the crystalline structure results in strong dipole moments that spontaneously align to each other [4]. Ferroelectric materials typically have very high

Figure 1.4 Schematic of magnetic domains within a material.

permittivities and thus can be useful at microwave frequencies for designing artificial dielectrics. They also are tuneable and can be useful for microwave phase shifter components.

1.3 Dispersion

Measurement of the dielectric and magnetic properties of a material specimen can be difficult. However, there are additional complications that make determining intrinsic properties even harder: dispersion and anisotropy. Anisotropy is discussed in the next section and this section describes dispersion, which is the property of a material to have a frequency dependent dielectric permittivity and/or magnetic permeability. This is related to the time-dependent behavior described in the previous sections, where dipole relaxation or magnetic spin reorientation only occurs when it has a time scale similar to or shorter than the period of the incident electric or magnetic fields.

A notional dispersive or frequency-dependent curve for dielectric permittivity is shown in Figure 1.5. Permittivity is a complex number, so the left side of Figure 1.5 shows a dispersion curve for the real permittivity, and the right side of Figure 1.5 shows the corresponding imaginary part. The behavior shown is typical of a wide range of materials, where the real permittivity, ε' undergoes a step decrease as frequency increases and the imaginary permittivity, ε'' shows a peak close to the maximum slope in ε'. These changes, called relaxations, are common in the dielectric permittivity of most materials. Relaxations occur in the permeability of magnetically active materials as well. For this reason, a number of analytical models exist to describe relaxations, including the Debye and Lorentz models for dielectric and magnetic relaxations and the Drude model for conductive materials [5–7].

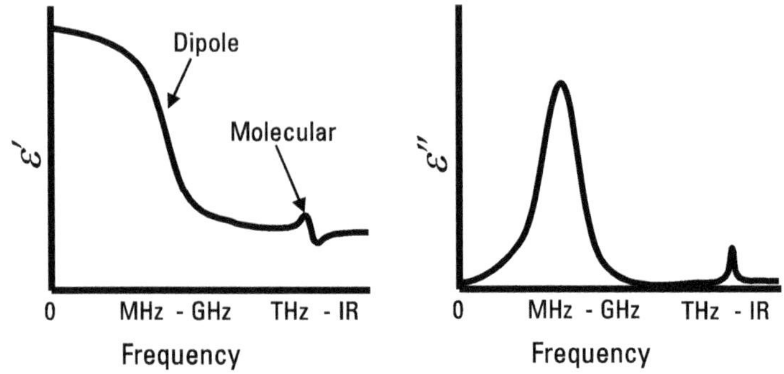

Figure 1.5 Notional real and imaginary relative permittivity for a dispersive material.

Relaxation theories such as these are based on physics approximations that describe dipole displacement or conductor transport in a medium. Fitting measured data to relaxation models and documenting the fitted parameters is a more concise way to compare different measurements. These models also provide a convenient means for defining dispersive materials in computational electromagnetics codes used in design and prediction. The most common models are summarized as follows.

- Debye:

$$\varepsilon = \varepsilon_U + \frac{\varepsilon_R - \varepsilon_U}{1 + i\omega\tau} \tag{1.1}$$

- Lorentz:

$$\varepsilon = \varepsilon_U + \varepsilon_R \frac{\omega_0^2}{\omega_0^2 - \omega^2 + 2i\omega\delta} \tag{1.2}$$

- Drude:

$$\varepsilon = \varepsilon_U - \frac{\omega_p^2}{\omega^2 - i\omega\delta} \tag{1.3}$$

where ε_U and ε_R are the high and low frequency limits (*unrelaxed* and *relaxed*) of the permittivity, τ is the characteristic relaxation time, ω_0 is a characteristic relaxation frequency, ω_p is another characteristic relaxation frequency called the plasma frequency, δ is a damping parameter, and ω is the angular frequency in radians/second. The parameter, τ, is sometimes also expressed as a characteristic relaxation frequency (i.e., $\tau = 1/\omega_0$). Note that while these equations are specified for permittivity, they can also be applied to magnetic permeability. Usually, the Lorentz dispersion model is most relevant to magnetic materials, and it can be applied by replacing the dielectric permittivity with magnetic permeability.

These dispersion models are ideal, and most materials do not exactly fit them. Typically, the relaxation phenomena in actual materials occur over a wider bandwidth than these models predict, and additional empirical parameters are incorporated to improve the fit. For example, Cole and Cole generalized the Debye model by including an empirical exponent, α, [8]

$$\varepsilon = \varepsilon_U + \frac{\varepsilon_R - \varepsilon_U}{1+(i\omega\tau)^{1-\alpha}} \tag{1.4}$$

This exponent broadens the relaxation over a wider range of frequencies. Figure 1.6 demonstrates the difference between the Debye function and the Cole-Cole variant by fitting these models to measured data for a graphite-filled polyimide material; the measured data are shown as a solid line in these plots, and the model fit data were calculated with a standard computational function minimization method. These results show the necessity for modifying the idealized dispersion models when trying to fit them to real data. Going further, Havriliak and Negami [9] defined an even more general variant to the Debye model,

$$\varepsilon = \varepsilon_U + \frac{\varepsilon_R - \varepsilon_U}{\left[1+(i\omega\tau)^{\alpha}\right]^{\beta}} \tag{1.5}$$

Havriliak and Negami added the two empirical parameters, α and β, which allow for both flatness and asymmetrical skew in the relaxation behavior. Numerous researchers have proposed other empirical variations on these equations as well [10–14].

Figure 1.6 shows data indicating a relatively low loss with an imaginary permittivity much smaller than the real part. This is characteristic of a material with bound charges, where dipole polarizability is the primary mechanism for dispersion. In some materials, however, there may be unbound electrons that conduct, which can lead to a different dispersion characteristic. For example, Figure 1.7 shows measured data from a carbon-loaded foam, where the carbon loading is sufficient to create long-range conduction pathways for unbound electrons. In a material such as this, the imaginary permittivity can be significantly greater than the real permittivity. More distinctly, the imaginary permittivity continues increasing as measurement frequency is decreased.

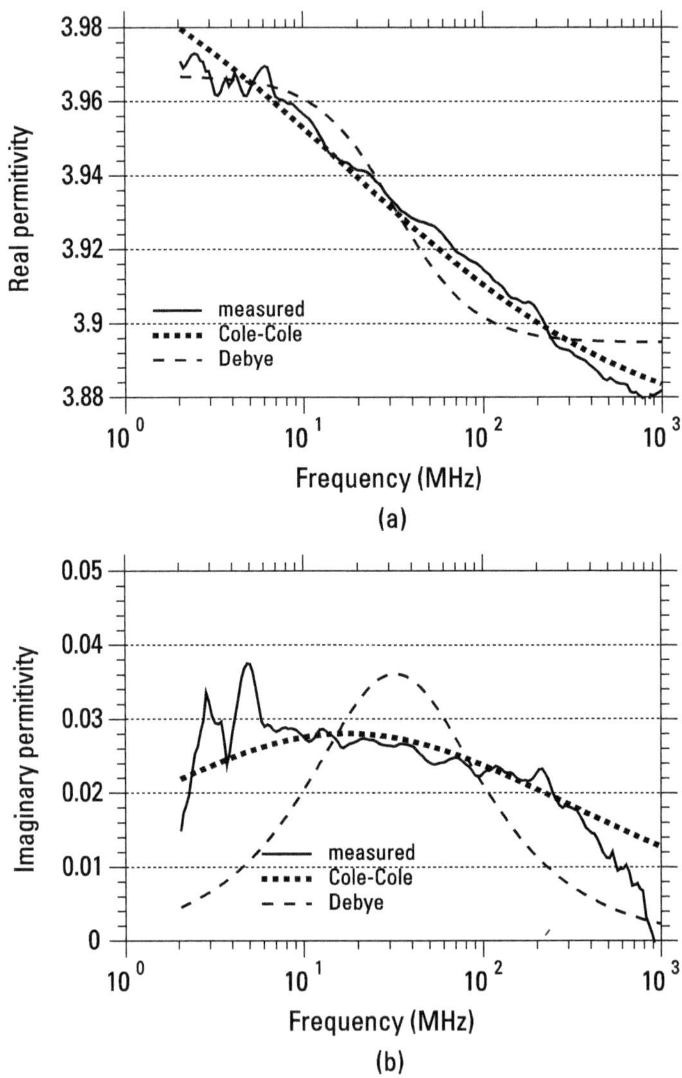

Figure 1.6 Comparison of Debye (1.1) and Cole-Cole (1.4) dispersion models to measured data.

In a conductive material such as the carbon-loaded foam of Figure 1.7, it may make sense to model the conductive dispersion with a Debye model plus an additional conduction term,

$$\varepsilon = \varepsilon_U + \frac{\varepsilon_R - \varepsilon_U}{1 + i\omega\tau} - i\frac{\sigma}{\omega\varepsilon_0} \tag{1.6}$$

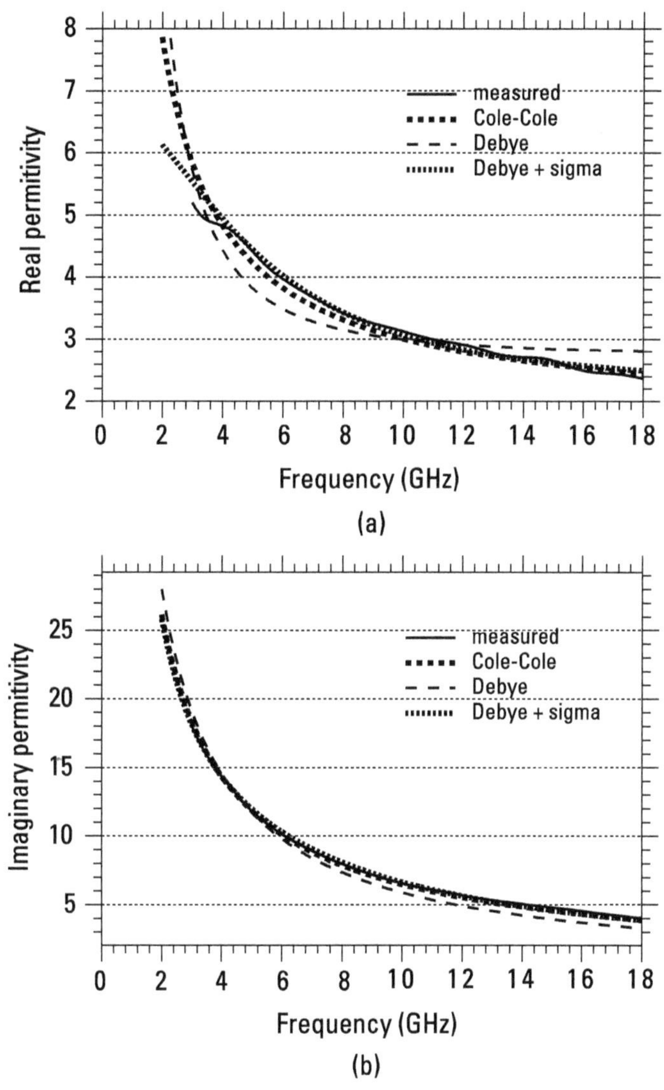

Figure 1.7 Comparison of measured data to three different dispersion models.

where σ is the conductivity in the limit of low frequency and ε_0 is the permittivity of free space. In addition to the measured real and imaginary permittivity of a carbon-loaded foam, Figure 1.7 shows several model fits to the data, including a simple Debye model (1.1), the Cole-Cole model (1.4), and the Debye-plus-conductivity term (1.6). The simple Debye model does not fit the data well, which is not surprising since the Debye was originally derived as a model for materials with bound charges. The Cole-Cole function and the Debye-plus-conductivity term models both do a significantly better job fitting the measured data.

The disadvantage of modified relaxation models such as Cole-Cole or Havriliak-Negami is that their modifications are empirical. Another way to broaden the basic models (e.g., Debye and Lorentz) is to realize that there may be a distribution of dipole moments that need to be described. For example, instead of the single relaxation term of the Debye model, it can be generalized as a summation of relaxations that reflect the distribution of dipoles within a material,

$$\varepsilon = \varepsilon_U + \sum_{n=1:N} \frac{\varepsilon_n - \varepsilon_U}{1 + i\omega\tau_n} \tag{1.7}$$

where there are N terms to describe the range of relaxations. Of course, this can significantly complicate the parameterization of the permittivity or permeability data. In a practical application, it is often sufficient to include just two terms of this summation to fit the dispersion behavior over the desired frequency range. A function that I have used with some success to describe frequency dispersion of broadband magnetic composites is a double-Lorentz function. This double-Lorentz simply adds a second dispersion term to the usual Lorentz,

$$\mu = \mu_U + \frac{\omega_1^2\left(\mu_R - \mu_{Int}\right)}{\omega_1^2 - \omega^2 + 2i\omega\delta_1} + \frac{\omega_1^2\left(\mu_{Int} - \mu_R\right)}{\omega_2^2 - \omega^2 + 2i\omega\delta_2} \tag{1.8}$$

where μ_U, μ_{Int}, and μ_R are the *unrelaxed* (high-frequency), *intermediate*, and *relaxed* (low-frequency) permeabilities, ω_1 and ω_2 are the relaxation frequencies for the first and second terms, and δ_1 and δ_2 are the damping factors for the first and second terms.

Understanding these dispersion models can provide useful insight about the measurement data acquired from broadband materials measurement systems. For example, dielectric relaxations that occur in the microwave range are usually slowly varying, or they follow a Debye-like frequency dependence. As such, the real part of the dielectric permittivity, ε', almost always decreases as frequency increases. If a specimen measurement shows an increasing ε' with frequency, then that measurement data is likely suspect or at least has some systematic measurement error associated with it. On the other hand, the Lorentz model followed by many magnetic materials does allow for a rising real permeability with frequency over a limited part of the relaxation band.

Another use for dispersion models is in extrapolation or interpolation. For example, if measurement equipment is only available in certain frequency bands, fitting the data to an appropriate dispersion model can provide an estimate of the material properties outside those measured frequencies or in gaps between measurement frequencies.

1.4 Anisotropy

Another complicating factor in electromagnetic materials measurements is anisotropy, which is the capability of some materials to have directionally dependent intrinsic properties. For many engineering composites, a single value of complex permittivity or permeability is insufficient as the intrinsic properties depend on the orientation of constituents within the material. For example, fiber-reinforced composites with uniaxially oriented fibers can have a permittivity along the fiber direction that is markedly different than orthogonal to the fiber direction. Particulate-filled materials may have in-plane properties that are different than out-of-plane properties when those particulates are shaped and aligned. Specifically in platelet-filled materials, where the platelets are oriented in-plane, the properties parallel to the plane of the platelets will be significantly different than the properties orthogonal to the platelets.

A material that is used in aerospace applications or sometimes anechoic chambers is honeycomb core, which has effective properties that vary in all three principal directions. Honeycomb consists of a tube-like geometry as shown in Figure 1.8. In this geometry there is a length, width, and thickness direction, each having different values of permittivity corresponding to the geometrical differences in each direction. The thickness direction is along the tubes, the width direction crosses some of the flats of the hexagonal tubes, and the length direction crosses the tubes but is parallel to some of the flats of the hexagons. Because of this three-dimensional geometry, there can be differences in the material response depending on the orientation of the electric field with respect to these directions

The prevalence of anisotropic engineered materials in various applications means we must generalize dielectric permittivity and magnetic permeability to be represented with three-by-three tensors,

$$\mu = \begin{bmatrix} \mu_{11} & \mu_{12} & \mu_{13} \\ \mu_{21} & \mu_{22} & \mu_{23} \\ \mu_{31} & \mu_{32} & \mu_{33} \end{bmatrix} \text{ and } \varepsilon = \begin{bmatrix} \varepsilon_{11} & \varepsilon_{12} & \varepsilon_{13} \\ \varepsilon_{21} & \varepsilon_{22} & \varepsilon_{23} \\ \varepsilon_{31} & \varepsilon_{32} & \varepsilon_{33} \end{bmatrix} \tag{1.9}$$

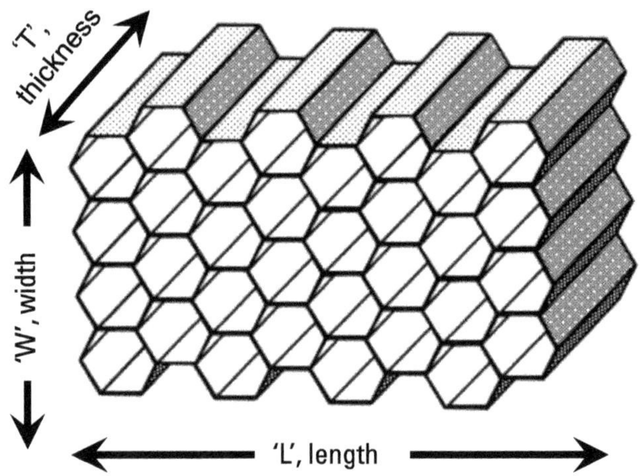

Figure 1.8 Geometry of honeycomb core, an anisotropic material with three principal directions.

where each element of these tensors is a complex number. In most composites, only the diagonal tensor elements are nonzero and the off-diagonal elements can be ignored. In this case, a magneto-dielectric material may have six complex constitutive parameters. However, it is possible to have nonzero off-diagonal elements as well. For example, some ferrite crystals and gyrotropic plasmas are known to have off-diagonal permeability tensor elements. However, the majority of practical engineered composites can be represented with diagonalized tensors.

Fortunately, when the permittivity and permeability tensors are diagonalized, the diagonal tensor components can be determined with independent measurements. For example, measurements of each component of ε and of μ are made by simply orienting the specimen so that the electric and magnetic fields of the incident wave correspond to the desired tensor components. This is possible in free-space measurement systems because free-space propagation can have a linearly polarized incident beam, with the E-fields and H-fields orthogonal to the propagation direction and orthogonal to each other. Other broadband methods may not have linearly polarized fields, and this can limit their utility on the measurement of anisotropic materials. Of course, it is important to know the principal directions of the tensor within a material system. Otherwise, if the specimen is oriented at an angle that does not line up with the polarization direction, the resulting data will be a combination of two or more tensor components, making interpretation more complicated.

1.5 Engineered Materials

Section 1.4 discussed dispersion and anisotropy as they are complicating factors that must be considered in dielectric or magnetic properties. Yet another complicating factor for material measurements is homogeneity. Homogeneity has to do with variation of material properties with position. Specifically, a homogeneous material is one in which the properties do not have any spatial variation, while an inhomogeneous material is one in which the properties do vary from location to location. For example, a composite is an engineered material that consists of two or more constituents mixed in some way. Each constituent may have different properties, so the composite can be considered inhomogeneous.

Homogeneity is a relative property that depends on the size scale of interest. If the spatial property variation of a composite happens over an electrically small size scale—one that is a small fraction of a wavelength—then the material may be treated as homogeneous for applications at that wavelength. However, if that same composite has manufacturing defects that cause larger-scale variations that are nearing or even larger than the application wavelength, then it is inhomogeneous, and those longer size-scale variations will affect the material performance. In a woven fiberglass composite, inhomogeneity could come in the form of density variations of the weave. In a carbon-ink–infused foam absorber, inhomogeneity could come in the form of a gradient of the carbon ink density.

Setting aside the effects from manufacturing defects, composite materials are commonly used in many applications, and under ideal conditions, broadband measurement techniques are applied to determine the average, or homogenized, set of dielectric and magnetic properties. For electromagnetic applications ranging from absorbers to radomes, composites are engineered specifically to control the dielectric or magnetic properties and to achieve a certain level of performance. Composites specifically engineered for electromagnetic purposes are sometimes called artificial-dielectric or artificial-magnetic materials. Since magnetic materials also have some dielectric permittivity associated with them, they may be referred to as artificial magneto-dielectrics. Of course, there are physics-based theories to estimate the expected homogenized properties of these composites, and measurements are often used to compare to theoretical models to establish their validity. This section briefly overviews some of these models, which are sometimes called effective medium theories (EMTs).

An EMT starts by approximating the field structure within a composite microstructure. A two-part composite can be understood as having a matrix and inclusions that make up the two different constituents. A dielectric or magnetic inclusion within a matrix has an internal field that sets up to partially cancel an external electric or magnetic field. This idea is represented in Figure 1.9, which shows an inclusion within a matrix. The differing dielectric permittivity of that inclusion results in a net field within the inclusion that is different from the externally imposed field. The average field of the composite then includes a weighted sum of the fields internal and external to the inclusions. For the case of the electric (E) and displacement (D) fields, these averaged fields are,

$$\langle E \rangle = \nu E_i + (1-\nu)E_0 \tag{1.10}$$

$$\langle D \rangle = \nu \varepsilon_i E_i + (1-\nu)\varepsilon_m E_0 \tag{1.11}$$

where v is the volume fraction of the inclusion, $(1 - v)$ is the volume fraction of the matrix, and E_0 is the externally imposed field. The subscript i designates the inclusions, the subscript m designates the matrix, and the dielectric permittivity is designated by ε. Similar equations can be written for the magnetic fields.

The field within the inclusion is a function of the polarizability or magnetization multiplied by a shape factor, N. For electric fields, N is called the depolarization factor, and for magnetic fields, N is called the demagnetizing factor. Since it is determined by only the shape of the inclusion, the same shape factor applies to both the electric and magnetic fields. By solving the boundary value problem on Laplace's potential equation, Stratton [15] derives an expression for the fields inside an ellipsoid within a matrix medium and

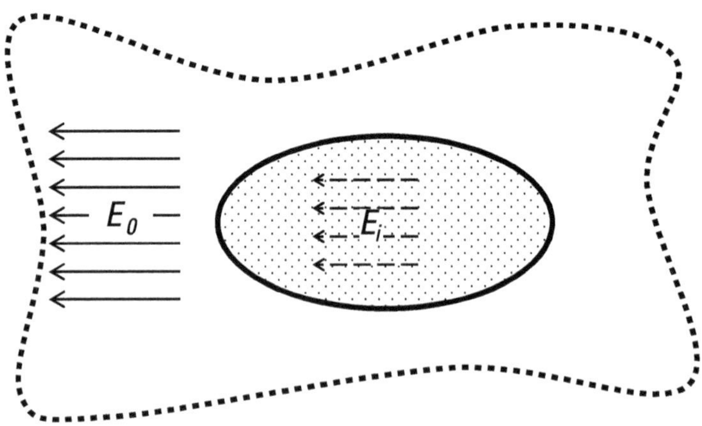

Figure 1.9 Representation of an inclusion within a matrix showing internal electric fields.

with a uniform external field applied. For the case of the electric fields in the direction along one of the principal planes, the internal E-field is given by,

$$E_i^a = \frac{E_0^a}{1 + \frac{(\varepsilon_i - \varepsilon_m)}{\varepsilon_m} N_a} \tag{1.12}$$

where the superscript a indicates that (1.12) is along one Cartesian direction. The expressions for the other orthogonal directions are similar. Combining (1.10) to (1.12) results in the well-known Maxwell Garnett theory [16], which was originally derived for spheres,

$$\varepsilon^a = \varepsilon_m + v\varepsilon_m \frac{\varepsilon_i - \varepsilon_m}{\varepsilon_m + (1 - v) N_a (\varepsilon_i - \varepsilon_m)} \tag{1.13}$$

Again, the superscript a denotes that this effective permittivity is directional—depending on the direction relative to the alignment of the ellipsoids. Note that when $N = 0$, (1.13) reduces to a simple rule of mixtures. The magnetic version of (1.13) is written by replacing the dielectric permittivity variables with magnetic permeabilities.

For an ellipsoid, the shape factor, N, is expressed in terms of an integral equation,

$$N_a = \frac{abc}{2} \int_0^\infty \frac{ds}{(s + a^2)\sqrt{(s + a^2)(s + b^2)(s + c^2)}} \tag{1.14}$$

where a, b, and c are the semiprincipal axes of the ellipsoid. The shape factors in the other principal directions of the ellipsoids are found by transposing a, b, and c. Explicit expressions of N exist for certain types of ellipsoids [17]. For prolate ellipsoids ($a > b = c$),

$$N_a = \frac{1 - e^2}{2e^3}\left(\ln\frac{1 + e}{2 - e} - 2e\right), N_b = N_c = \frac{1}{2}(1 - N_a), \text{ and } e = \sqrt{1 - \frac{b^2}{a^2}} \tag{1.15}$$

For oblate ellipsoids ($a = b > c$),

$$N_c = \frac{1 + e^2}{e^3}(e - \tan^{-1} e), N_a = N_b = \frac{1}{2}(1 - N_c), \text{ and } e = \sqrt{\frac{a^2}{c^2} - 1} \tag{1.16}$$

For spheres,

$$N_a = N_b = N_c = \frac{1}{3} \tag{1.17}$$

There is even an analytical expression for the case when the inclusions are prism-shaped [18].

The Maxwell Garnett theory assumes that the inclusions do not interact with each other. This assumption is good for low-volume fractions of inclusion and applies reasonably well in some types of composites, such as engineered metamaterials, where the inclusions are regularly spaced. Other artificial dielectric or magnetic materials may have inclusions that are randomly spaced so that they do interact with each other. In this case, the Bruggeman EMT is a more appropriate model for characterizing this behavior [19]. One way to derive the Bruggeman EMT is to first rearrange the Maxwell Garnett model of (1.13) into the following form,

$$\frac{\varepsilon^a - \varepsilon_m}{\varepsilon_m + N_a\left(\varepsilon^a - \varepsilon_m\right)} = \nu \frac{\varepsilon_i - \varepsilon_m}{\varepsilon_m + N_a\left(\varepsilon_i - \varepsilon_m\right)} \tag{1.18}$$

In this format, we can generalize the Maxwell Garnett model to account for M different inclusions,

$$\frac{\varepsilon^a - \varepsilon_m}{\varepsilon_m + N_a\left(\varepsilon^a - \varepsilon_m\right)} = \sum_{n=1}^{M} \nu_n \frac{\varepsilon_n - \varepsilon_m}{\varepsilon_m + N_a\left(\varepsilon_n - \varepsilon_m\right)} \tag{1.19}$$

The goal of our derivation is to extend the EMT so that it accounts for inclusion fractions that are high enough to violate the assumption of noninteraction between the inclusions. We can approximate these inclusion interactions by assuming that our matrix properties are approximately equivalent to the effective properties of the composite (i.e., that $\varepsilon_m \rightarrow \varepsilon^a$). Substituting this assumption into (1.19) results in,

$$0 = \sum_{n=1}^{M} \nu_n \frac{\varepsilon_n - \varepsilon^a}{\varepsilon^a + N_a\left(\varepsilon_n - \varepsilon^a\right)} \tag{1.20}$$

For a two-component mixture, the Bruggeman EMT is therefore written as,

$$0 = \nu \frac{\varepsilon_1 - \varepsilon^a}{\varepsilon^a + N_a\left(\varepsilon_1 - \varepsilon^a\right)} + (1-\nu) \frac{\varepsilon_2 - \varepsilon^a}{\varepsilon^a + N_a\left(\varepsilon_2 - \varepsilon^a\right)} \tag{1.21}$$

As with the Maxwell Garnett model, the Bruggeman equation can also be used for magnetic permeability by replacing the dielectric permittivity variables with permeability. Finally, (1.21) can also be rearranged to provide a direct expression for the effective properties in terms of the constituents,

$$\varepsilon^a = \frac{\varepsilon_1\left(v - N_a\right) + \varepsilon_2\left(1 - v - N_a\right) + \sqrt{\left(\varepsilon_1\left(v - N_a\right) + \varepsilon_2\left(1 - v - N_a\right)\right)^2 - 4\varepsilon_1\varepsilon_2 N_a\left(N_a - 2\right)}}{2\left(1 - N_a\right)} \tag{1.22}$$

The Bruggeman and Maxwell Garnett models have very different behaviors as illustrated in Figure 1.10, which shows the calculated real and imaginary permittivity for mixtures of conductive spheres in a nonconductive matrix as a function of volume fraction of the conductive constituent. The matrix permittivity is 3 – 0.01i, which is typical of many simple polymers. The inclusions are modeled as conductive spheres ($N = 1/3$) with a real permittivity of 1 and an imaginary permittivity modeled by a simple conductivity, $i\sigma/\omega\varepsilon_0$. Figure 1.10 includes curves from three different conductivities: 10, 100, and 1,000 S/m. These conductivities are in the range typical for carbon black, a commonly used conductive additive in microwave absorber materials.

As the bottom of Figure 1.10 shows, the imaginary permittivity for a Maxwell Garnett mixture is lower than that predicted by Bruggeman. A more distinct difference is that the Bruggeman mixtures undergo a rapid transition right around a volume fraction of 33%. This transition is called the percolation threshold, and a physical interpretation of this effect is that the assembly of conductive particles touch each other, creating long-range connectivity above a certain concentration. The real permittivity shows that the Bruggeman mixture has a peak corresponding to this threshold behavior, while the Maxwell Garnett mixture gradually transitions from low to high over most of the volume fraction range. Only at the highest volume fraction does it take on the dielectric properties of the spheres. The Bruggeman and Maxwell Garnett behaviors represent the two extremes of binary mixtures. Most actual mixtures may exhibit aspects of both. Furthermore, many mixtures have inclusions that are not spherical, but fibers or platelets. For these other shapes the percolation volume fraction may be different. For example, high-aspect ratio fibers tend to have percolation thresholds of just a few percent or lower. Because of the complexity of real mixtures, these mixture theories may be more qualitative than quantitative. However, they are still useful for understanding trends and for providing insights about the dielectric and magnetic properties of engineering composites.

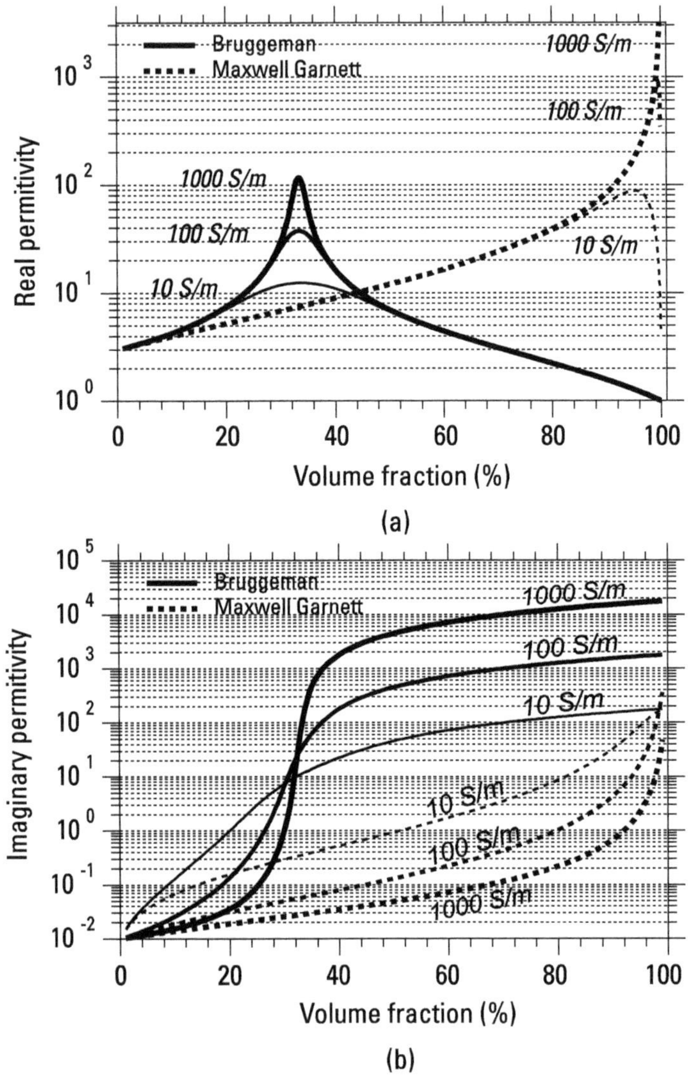

Figure 1.10 (a) Real and (b) imaginary permittivity predicted by the Bruggeman and Maxwell Garnett models for conductive particles in a nonconductive matrix.

References

[1] Pauling, L., and E. Bright Wilson, *Introduction to Quantum Mechanics with Applications to Chemistry*, New York, NY: McGraw-Hill, 1935.

[2] Soohoo, R. F., *Microwave Magnetics*, New York, NY: Harper & Row, 1985.

[3] O'Handley, R. C., *Modern Magnetic Materials Principles and Applications*, Hoboken, NJ: Wiley-Interscience, 2000.

[4] Robert, P., *Electrical and Magnetic Properties of Matter*, Norwood, MA: Artech House, 1988.

[5] Sihvola, A., *Electromagnetic Mixing Formulas and Applications*, London: IEE, 1999.

[6] Debye, P., *Polar Molecules*, Mineola, NY: Dover, 1929.

[7] Drude, P., *The Theory of Optics* (translated by C. Riborg Man and R. A. Millikan), Mineola, NY: Dover, 1902.

[8] Cole, K. S., and R. H. Cole, "Dispersion and Absorption in Dielectrics," *J. Chem. Phys.*, Vol. 9, 1941, pp. 341–351.

[9] Havriliak, S., Jr., and S. J. Havriliak, "Unbiased Modeling of Dielectric Dispersions," Chapter 6 in *Dielectric Spectroscopy of Polymeric Materials* (ed. by J. P. Runt and J. J. Fitzgerald), American Chemical Society, 1997, pp. 175–200.

[10] Fuoss, R. M., and J. G. Kirkwood, "Electrical Properties of Solids. VIII. Dipole Moments in Polyvinyl Chloride-Diphenyl Systems," *J. Am. Chem. Soc.*, Vol. 63, 1941, pp. 385–401.

[11] Davidson, D. W., and R. H. Cole, "Dielectric Relaxation in Glycerine," *J. Chem. Phys.*, Vol. 18, 1950, p. 1417.

[12] Kohlrausch, F., "Ueber die elastische Nachwirkung bei der Torsion," *Pogg. Ann. Phys. Chem.*, Vol. 119, 1863, pp. 337–368.

[13] Williams, W., and D. G. Watts, "Non-Symmetrical Dielectric Relaxation Behavior Arising from a Simple Empirical Decay Function," *Trans. Faraday Soc.*, Vol. 66, 1970, pp. 80–85.

[14] Jonscher, A. K., *Dielectric Relaxation in Solids*, London: Chelsea Dielectric Press, 1983.

[15] Stratton, J. A., *Electromagnetic Theory*, New York, NY: McGraw-Hill, 1941.

[16] Maxwell Garnett, J. C., "Colours in Metal Glasses and in Metallic Films," *Philosophical Trans. Royal Soc. London A*, 1904, pp. 385–420.

[17] Landau, L. D., and E. M. Lifshitz, *Electrodynamics of Continuous Media* (Second Ed.), Amsterdam, Netherlands: Elsevier, 1984.

[18] Aharoni, A., "Demagnetizing Factors for Rectangular Ferromagnetic Prisms," *J. Appl. Phys.*, Vol. 83, No. 6, 1998, pp. 3432–3434.

[19] Bruggeman, D. A. G., "Berechnung verschiedener physikalis- cher Konstanten von heterogenen Substanzen. I. Dielek- trizitätskonstanten und Leitfähigkeiten der Mischkörper aus isotropen Substanzen," *Annalen der Physik*, Vol. 416, No. 7, 1935, pp. 636–664.

2

Free-Space Methods

2.1 Historical Perspective

With the unit of frequency named after him, Heinrich Hertz is considered one of the founders of modern electromagnetics. Hertz's seminal work in the 1880s confirmed theories about the wave nature of electromagnetic energy posed by Maxwell and others [1]. From that foundation, some of the earliest known experimental research in the interaction of electromagnetic energy with materials was conducted by J. C. Bose in the 1890s [2]. During this time, Bose invented horn antennas, including waveguide-lens antennas, along with polarizers, prisms, and other basic components for manipulating microwave- and millimeter-wave energy. Bose's work was groundbreaking, but radio frequency (RF) materials characterization did not occur in earnest until the advent of the radar in World War II. Continued development of radar equipment and related RF and microwave technologies drove the need for understanding material properties at these frequencies, whether they absorb, reflect, or transmit RF energy. This importance stemmed from the need to incorporate materials in microwave components and antennas, as well as the desire to use materials for reducing the radar signatures of military vehicles.

An early treatise on broadband measurement methods is found in the MIT Radiation Laboratory series published in the late 1940s [3]. This extensive reference documents the state of the art in radar-related technologies from that time. In terms of free-space material characterization, [3] describes methods for characterizing transmission and reflection of planar dielectric sheets illuminated by antennas in various configurations and offers some techniques for inverting dielectric permittivity. Figure 2.1 shows drawings of fixtures developed prior to network analyzers, for measuring the free-space transmission phase (a) and amplitude (b) through a material specimen. Without the benefit of modern vector network analyzers, phase measurement required manual adjustments with a micrometer for mechanically determining phase shifts. Additional details of these free-space methods for obtaining permittivity can be found in [4], which includes research dating back to 1942. These references are restricted to dielectric properties of materials. However, less

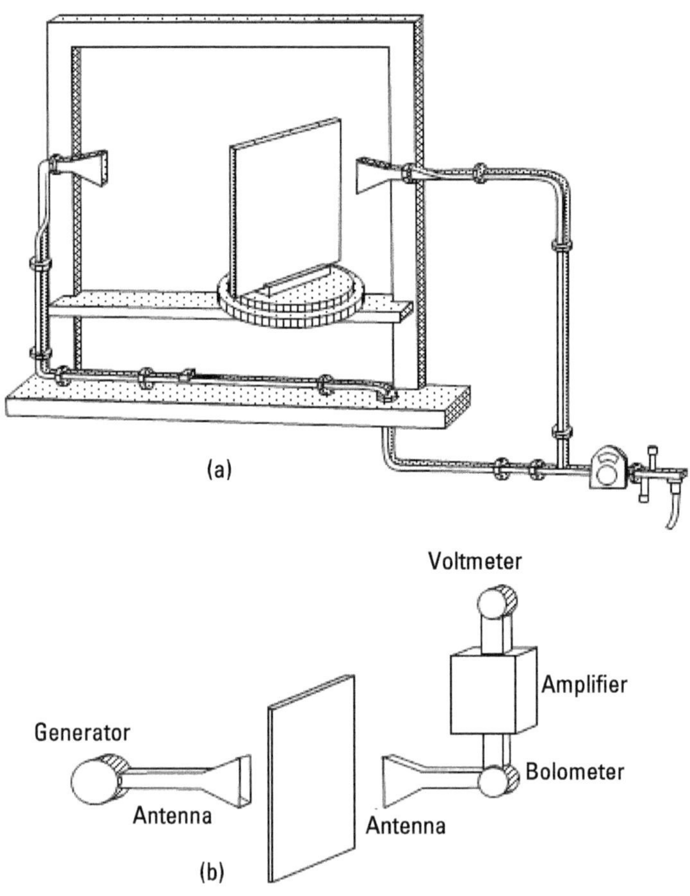

Figure 2.1 Early configurations for free-space transmission from [3], including (a) phase and (b) amplitude.

than a decade later, methods were generalized to include magnetic property determination [5].

Figure 2.2 shows different configurations used in the 1940s for measuring material reflectivity. These geometries enabled both normal incidence and oblique angle measurements of specimens. The geometry in Figure 2.2(b) was pioneered at the U.S. Naval Research Laboratory (NRL) and is also known as the NRL arch method [6], implemented on or about 1945 [7]. With a relatively simple setup and modest cost, the NRL arch measurement method remains in common practice today.

The configuration in Figure 2.1 has also seen continued use in the form of an admittance tunnel. Unlike the fixture in Figure 2.1, a more modern version of the admittance tunnel contains the specimen and antenna(s) within an absorber-lined box. It derives its name from its original use in measuring the sheet admittance or impedance of thin conductive materials. In one version of the admittance tunnel, pictured in Figure 2.3 [8], the specimen is mounted at the end of the absorber-lined box, and a moveable metal plate is placed behind the specimen. The metal plate position is varied to find the position producing the maximum and minimum in the reflection amplitude. Phase was also measured so that both the resistivity and capacitance (or dielectric substrate permittivity) could be determined. A more prevalent admittance tunnel configuration places the specimen midway between transmit and receive

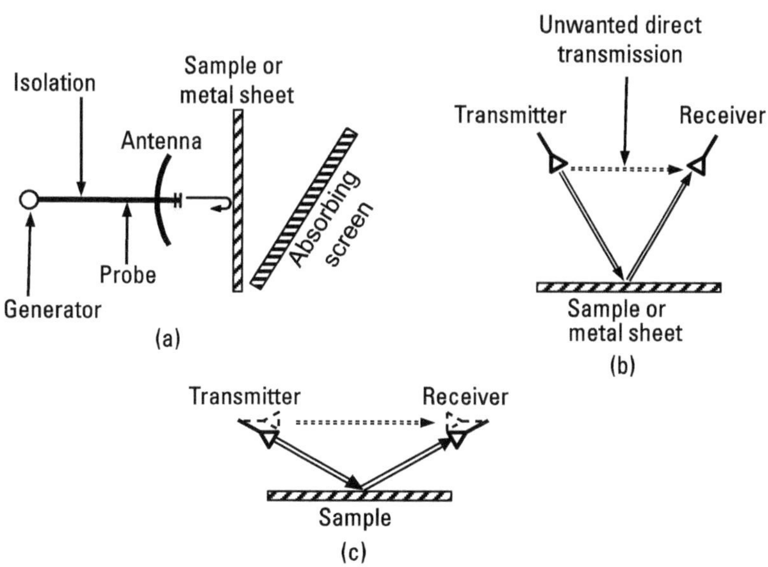

Figure 2.2 Early configurations for measuring free-space reflection from materials [3].

horn antennas. This type of tunnel, still used today, is also called a transmission tunnel. It can characterize thin sheet materials as well as dielectric and magnetic slab specimens.

These early free-space tunnels and arches have two primary disadvantages. The first is that practical limitations place the specimen near enough to the antennas to have significant phase taper across the specimen. This near-field illumination can significantly deviate from an ideal far-field plane wave, resulting in small but nonnegligible errors in the measured transmission and reflection coefficients. A more significant disadvantage, however, is the potentially large diameter of the illuminating beam. If a specimen is moved farther away from the transmit or receive antennas to compensate for near-field effects, the resulting beam diameter grows large enough to illuminate the edges of the specimen. The subsequent edge diffraction then interferes with the specular reflection or direct transmission, resulting in amplitude and phase errors in the measured characteristics.

While increased specimen size is a way to minimize edge-diffraction errors, this is often impractical. Another method, employed with some success in modern admittance tunnels, is the use of tapered apertures. A tapered aperture imposes a controlled amplitude taper on the illuminating microwave energy; however, it does not address the issue of phase curvature due to near-field effects. That said, in many cases this phase taper is relatively small and can be ignored.

As early as 1950, researchers at NRL addressed these errors by incorporating dielectric lenses into horn antennas [9]. The need in this case was to characterize the microwave properties of engine exhaust plumes. The illumination pattern from standard horn antennas was too broad, resulting in significant

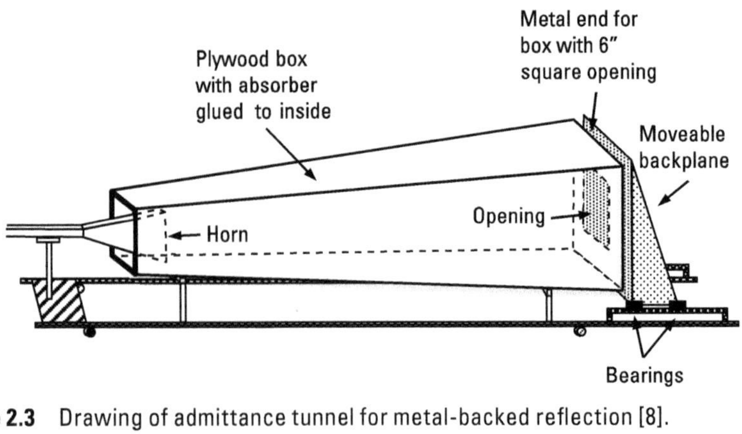

Figure 2.3 Drawing of admittance tunnel for metal-backed reflection [8].

over-illumination of the plume. Dielectric lenses were designed and built which successfully demonstrated the use of a focused beam to interrogate a finite-sized specimen [9]. Similar lens-focusing elements were used later to examine microwave properties of ionized trails behind hypersonic projectiles [10, 11].

In the late 1940s and early 1950s, another parallel effort was under way to achieve the same result [12]. In this case, horn antennas combined with metal artificial dielectric lenses were used to correct the phase front for improved accuracy of moisture content in bales of hay and other agricultural specimens. Also in the same period, Culshaw developed a free-space measurement system that determined permittivity from bistatic reflectivity measurements at oblique incidence [13]. In this work, Culshaw used lenses to either collimate or focus the incident beam on smaller samples. His work demonstrated the improvement in accuracy that occurred by reducing edge diffraction errors.

With optical lenses as a design inspiration, it is not surprising that early focused-beam devices incorporated the direct dielectric analog. However, subsequent researchers also developed alternate methods for focusing energy from a feed antenna. In 1961, Goubaou patented the idea of a beam-waveguide system that could use either dielectric lenses or parabolic reflectors to achieve beam focusing [14]. In 1966, Datlov, Musil, and Zacek applied focusing reflectors with horn antenna feeds to focus a beam used to characterize the microwave properties of a confined plasma [15]. A few years later, Bassett at Georgia Tech used four-foot diameter horn-fed ellipsoidal reflectors to focus microwave energy onto specimens that were heated to extremely high temperatures (2,000°C) [16–18].

In the 1970s, Musil et al. developed a unique alternative to a standard lens: a dielectric rod antenna [19]. This consisted of a dielectric rod inserted into the end of a horn antenna, which confined the radiated energy to a smaller area, much like a focusing lens. They successfully measured dielectric properties of materials with this device through direct contact of the dielectric rods with the specimen. This rod antenna concept has more recently been revisited with computational tools to improve the impedance match, and the rods are held a small distance away from the specimen rather than in direct contact [20]. The disadvantage of such an approach is that the proximity of the probe to the specimen reduces the convenience of high-temperature measurements or other applications where a longer standoff distance between the fixture and the specimen is required. In both the dielectric-rod or spot-probe antenna and in small–focal length lenses, excessive focusing errors can occur with a beam diameter that is too small to satisfy the paraxial approximation. However, the advantages of such spot probes include their very small illumination area, similar to that of a small focal length lens. Being small and lightweight, they can also be easily deployed in manufacturing situations or as handheld or on robotic automation systems.

While there continues to be active development of systems with alternate focusing mechanisms for controlling the illumination area, the dielectric lens remains a broadly used method for laboratory-based microwave-focused beam measurements. The lens' advantages include a relatively simple concept of operation, a sizeable standoff distance between the lens and specimen, and flexibility in accommodating a varied range of illumination areas (or focal lengths). That said, the lower cost and smaller size and weight of nonfocused methods make them prevalent for applications outside the laboratory, such as in manufacturing quality assurance. This chapter, along with Chapters 3 and 4, discusses the design and use of wide-bandwidth free-space methods including spot probes and focused beam systems. Table 2.1 provides a brief summary of the different free-space configurations commonly in use today.

Table 2.1
Typical Configurations of Modern Free-Space Measurements

Method	Description
NRL arch	Two horns on an arch aimed at a specimen
	Pros: Simple, low-cost setup
	Cons: Over-illumination of the specimen leads to edge-diffraction errors; reflection-only measurement
Admittance tunnel	Specimen placed in middle of an absorber tunnel with a horn antenna on each side
	Pros: Absorber reduces room reflection effects
	Cons: Absorber limits access to the specimen for angle measurements or environmental control; antennas need to be far away to simulate plane-wave illumination
Spot probes	Two specially designed antennas on one or both sides of a specimen
	Cons: Compact and can be made rugged for nonlaboratory environments
	Pros: Non-plane-wave illumination reduces accuracy
Focused beam	Two focusing elements (lenses or shaped reflectors) combined with antennas on one or both sides of a material specimen
	Pros: Focusing elements control illumination area and phase to better simulate plane-wave illumination and improve accuracy
	Cons: Lenses or reflectors add cost and complexity to the setup

2.2 Calibration

Determining intrinsic dielectric or magnetic properties of a material specimen with a free-space method involves several steps: calibration, time-domain filtering, and property inversion. Calibration establishes the quantitative scattering parameters of a specimen by comparing measured data to appropriate reference standards. Time-domain filtering minimizes systematic errors caused by multipath reflections within the fixture hardware. Property inversion determines the desired dielectric and magnetic properties of the specimen, usually through numerical solution of equations that relate scattering parameters to the intrinsic properties. These various procedures are described in this chapter.

The first step to characterizing materials in a free-space system involves quantitative measurement of the network scattering parameters. A free-space system works by approximating a plane wave that interacts with the specimen under test, which is usually a planar slab of material. For homogeneous materials, this planar slab can be considered as a two-port network, defined by a two-by-two matrix of scattering parameters to be measured. Modern free-space measurement systems utilize automated vector network analyzers (VNAs) to acquire scattering parameter data. Scattering parameters are discussed in more detail in Section 2.4.1. For the present, it is sufficient to know that the scattering parameters represent the forward and backward reflection (S_{11} and S_{22}) and forward and backward transmission (S_{21} or S_{12}) of a material specimen.

Raw data from the microwave receiver must be calibrated to obtain quantitative scattering parameters. This data is complex with both an amplitude and phase or alternately real and imaginary components associated with each measured frequency. When expressed as real and imaginary pairs, the units of the scattering parameters (S-parameters) are in relative volts. More traditionally S-parameters are reported in terms of amplitude and phase. S-parameter amplitude is also typically provided in terms of decibels, where the S-parameters are squared first before converting to a logarithmic scale so that they represent power: $S_{\mathrm{dB}} = 10\ \log_{10}|S_{\mathrm{volts}}|^2 = 20\log_{10}|S_{\mathrm{volts}}|$. The measured phase is the angle in the complex plane formed by the real and imaginary pairs, and in a material measurement, phase corresponds to the time delays of the free-space system with specimen or calibration materials.

2.2.1 One-Parameter Calibration

For reflection data, a simple calibration is achieved by dividing the data from the specimen under test with a separate measurement of an ideal reflector, typically represented by a conductive metal plate. Such a standard is analogous to the circuit concept of an electrical short, where a conductor is placed across two points of a circuit to short it out. For transmission data, a simple calibration is made by dividing the specimen data with a measurement of a thru, which is the transmission of the fixture with no specimen. A thru is so named because the energy travels through the fixture without a specimen to impede it; it is sometimes also called a clear site.

Because of the thickness of the specimen, additional phase corrections must also be applied to the measured signals. These phase corrections correspond to the signal path in the free-space system displaced by the specimen when it is inserted. For example, with S_{11} reflection measurements, the front face of the conductive calibration plate can be mounted in the same position as that of the material specimen. Here, front refers to the side of the specimen or shorting plate facing the incident energy. When these are in the same physical location, no phase correction is needed. Hence, the calibrated S_{11} is just the ratio of the specimen reflection (S_{11}^{specimen}) to the metal plate or short reflection (S_{11}^{short}),

$$S_{11}^{\text{cal}} = \frac{S_{11}^{\text{specimen}}}{S_{11}^{\text{short}}} \tag{2.1}$$

The position of the front face of the metal calibration plate is called the reference plane. Alternatively, if the specimen is placed in front of the shorting plate position so that the back of the specimen is in the same position as the front of the short plate, then (2.1) is multiplied by a phase correction, e^{-2ik_0t}, where k_0 is the wave number in free space, and t is the specimen thickness. The factor of 2 in the exponent of this phase correction is due to the two-way travel of the reflected energy. This correction corresponds to the path length that is displaced by the specimen's thickness. For S_{21} transmission measurements, the signal path without a specimen has an extra length corresponding to the thickness of the specimen, t. Hence a phase correction is included that corresponds to this extra length,

$$S_{21}^{\text{cal}} = e^{-ik_0t} \frac{S_{21}^{\text{specimen}}}{S_{21}^{\text{thru}}} \tag{2.2}$$

The simple calibration equations, (2.1) and (2.2), use only a single calibration standard for each S-parameter, and are termed response calibrations. A more rigorous calibration method that uses two different calibration standards is called a response-and-isolation calibration. The calibrated scattering parameters for this method are obtained by,

$$S_{ij}^{\text{cal}} = \frac{S_{ij}^{\text{specimen}} - S_{ij}^{\text{isolation}}}{S_{ij}^{\text{response}} - S_{ij}^{\text{isolation}}} \tag{2.3}$$

where ij is the set of scattering parameters (11, 12, 21, 22). S_{ij}^{response} are the same response standards already discussed—metal plate for reflection and no specimen for transmission. In addition, isolation standards, $S_{ij}^{\text{isolation}}$, are used consisting of an opaque metal plate for transmission and a matched load or clear site for reflection. Like (2.1) and (2.2), (2.3) should include a phase correction to account for sample thickness displacement as appropriate.

In the case of reflection, this response and isolation calibration method is analogous to radar cross-section (RCS) measurements. In RCS measurement ranges, a radar illuminates a target, and the backscatter from that target is measured. Whether RCS measurements are done in an outdoor facility or in an indoor anechoic chamber, there are always undesired clutter sources in the vicinity of the target under test, contributing to the total measured signal. Therefore, an RCS measurement calibration utilizes a background measurement that occurs in the absence of the target under test, which is subtracted from the target data to minimize unwanted clutter signals. In a free-space system, the isolation calibration is akin to this RCS background subtraction, where the free-space background consists of reflections from discontinuities in the transmission path, such as imperfect feed antennas and network analyzer ports.

This RCS measurement analogy applies to the response calibration standard as well. In an RCS range, a measurement of a known standard target (such as a sphere or cylinder) is also conducted to quantify the unknown target scatter by accounting for the power levels of the transmitter, the sensitivity of the receiver, and geometrical considerations in a measurement facility. This known target measurement is akin to the response standard in the free-space calibration, which is used to normalize the unknown specimen response to account for transmitter power, receiver sensitivity, and transmission line losses in the free-space system.

2.2.2 Four-Parameter Calibration

When accurate reflection phase is needed, the position of the specimen should exactly correspond to the position of the reflection response standard or metal plate. Any deviation between the positions of the two will result in a phase error. These errors can also arise from warped specimen geometries, where curvature may occur due to material flexibility or internal material stresses. Specimen curvature results in a net position offset of the middle of the specimen from the reflection calibration plane, causing a phase error in the measured reflection coefficient. However, additional procedures can be used to account for these positional phase errors.

For ideal homogeneous materials, the reflection and transmission should be identical in both directions. Historically, only one transmission scattering parameter and one reflection scattering parameter (e.g., S_{21} and S_{11}) were measured to determine the complex permittivity and permeability. Measuring the scattering parameters in the other directions provides theoretically redundant data. However, when the specimen location deviates from the reference plane position, equal but opposite phase offsets occur in the forward and backward reflections (S_{11} and S_{22}). Therefore, measuring all scattering parameters provides information on this phase error, which can then be corrected in the inversion algorithm. This four-parameter method has the advantage of not needing a precise placement of the specimen relative to the calibration reference plane. In particular, the phase error associated with specimen placement is eliminated, resulting in greater accuracy for the permittivity and permeability calculations. The procedure outlined in the following can be used to obtain the response and isolation calibration coefficients in this method:

1. Measure S_{21} and S_{12} of a thru (or fixture with no specimen) to obtain transmission response coefficients (S_{21}^{response} and S_{12}^{response}).
2. Insert a flat metal plate with known thickness t_m and measure S_{11} and S_{22} (S_{11}^{response} and S_{22}^{response}) *without moving the plate* to obtain reflection response data.
3. Leave the metal plate in place and measure S_{21} and S_{12} to obtain transmission isolation coefficients ($S_{21}^{\text{isolation}}$ and $S_{12}^{\text{isolation}}$).
4. Remove the metal plate and either measure no specimen in the two directions, or optionally insert a broadband foam absorber, offset from the specimen position in each direction so that it is as far away as possible from the specimen location.

Once these calibration measurements are complete, the specimen is inserted, and all four scattering parameters are measured without moving the specimen. The calibrated scattering parameters for the specimen are then obtained with (2.3). The calibration must also account for the transmission line displaced with the metal calibration plate by including the calibration plate thickness, t_m, as well as the specimen thickness, t_s, in phase corrections for the inversion algorithm as appropriate.

For most applications, the four-parameter response-and-isolation calibration method provides sufficient accuracy. However, there may be occasions when a more extensive calibration method is desired. One such method that has been implemented [21] is thru-reflect-line (TRL) calibration. TRL calibration also uses the metal plate and thru standards of the simple response method. However, instead of the isolation standards of the response-and-isolation method, TRL uses line standards. The line standards consist of extra sections of transmission line, which impose a phase shift on the measured transmission. Ideally the phase shift should be approximately 90 degrees. When wideband feed antennas are used, this dictates that multiple line standards are needed to avoid the ambiguity of a 180-degree phase shift somewhere within the measurement band. In a free-space measurement, the line standard is implemented by physically increasing the separation between the two sides of the free-space system.

While the TRL calibration method can theoretically give better results than a response-and-isolation method, it requires moving the feeds when used in a free-space system, which requires additional hardware such as an accurate linear-translation mechanism. Furthermore, even with a TRL calibration, the long cable lengths and their sensitivity to temperature can cause residual errors due to multipath reflections, and time-domain filtering is usually applied to further reduce measurement ripple. This time-domain filtering or gating is described in Section 2.3. When a time-domain gate is used, the accuracy difference between TRL and response-and-isolation calibration methods becomes relatively small. Finally, unlike waveguides, free-space systems have some degree of divergence of the propagating beam, also known as space loss. In a TRL calibration, there is an inherent assumption of no space loss. Therefore, in practice a TRL calibration may be less accurate than the simpler response-and-isolation calibration plus time-domain gating. For this reason, TRL calibration is not recommended for most free-space measurement setups.

2.3 Time-Domain Processing

When data is acquired over a reasonably wide range of frequencies, a Fourier transform can be applied to convert frequency-domain data into time-domain data. This enables the separation of the measured signals into different components from the unknown specimen and from other discontinuities within the measurement system. Data measured with a network analyzer is discrete in frequency, so a discrete Fourier transform is required to view the data in the time domain. The resolution in the time domain is proportional to the bandwidth of the frequency domain, $\Delta t = 1/(f_{max} - f_{min})$. The unambiguous range of the time domain is proportional to the number of points, N, in the frequency domain, $t_{unambiguous} = N\Delta t$. For free-space measurements, this unambiguous time can be converted to the unambiguous distance, $R_{unambiguous} = cN/(f_{max} - f_{min})$, where c is the speed of light. Therefore, using a high number of frequency points is preferred so that the unambiguous range is large enough to avoid aliasing (overlap) with other undesired signals.

Examples of time-domain data with different feed horn antennas are in Figure 2.4, which shows reflection amplitude measured with a free-space focused-beam system after a Fourier transform. In the ridged horn plot in Figure 2.4(a), the Fourier transform was performed over the 2–18-GHz band, while the standard gain horn data in Figure 2.4(b) was limited to 8–13 GHz because of the limited bandwidth of that horn. The dashed line of each plot shows the measured response when a tilted absorber is placed just beyond the specimen position, and the black solid lines show the response when a normal incidence metal plate is placed at the specimen position. Peaks are labeled according to their source. The initial peak at 0 ns is from the port of the network analyzer. The next major peak in these plots is from the mismatch of the feed antennas. It is at different times for the two plots because different length cables were used to connect the horns to the network analyzer. The peak from the standard gain horn is lower than that of the ridged horn because it has a better impedance match. However, the additional bandwidth of the ridged horn makes all the peaks narrower because of the additional time resolution afforded by the wider bandwidth. Additional multipath reflections are evident at multiples of the antenna-specimen separation.

Calibration is effective in reducing the mismatch peaks from the cable and feed antenna mismatches. However, the multipath reflections depend on the amplitude attenuation and phase delay introduced with the specimen, so they can still be significant, even after calibration. Thus, a filter is usually applied to selectively minimize all the time-domain peaks except for the specimen signal. In time-domain, this filter is a window, or gate, that is multiplied with the data to preserve the desired signal while minimizing other signals at other times. Different window shapes exist for filtering discrete signals, and a window that works particularly well for the signals of interest here is the Kaiser-Bessel window [22]. Filtering measured data with a time-domain gate is effective in cleaning up the measured data. Gate width is typically specified in nanoseconds, and the choice of what width to use depends on the measurement frequency range and the characteristics of the measured specimen. Higher-index or thick specimens require increased gate widths compared to electrically thinner specimens.

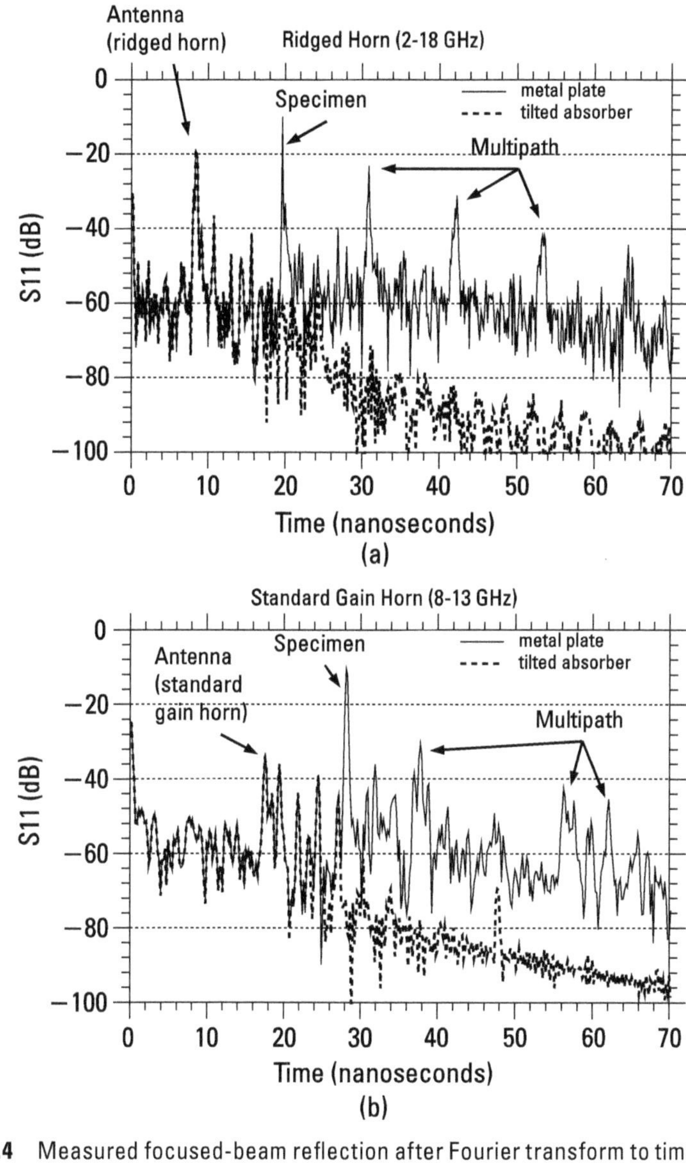

Figure 2.4 Measured focused-beam reflection after Fourier transform to time domain: (a) wideband ridged horn feed and (b) X-band standard gain horn feed.

Minimizing the width of the gate function also minimizes the ripple from undesired multipath signals. However, if the gate width is less than the width of the desired signal, then this filtering will induce systematic errors in the result. Care should be used for gating resonant specimens such as frequency-selective surfaces, radomes, or metal-backed narrowband absorbers because they may ring for a significant amount of time. A gate width that is used for a simple slab of material may be too narrow for these more resonant structures, even when the physical specimen thickness is the same. Thus, the choice of the gate width will be a compromise between minimizing noise and ensuring that the full response of the specimen is captured.

2.4 Inverting Intrinsic Properties

Once calibrated scattering parameters are obtained, then an inversion algorithm is applied to convert the measured observables into intrinsic properties such as dielectric permittivity or magnetic permeability. For free-space techniques, inversion methods are derived by solving the boundary value problem of a plane wave interacting with a planar slab of material. In particular, the intrinsic properties can be determined by directly applying Maxwell's equations to the various boundaries of a given geometry and solving for the entire system. This approach is straightforward for a single boundary, however as the complexity of a specimen geometry grows with two or more boundaries, such as in multilayer systems, property inversion becomes more complicated. As a result, boundary-value problems in microwave systems are often framed in terms of microwave network analysis. Section 2.4.1 describes network analysis and then subsequent sections apply it to develop equations for calculating intrinsic properties from free-space reflection and transmission.

2.4.1 Microwave Network Analysis

Microwave network theory is a formalism that enables a more complicated transmission line problem to be solved by breaking it up into smaller pieces, which can later be recombined with matrix multiplication to solve the entire problem. In the most general case, a microwave network is a region of space having an arbitrary shape and some number of waveguide or transmission line inputs and outputs. The inputs and outputs are also called ports. A simple free-space measurement of a single-layer dielectric or magnetic slab is an example of a two-port network, and we can evaluate the characteristics of that slab in terms of the fields at the two ports or faces of the material. Figure 2.5 shows a schematic representation of a two-port network with input voltages a_i and output voltages b_i. The input and output voltages are related to each other by a matrix formulation known as a scattering matrix,

$$\begin{bmatrix} b_1 \\ b_2 \end{bmatrix} = \begin{bmatrix} S_{11} & S_{12} \\ S_{21} & S_{22} \end{bmatrix} \begin{bmatrix} a_1 \\ a_2 \end{bmatrix} \tag{2.4}$$

where S_{ij} are the elements of the scattering matrix, or S-parameters. Another formulation that is more convenient for cascading multiple networks together is the R-matrix form,

$$\begin{bmatrix} b_1 \\ a_1 \end{bmatrix} = \begin{bmatrix} R_{11} & R_{12} \\ R_{21} & R_{22} \end{bmatrix} \begin{bmatrix} a_2 \\ b_2 \end{bmatrix} \tag{2.5}$$

Thus, it is necessary to convert between these two forms. The following formulas convert back and forth between the R-matrix and S-matrix.

$$R = \frac{1}{S_{21}} \begin{bmatrix} S_{12}S_{21} - S_{11}S_{22} & S_{11} \\ -S_{22} & 1 \end{bmatrix} \tag{2.6}$$

$$S = \frac{1}{R_{22}} \begin{bmatrix} R_{12} & R_{11}R_{22} - R_{12}R_{21} \\ 1 & -R_{21} \end{bmatrix} \tag{2.7}$$

In measurements of transmission and reflection from a material specimen, this network analysis formalism is used to derive the necessary relationships between measured scattering parameters and the intrinsic properties of that specimen. To obtain the intrinsic properties from the network scattering parameters, it is first necessary to segment the problem into simpler parts. Specifically, the slab can be defined in terms of three different cascaded two-port networks that represent the front surface (i), the region between the front and back (ii), and the back surface (iii), as shown schematically in Figure 2.6.

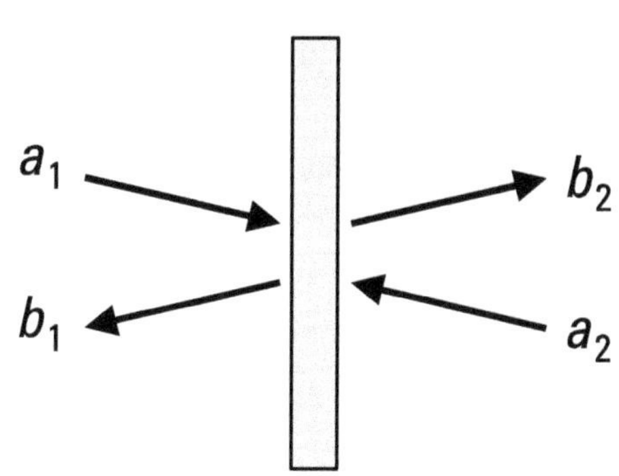

Figure 2.5 Simple two-port network showing input and output voltages.

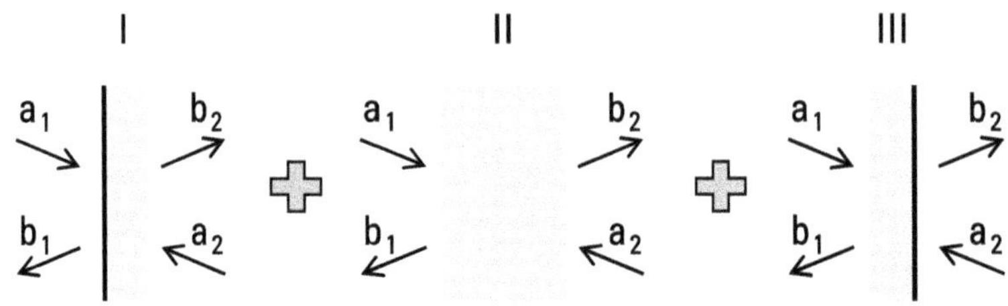

Figure 2.6 Two-port network showing input and output voltages for three different regions of a homogeneous slab specimen.

In the following derivation, we calculate the R-matrix of each of the three networks by analyzing the voltages in terms of transmission coefficient, τ, or reflection coefficient, Γ; and in the case of region II, we analyze the R-matrix in terms of the wave propagation through the medium, T. The transmission coefficient is the ratio of the transmitted wave to the incident wave, and the reflection coefficient is the ratio of the reflected wave to the incident. The wave propagation factor, T, is the phase and amplitude change in the wave as it traverses through the thickness of the material specimen.

For this derivation, we first assume an incoming wave from either the left side traveling toward the right or from the right side traveling toward the left and with a voltage, V_0. Region I in Figure 2.6 represents the leftmost boundary between free space and the material specimen. So, we can assign the network voltages of region I in terms of this input voltage, as shown in Table 2.2.

Inserting these voltages into (2.5) and recognizing that $\Gamma^+ = -\Gamma^-$, the R-matrix elements can then be determined:

$$R^{\mathrm{I}} = \begin{bmatrix} \tau^- + \dfrac{\Gamma^{+2}}{\tau^+} & \dfrac{\Gamma^+}{\tau^+} \\ \dfrac{\Gamma^+}{\tau^+} & \dfrac{1}{\tau^+} \end{bmatrix} \tag{2.8}$$

Table 2.2
Voltages at Air/Material Interface

Right Traveling Wave	**Left Traveling Wave**
$a_1 = V_0$	$a_1 = 0$
$b_1 = \Gamma^+ V_0$	$b_1 = \tau^- V_0$
$a_2 = 0$	$a_2 = V_0$
$b_2 = \tau^+ V_0$	$b_2 = \Gamma^- V_0$

At the boundary, the tangential E-fields of the incident and reflected waves must match that of the transmitted. Thus, we write equations that relate the transmission and reflection coefficients: $1 + \Gamma^{+} = \tau^{+}$ and $1 + \Gamma^{-} = \tau^{-}$. Plugging these relationships into (2.8) then provides the R-matrix of region I in terms of Γ^{+},

$$R^{\mathrm{I}} = \frac{1}{\tau}\begin{bmatrix} 1 & \Gamma \\ \Gamma & 1 \end{bmatrix} = \frac{1}{1+\Gamma}\begin{bmatrix} 1 & \Gamma \\ \Gamma & 1 \end{bmatrix} \tag{2.9}$$

where the superscript + has been dropped for brevity. Additionally, the reflection coefficient can be expressed in terms of the intrinsic permittivity and permeability [23],

$$\Gamma = \frac{\sqrt{\frac{\mu}{\varepsilon}} - 1}{\sqrt{\frac{\mu}{\varepsilon}} + 1} \tag{2.10}$$

where normal incidence transmission through the specimen is assumed.

Region II in Figure 2.6 represents the sample medium between the two air-material interfaces. For an incident wave with a voltage of V_0 we can assign the network voltages of region II in terms of this input voltage, as shown in Table 2.3, where T is the propagation factor of the wave through region II, $T = e^{-ikt}$; $k = k_0\sqrt{\varepsilon_r\mu_r}$ is the wave number in the sample medium; and t is the distance traveled between the two ports. For a simple slab with a normal incidence-illuminating wave, t is simply the thickness of the slab. Plugging these boundary conditions into (2.5) then gives the R-matrix for region II,

$$R^{\mathrm{II}} = \begin{bmatrix} T & 0 \\ 0 & \frac{1}{T} \end{bmatrix} \tag{2.11}$$

Table 2.3
Voltages Going Through a Medium (Region II in Figure 2.6)

Right-Traveling Wave	Left-Traveling Wave
$a_1 = V_0$	$a_1 = 0$
$b_1 = 0$	$b_1 = TV_0$
$a_2 = 0$	$a_2 = V_0$
$b_2 = TV_0$	$b_2 = 0$

Finally, the R-matrix for the back face of the material slab, region III, follows a derivation similar to the front face. Because it is the reverse of the front face, it is essentially the same matrix, but with the reflection coefficients replaced by their negatives,

$$R^{\mathrm{III}} = \frac{1}{1-\Gamma}\begin{bmatrix} 1 & -\Gamma \\ -\Gamma & 1 \end{bmatrix} \tag{2.12}$$

The equivalent R-matrix for all three regions put together is then calculated by matrix multiplication,

$$R = R^{\mathrm{I}}R^{\mathrm{II}}R^{\mathrm{III}} = \frac{1}{T\left(1-\Gamma^2\right)}\begin{bmatrix} T^2-\Gamma^2 & \Gamma-\Gamma T^2 \\ \Gamma T^2-\Gamma & 1-\Gamma^2T^2 \end{bmatrix} \tag{2.13}$$

Then with (2.7), we convert this R-matrix to the elements of the scattering matrix for the full material slab,

$$S = \frac{1}{1-\Gamma^2T^2}\begin{bmatrix} \Gamma\left(1-T^2\right) & T\left(1-\Gamma^2\right) \\ T\left(1-\Gamma^2\right) & \Gamma\left(1-T^2\right) \end{bmatrix} \tag{2.14}$$

This gives us the relationship between the experimentally measured scattering parameters and the intrinsic permittivity and permeability of the material under test. In one case, (2.14) can be written such that the permittivity and/or permeability are expressed as a direct function of the measured scattering parameters. More often, an iterative root finding method must be used to determine intrinsic properties.

The following sections present common inversion algorithms based on (2.14) or other similar equations, which can be applied based on the needs of a given specimen measurement scenario. Note that in the following discussion, μ and ε are assumed to refer to the relative permeability and permittivity and the subscript r is dropped for brevity. The presented inversion algorithms are summarized in Table 2.4.

2.4.2 Nicolson-Ross-Weir Algorithm

The Nicolson-Ross-Weir (NRW) algorithm is a well-known algorithm for inverting permittivity and permeability from the S_{11} and S_{21} network scattering parameters [24, 25]. From a specimen's measured S_{11} and S_{21}, the reflection Γ at the air/material boundary and transmission T through the material are calculated,

Table 2.4
Summary of Inversion Algorithms

Inversion Algorithm	Section	Extracted Parameters
Nicolson-Ross-Weir	2.4.2	μ and ε
Iterative S_{11} or S_{21}	2.4.3	ε
Iterative S_{11} and S_{21}	2.4.4	μ and ε
Iterative shorted S_{11}	2.4.5	ε
Iterative shorted S_{11} and S_{21}	2.4.6	μ and ε
Iterative 4-parameter	2.4.7	μ and ε
Sheet Impedance	2.4.8	Z
Iterative N-layer	2.5.1	μ and ε
Iterative 2-thickness	2.5.2	μ and ε
Model-based	2.5.3	μ and ε

$$\Gamma = X \pm \sqrt{X^2 - 1}, \text{ where } X = \frac{S_{11}^2 - S_{21}^2 + 1}{2S_{11}} \tag{2.15}$$

$$T = \frac{S_{11} + S_{21} - \Gamma}{1 - \left(S_{11} + S_{21}\right)\Gamma} \tag{2.16}$$

By defining a third parameter, Λ, such that,

$$\frac{1}{\Lambda^2} = -\left(\frac{1}{2\pi t}\ln T\right)^2 \tag{2.17}$$

then the permeability and permittivity can be solved explicitly,

$$\mu = \frac{2\pi}{\Lambda k_0}\left(\frac{1+\Gamma}{1-\Gamma}\right) \tag{2.18}$$

$$\varepsilon = \frac{1}{\mu k_0^2}\left(\frac{4\pi^2}{\Lambda^2}\right) \tag{2.19}$$

where $k_0 = 2\pi/\lambda$ is the wave number in air. As Baker-Jarvis [26] points out, the NRW algorithm outlined above suffers from a numeric instability when the frequency corresponds to a multiple of one-half wavelength in the specimen.

This instability is caused in part by phase uncertainties and is more likely in low-loss materials. Boughriet et al. [27] noted that the instability arises from the computation of the factor, $(1+\Gamma)/(1-\Gamma)$. Therefore, if μ is already known, this instability can be circumvented since this factor affects the computation of ε only by its influence on μ.

Another difficulty with this algorithm stems from the fact that the logarithm of a complex number is multivalued. In other words, there is a phase ambiguity in (2.17) such that $\ln T = \ln T + i2\pi n$, where n is an integer. In most "nice" specimens (i.e., when the permittivity and permeability are moderate to small, and the sample electrical thickness is less than a half wavelength), the $n = 0$ solution is the correct one. However, in electrically thick cases where $n \neq 0$, additional information is needed to identify the correct solution. For example, Baker-Jarvis et al. [26] outline a procedure comparing measured and calculated group delays to identify the correct solution. Alternatively, a priori knowledge of the expected range of permittivity and permeability values can be used to identify the correct root.

2.4.3 Iterative Algorithm: S_{11} or S_{21}

While the NRW algorithm is useful for some situations, there are times where both S_{11} and S_{21} cannot be accurately measured. For example, in-situ characterization of components may prevent access to both sides of the material so only S_{11} can be obtained. More often, materials may be difficult to make perfectly flat, which causes focusing or defocusing of reflected energy as well as position uncertainty that results in phase errors of the measured S_{11}. In these situations, a calculation based solely on S_{21} is preferable. That said, for dielectric-only materials where the magnetic permeability is already known to be the same as free space, an iterative calculation based on either S_{11} or S_{21} can be performed.

All iterative algorithms begin with initial estimates at all frequencies for permittivity (and permeability when both parameters are unknown). The algorithm then refines the initial estimates with a numerical root-finding or function-fitting technique. There are often multiple solutions to these equations, and an initial estimate is necessary to select the proper root. The suitability of the initial estimates is determined by the stability of permittivity and permeability results, or by comparison with other measurement methods. If the initial estimates start the iterative calculation on the wrong root, the calculated results tend to have poor convergence and may jump to another root when plotted as a function of frequency. Improper solutions may also violate energy conservation by having negative loss.

For a two-port network, the abovementioned reflection and transmission coefficients are related to the scattering parameters, S_{11} and S_{21} by [26],

$$S_{11} = \frac{\Gamma(1-T^2)}{1-\Gamma^2 T^2} \quad \text{and} \quad S_{21} = \frac{T(1-\Gamma^2)}{1-\Gamma^2 T^2} \tag{2.20}$$

where S_{11} and S_{21} are the calibrated S-parameters. In free-space measurements a specimen is typically inserted into an existing transmission path. So, these equations also assume that a phase correction was already applied to the S-parameters for the displacement of the transmission path by the specimen. The general relations for the reflection and transmission coefficients at the arbitrary incidence angle are

$$\Gamma_{\text{TE}} = \frac{\mu\cos\theta - \sqrt{\mu\varepsilon - \sin^2\theta}}{\mu\cos\theta + \sqrt{\mu\varepsilon - \sin^2\theta}} \tag{2.21}$$

$$\Gamma_{\text{TM}} = \frac{\sqrt{\mu\varepsilon - \sin^2\theta} - \varepsilon\cos\theta}{\sqrt{\mu\varepsilon - \sin^2\theta} + \varepsilon\cos\theta} \tag{2.22}$$

$$T = e^{-ik_0 t\sqrt{\mu\varepsilon - \sin^2\theta}} \tag{2.23}$$

where t is the thickness of the specimen, and θ is the angle between the direction of propagation and the specimen normal. Equations (2.21) and (2.22) are derived by the matching boundary conditions at the specimen/air interface. Equation (2.23) is the propagation through a specimen of thickness t and at an angle θ. Note that the propagation angle inside the specimen is different than the incidence angle. However, the angle within the specimen was eliminated from the above equations by Snell's law combined with a geometric identity. Typical measurements orient the specimen at normal incidence ($\theta = 0$), and the above equations reduce to

$$\Gamma = \frac{\sqrt{\frac{\mu}{\varepsilon}} - 1}{\sqrt{\frac{\mu}{\varepsilon}} + 1} \quad \text{and} \quad T = e^{-ik_0 t\sqrt{\mu\varepsilon}} \tag{2.24}$$

Either of the equations in (2.20) along with the reflection or transmission coefficient as a function of frequency can then be solved via Newton's method using ε or μ as the unknown variable. Newton's method is recommended since it solves for complex valued roots [28]. It also requires the evaluation of the first derivative of the function. The reformulated functions and their derivatives for solving S_{11} and S_{21} are outlined as follows.

- S_{11}:

$$f = 0 = \left(1 - \Gamma^2 T^2\right) S_{11} - \Gamma\left(1 - T^2\right) \tag{2.25}$$

$$f' = \left[T^2\left(1 - 2S_{11}\Gamma\right) - 1\right]\Gamma' - 2T\Gamma\left(1 - S_{11}\Gamma\right)T' \tag{2.26}$$

- S_{21}:

$$g = 0 = \left(1 - \Gamma^2 T^2\right) S_{21} - T\left(1 - \Gamma^2\right) \tag{2.27}$$

$$g' = \left[\Gamma^2\left(1 - 2S_{21}T\right) - 1\right]T' + 2T\Gamma\left(1 - S_{21}T\right)\Gamma' \tag{2.28}$$

where the superscript prime indicates the first derivative.

While this iterative technique will work for any value of μ, it is usually only applied to dielectric materials, where $\mu = 1$, since a separate measurement of μ would otherwise be required. The derivatives of the reflection and transmission coefficients with respect to permittivity are

$$\frac{\partial \Gamma_{\text{TE}}}{\partial \varepsilon} = \frac{-\mu^2 \cos\theta}{\sqrt{\mu\varepsilon - \sin^2\theta}\left(\mu\cos\theta + \sqrt{\mu\varepsilon - \sin^2\theta}\right)^2} \tag{2.29}$$

$$\frac{\partial \Gamma_{\text{TM}}}{\partial \varepsilon} = \frac{\cos\theta\left(\dfrac{\mu\varepsilon}{\sqrt{\mu\varepsilon - \sin^2\theta}} - 2\sqrt{\mu\varepsilon - \sin^2\theta}\right)}{\left(\varepsilon\cos\theta + \sqrt{\mu\varepsilon - \sin^2\theta}\right)^2} \tag{2.30}$$

$$\frac{\partial T}{\partial \varepsilon} = -\frac{ik_0\mu t}{2\sqrt{\mu\varepsilon - \sin^2\theta}}T \tag{2.31}$$

For S_{21} measurements, Newton's method solves these equations by iteratively calculating $\varepsilon_{n+1} = \varepsilon_n - g/g'$ until $\varepsilon_{n+1} - \varepsilon_n$ is sufficiently small.

Since iterative methods rely on starting guess values, it is beneficial in some cases to iterate serially—one frequency at a time. For a material that is not very dispersive with a permittivity (or permeability) that is stable across the measured frequencies, the same starting guess value can work for all the frequencies. However, for a material that is dispersive (i.e., has a frequency-dependent permittivity or permeability), conducting the iterative procedure in series requires a guess value only for the first frequency point. The converged solution for that frequency then becomes the guess value for the next frequency point, and so on. Otherwise, a frequency-dependent set of guess values is needed to get good convergence for the whole bandwidth, which could be especially difficult to estimate for a material that is undergoing a relaxation within the measurement bandwidth. Instead, it is easier to start at one end or the other of the band where the material permittivity or permeability may be better known. Often it is easiest to start at the lowest frequency and work upward, since the material is electrically thinner at lower frequencies, and the multiple solution branches are spaced further apart.

2.4.4 Iterative Algorithm: S_{11} and S_{21}

An iterative algorithm to solve for μ and ε simultaneously can also be implemented based on measurements of both S_{11} and S_{21}. Because both S_{11} and S_{21} equations must be solved, Newton's iteration for a system of equations is used. In the case of more than one variable to be solved, the algorithm is most easily expressed in matrix form [29] where the following linear system is solved.

$$\begin{bmatrix} \dfrac{\partial f}{\partial \varepsilon} & \dfrac{\partial f}{\partial \mu} \\ \dfrac{\partial g}{\partial \varepsilon} & \dfrac{\partial g}{\partial \mu} \end{bmatrix} \cdot \begin{bmatrix} \Delta\varepsilon \\ \Delta\mu \end{bmatrix} = \begin{bmatrix} f \\ g \end{bmatrix} \tag{2.32}$$

and where f and g and their derivatives are defined previously by (2.25) to (2.31). Since μ is no longer assumed fixed, this algorithm also requires the derivatives of the reflection and transmission coefficients with respect to μ,

$$\frac{\partial \Gamma_{\mathrm{TE}}}{\partial \mu} = \frac{\cos\theta\left(-\dfrac{\mu\varepsilon}{\sqrt{\mu\varepsilon - \sin^2\theta}} + 2\sqrt{\mu\varepsilon - \sin^2\theta}\right)}{\left(\mu\cos\theta + \sqrt{\mu\varepsilon - \sin^2\theta}\right)^2} \tag{2.33}$$

$$\frac{\partial \Gamma_{\mathrm{TM}}}{\partial \mu} = \frac{\varepsilon^2\cos\theta}{\sqrt{\mu\varepsilon - \sin^2\theta}\left(\varepsilon\cos\theta + \sqrt{\mu\varepsilon - \sin^2\theta}\right)^2} \tag{2.34}$$

$$\frac{\partial T}{\partial \mu} = -\frac{ik_0\varepsilon t}{2\sqrt{\mu\varepsilon - \sin^2\theta}} T \tag{2.35}$$

The two-by-two matrix of (2.32) is the Jacobian of the system of two complex equations. In matrix notation the discretized system of equations can be represented as follows,

$$J\Delta X = Y \tag{2.36}$$

where the vector X contains the estimated values of permittivity and permeability, and the vector Y contains the zero-valued functions f and g given above. Solving for ΔX gives $\Delta X = J^{-1}Y$. Like the single equation iteration algorithm discussed above, this algorithm starts with an initial estimate of X and calculates ΔX, which is a function of both permittivity and permeability. The functional iteration procedure is then $X^{new} = X^{old} - J^{-1}Y$ and is repeated until $X^{new} \approx X^{old}$, at which point the converged values of permittivity and permeability have been found. This iterative algorithm should give the same results as the more direct NRW method, so it is of limited usefulness. It also suffers from the same $\lambda/2$ wavelength instability as the NRW algorithm.

2.4.5 Iterative Algorithm: Shorted S_{11}

The iterative algorithm for S_{11} outlined above assumes that there is air behind the specimen. It is also possible to calculate permittivity when there is a conductive plate behind the specimen. This is useful for materials that are formed by being deposited onto a metal substrate and where separating the specimen from the substrate is impractical. This method was originally developed by Roberts and von Hippel [30] and has been reviewed by Baker-Jarvis [26]. When the specimen is flush against the short, the permittivity and permeability are related to S_{11} by

$$S_{11} = \frac{\mu \tanh\left(i\sqrt{\mu\varepsilon}k_o t\right) - \sqrt{\mu\varepsilon}}{\mu \tanh\left(i\sqrt{\mu\varepsilon}k_o t\right) + \sqrt{\mu\varepsilon}} \tag{2.37}$$

Like the previous iteration methods, (2.37) is reformulated to apply Newton's root finding method and iteratively solve for ε. The iterated function and its derivative are as follows:

$$f(\varepsilon) = 0 = \left(\tanh\left(i\sqrt{\varepsilon}k_o t\right) + \sqrt{\varepsilon}\right)S_{11} - \left(\tanh\left(i\sqrt{\varepsilon}k_o t\right) - \sqrt{\varepsilon}\right) \tag{2.38}$$

$$f'^{(\varepsilon)} = \frac{k_0}{2\sqrt{\varepsilon}}\begin{bmatrix} \left(k_0 \tanh^2\left(i\sqrt{\varepsilon}k_o t\right) + k_0 t + i\right)S_{11} \\ - k_0 \tanh^2\left(i\sqrt{\varepsilon}k_o t\right) + k_0 t + i \end{bmatrix} \tag{2.39}$$

While this method does make it possible to determine the dielectric permittivity of a material adjacent to a conductive plate, it is not a preferred method if the material can be removed from the plate. In particular, the tangential electric field next to a conductive surface goes to zero at that surface. Dielectric permittivity is about the interaction of the material with the electric field, so measurement accuracy declines for a specimen next to a metal plate compared to one that is suspended in air.

2.4.6 Iterative Algorithm: Shorted S_{11} and S_{21}

The shorted S_{11} iteration equations can be combined with the equations for S_{21} so that magnetic specimens can be measured. As in the nonshorted S_{11} and S_{21} iteration, this algorithm uses Newton's method for a system of equations as outlined in (2.32). In this case, the iterated function, f, is obtained from (2.37), and the second iterated function, g, is obtained from (2.20).

$$f(\varepsilon,\mu)=\left(\mu\tanh\left(i\sqrt{\mu\varepsilon}k_o t\right)+\sqrt{\mu\varepsilon}\right)S_{11}-\left(\mu\tanh\left(i\sqrt{\mu\varepsilon}k_o t\right)-\sqrt{\mu\varepsilon}\right) \tag{2.40}$$

$$g(\varepsilon,\mu)=\left(1-\Gamma^2T^2\right)S_{21}-T\left(1-\Gamma^2\right) \tag{2.41}$$

The derivatives of these two functions for solving via Newton's algorithm are

$$\frac{\partial f}{\partial \varepsilon}=\frac{k_0^2\mu^2 t}{2\sqrt{\mu\varepsilon}}\left[\begin{array}{c}\left(\tanh^2\left(i\sqrt{\mu\varepsilon}k_o t\right)-1+\dfrac{i}{\mu k_0 t}\right)S_{11}\\ -\tanh^2\left(i\sqrt{\mu\varepsilon}k_o t\right)+1+\dfrac{i}{\mu k_0 t}\end{array}\right] \tag{2.42}$$

$$\frac{\partial f}{\partial \mu}=\frac{k_0}{2\sqrt{\mu\varepsilon}}\left[\begin{array}{c}\left(\begin{array}{c}\mu\varepsilon t k_0\tanh^2\left(i\sqrt{\mu\varepsilon}k_o t\right)\\ -2i\sqrt{\mu\varepsilon}\tanh\left(i\sqrt{\mu\varepsilon}k_o t\right)-k_0\mu\varepsilon t+i\varepsilon\end{array}\right)S_{11}\\ -\mu\varepsilon t k_0\tanh^2\left(i\sqrt{\mu\varepsilon}k_o t\right)\\ +2i\sqrt{\mu\varepsilon}\tanh\left(i\sqrt{\mu\varepsilon}k_o t\right)+k_0\mu\varepsilon t+i\varepsilon\end{array}\right] \tag{2.43}$$

$$\frac{\partial g}{\partial \varepsilon}=\left[\Gamma^2\left(1-2S_{21}T\right)-1\right]\frac{\partial T}{\partial \varepsilon}+2T\Gamma\left(1-S_{21}T\right)\frac{\partial \Gamma}{\partial \varepsilon} \tag{2.44}$$

$$\frac{\partial g}{\partial \mu}=\left[\Gamma^2\left(1-2S_{21}T\right)-1\right]\frac{\partial T}{\partial \mu}+2T\Gamma\left(1-S_{21}T\right)\frac{\partial \Gamma}{\partial \mu} \tag{2.45}$$

These equations along with the derivatives for Γ and T given above are solved iteratively via (2.32). While this method does have good accuracy, it is less convenient than other methods since it requires separate measurements of the specimen with and without a conductive plate adjacent to it.

2.4.7 Iterative Algorithm: Four-Parameter

A disadvantage of the previous algorithms utilizing S_{11} is that they can have significant measurement uncertainties. These errors arise from warped specimen geometries, where slight or not-so-slight curvature occurs due to material flexibility or internal material stresses. Specimen curvature results in a net position offset of the specimen from the reflection calibration plane, causing a phase error in the measured reflection. The waveguide and coaxial airline methods discussed later in this book use relatively small specimens that can be maintained flat and more accurately positioned to effectively use the NRW or iterative S_{11} and S_{22} inversions. However, these inversions are not recommended for free-space material measurements. The larger specimen size inevitably results in too much position uncertainty. Instead, free-space measurements of magneto-dielectric materials are more accurately done with a four-parameter method that uses all four scattering parameters (S_{11}, S_{22}, S_{21}, S_{12}) to determine permittivity and permeability. This four-parameter method has the advantage of not needing precise specimen placement relative to the calibration reference plane. It eliminates the phase error associated with specimen placement, resulting in greater accuracy for the permittivity and permeability calculations.

At first glance, using forward and backward scattering coefficients overspecifies the inversion problem since only two complex intrinsic properties are calculated. By reciprocity, the transmission amplitudes and phases should all be the same. For homogeneous specimens the reflection amplitudes should also be the same. Only the reflection phases should differ, depending on specimen position. Thus, the four-parameter inversion includes additional information that can determine specimen position without a physical distance measurement. These additional parameters are used to algebraically eliminate the specimen position dependence.

The four-parameter inversion for free-space measurements is similar to a procedure used in coaxial airline fixtures [29]. Equations (2.46) and (2.47) relate all four scattering parameters to specimen thickness, permittivity, and permeability:

$$S_{11}^{\text{cal}} S_{22}^{\text{cal}} e^{-2ik_0\left(t_s - t_m\right)} - S_{21}^{\text{cal}} S_{12}^{\text{cal}} e^{-2ik_0\left(t_s\right)} = \frac{\Gamma^2 - T^2}{1 - \Gamma^2 T^2} \tag{2.46}$$

$$e^{-ik_0 t_s}\frac{S_{21}^{\text{cal}}+S_{12}^{\text{cal}}}{2}=\frac{T\left(1-\Gamma^2\right)}{1-\Gamma^2T^2} \tag{2.47}$$

where $\Gamma = (\mu - \sqrt{\mu\varepsilon})/(\mu + \sqrt{\mu\varepsilon})$ and $T = e^{-ik_0 t_s\sqrt{\mu\varepsilon}}$. The *cal* superscript designates the use of already calibrated scattering parameters. Equations (2.46) and (2.47) also explicitly include the phase correction for the transmission line displaced with the metal calibration plate by including the calibration plate thickness, t_m as well as the specimen thickness, t_s. This algorithm requires the use of the four-parameter calibration method described above, where S_{11} and S_{22} response calibration factors are measured simultaneously without moving the metal calibration standard.

These equations are then solved to invert permittivity and permeability using the system of equations defined by (2.32). The zero-valued functions whose roots must be obtained are

$$f=\left(1-\Gamma^2T^2\right)\left(S_{11}^{\text{cal}}S_{22}^{\text{cal}}e^{-2ik_0\left(t_s-t_m\right)}-S_{21}^{\text{cal}}S_{12}^{\text{cal}}e^{-2ik_0\left(t_s\right)}\right)-\left(\Gamma^2-T^2\right) \tag{2.48}$$

$$g=\left(1-\Gamma^2T^2\right)\left(e^{-ik_0 t_s}\frac{S_{21}^{\text{cal}}+S_{12}^{\text{cal}}}{2}\right)-T\left(1-\Gamma^2\right) \tag{2.49}$$

2.4.8 Inverting Sheet Impedance

It is sometimes convenient to represent a material as a shunt impedance. Certain classical absorber materials such as the Salisbury screen and Jaumann absorber [31] are constructed using resistive sheets where the shunt impedance is expressed in units of ohms/square [32]. In this case the sheet impedance is directly related to the scattering parameters and can be derived using the same network analysis formalism shown earlier in this chapter. The shunt resistance of the sheet can be treated as a simple two-port network. When an incident wave, with a voltage, V_0, interacts with the sheet, the network voltages are as shown in Table 2.5.

Unlike the interface between air and a dielectric material discussed previously, the discontinuity represented by the sheet is symmetrical so that $\Gamma^+ = \Gamma^-$ and $\tau^+ = \tau^-$. Additionally, the tangential E-fields should match so the relationship, $1 + \Gamma = \tau$ applies. Finally, with straightforward circuit analysis [32], the reflection coefficient from a shunt impedance Z, in a transmission line with a characteristic impedance Z_0 is given by $\Gamma = (Z - Z_0)/(Z + Z_0)$. These relationships then lead to the following R-matrix for a resistive sheet,

Table 2.5
Voltages at a Resistive Sheet in Air

Right Traveling Wave	**Left Traveling Wave**
$a_1 = V_0$	$a_1 = 0$
$b_1 = \Gamma^+ V_0$	$b_1 = \tau^- V_0$
$a_2 = 0$	$a_2 = V_0$
$b_2 = \tau^+ V_0$	$b_2 = \Gamma^- V_0$

$$R = \frac{1}{2Z_s}\begin{bmatrix} 2Z_s - Z_0 & -Z_0 \\ Z_0 & 2Z_s + Z_0 \end{bmatrix} \tag{2.50}$$

With (2.7), this R-matrix can then be converted into scattering parameters. For transmission,

$$Z_s = \frac{Z_0 S_{21}}{2\left(1 - S_{21}\right)} \tag{2.51}$$

and for reflection,

$$Z_s = \frac{-Z_0\left(1 + S_{11}\right)}{2S_{11}} \tag{2.52}$$

where Z_0 is the bulk impedance of free space (377Ω). Unlike most of the permittivity and permeability inversions, sheet impedance is calculated directly from the scattering parameters and an iterative inversion method is not needed.

Sometimes resistive sheet material is mounted onto a thick dielectric substrate. The effect of an additional substrate is to provide additional capacitance to the complex impedance, resulting in a more negative imaginary impedance. When this is the case, it may be desirable to extract the impedance of just the resistive sheet while excluding the effect of the supporting substrate. Using the cascade matrix theory, the R-matrix of the sheet is multiplied with the R-matrix of the dielectric slab to obtain an R-matrix, (and corresponding S-matrix) for the multilayer stack. If the substrate permittivity, ε, and thickness, t, are accurately known, then the sheet impedance of just the resistive sheet by itself can be calculated from the transmission scattering parameter by

$$Z_s = S_{21} Z_0 \frac{\Gamma^2 T^2 - 1 + \Gamma\left(T^2 - 1\right)}{2\left(\Gamma^2 - 1\right)T - 2S_{21}\left(\Gamma^2 T^2 - 1\right)} \tag{2.53}$$

where $\Gamma = (\mu - \sqrt{\mu\varepsilon})/(\mu + \sqrt{\mu\varepsilon})$ and $T = e^{-ik_0 t\sqrt{\mu\varepsilon}}$ are of the substrate measured by itself (without a conductive layer). Thus, this two-layer inversion requires that a specimen of the substrate be measured first without any resistive sheet.

2.5 Advanced Material Inversions

Most of the above inversions apply for the simple case when there is a single layer of material. Sometimes, materials exist in a layered configuration, such as a painted coating applied to a substrate. That substrate can be another dielectric or magneto-dielectric material, or it can be multiple layers of different materials. In some cases, an unknown layer cannot be physically separated from the other layer(s) and a more sophisticated inversion algorithm must be employed, especially if it is magneto-dielectric material on a conductive substrate, and only one side of the material is accessible. In this case there are four unknowns (real and imaginary permittivity and permeability), but only two physical observables are possible at each frequency for a single material (amplitude and phase of reflection coefficient). Additional information or assumptions must then be made to extract all four unknown intrinsic parameters. Sections 2.5.1–2.5.3 describe inversion algorithms that can be used for these more complicated cases.

2.5.1 N-Layer Inversion

As shown in (2.53), it is possible to invert an unknown resistive layer that is deposited onto a known substrate. Similarly, unknown dielectric or magneto-dielectric layers can be inverted when they are on or in a multilayer stack. A notional multilayer stack is shown in Figure 2.7, which contains an unknown layer surrounded on top and bottom by other known layers.

For these more general cases, an inversion equation can be derived by determining the R-matrixes of each of the known layers and cascading them together with the unknown layer. For example, individual layers within a stack can be constructed using (2.13), where Γ for each layer is determined by (2.10) and T for each layer is calculated by $T = e^{-ikt}$ and $k = k_0\sqrt{\varepsilon_r \mu_r}$. Each known layer will have a permittivity, permeability, and thickness specific to that layer as indicated in Figure 2.7. Usually, the unknown layer will have a known thickness, t, but unknown permittivity and/or permeability. The R-matrixes for these layers are then multiplied together to construct a single R-matrix for the stack,

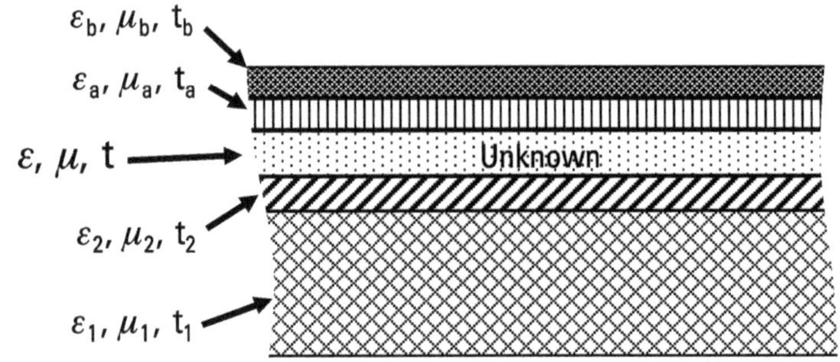

Figure 2.7 Notional multilayer stack of materials with thicknesses and intrinsic properties known for all but one layer.

$$R_{\text{stack}} = R_1 R_2 R_3 \ldots R_{\text{unknown}} R_a R_b R_c \ldots = R_{\text{bottom}} R_{\text{unknown}} R_{\text{top}} \tag{2.54}$$

In (2.54), R_1, R_2, R_3, and so on, are the known layers on one side of the unknown layer, while R_a, R_b, R_c, and so on, are the known layers on the other side of the unknown layer. Since the inversion will be iterative, multiple layers on one or the other side of the unknown layer should be multiplied together first to make bottom- and top-side R-matrixes. This avoids extra matrix multiplications during each step of the iteration. The R-matrix of the stack can then be converted to an S-matrix using (2.14) for comparison to the measured S-parameters, and a standard function minimization routine iteratively finds the best fit by varying the permittivity and/or permeability of the unknown layer within the stack [28]. In theory, a Newton's iteration method could also be used for solving these multilayer inversions, but the calculation of derivatives becomes more complicated.

Depending on what the unknown material is in the stack, the inversion can be based on S_{21} alone for dielectric materials or all four S-parameters for magneto-dielectric materials. For the S_{21} inversion, the R-matrix for the stack is used to calculate a model S_{21}, which is compared to the measured S_{21}. For the four-parameter inversion, two simultaneous equations of S-parameters are constructed from the R-matrix for the stack, similar to (2.46) and (2.47),

$$X = S_{11}^{\text{cal}} S_{22}^{\text{cal}} e^{2ik_0(t_m)} - S_{21}^{\text{cal}} S_{12}^{\text{cal}} \tag{2.55}$$

$$Y = \frac{S_{21}^{\text{cal}} + S_{12}^{\text{cal}}}{2} \tag{2.56}$$

where t_m is the thickness of the metal calibration plate. The model calculated X and Y are then compared to the measured X and Y in an iterative loop until the best fit solution is found.

2.5.2 Two-Thickness Inversion

Another situation where the inversions in Section 2.4 may not be applicable is when a magneto-dielectric material is deposited onto a conductive substrate and both μ and ε are unknown. In this case, only the amplitude and phase of the reflection S-parameter can be experimentally measured. Since the deposited material has both unknown permittivity and permeability, then there are not enough known measurement variables to uniquely solve for the four unknown parameters—real permittivity, imaginary permittivity, real permeability, and imaginary permeability.

One method for this conductor-backed scenario is a two-thickness method, where two different specimens are measured with different thicknesses to provide the necessary four independent measured quantities. The two-thickness inversion method starts with (2.37), which relates S_{11} to the permittivity and permeability of a metal-baked slab. After some rearrangement, and noting that $\sqrt{\mu\varepsilon}$ is the refractive index, n, and that $\sqrt{\mu/\varepsilon}$ is the bulk impedance, Z_m, we obtain the apparent normalized impedance of the conductor-backed slab,

$$\frac{1-S_{11}}{1+S_{11}} = Z = Z_m \tanh\left(itk_0 n\right) \tag{2.57}$$

Since the material's impedance does not vary with thickness, we rearrange (2.57) to solve for Z_m for each specimen measurement, and then set these equations equal for the two slabs such that

$$\frac{Z_1}{\tanh\left(it_1 k_0 n\right)} = \frac{Z_2}{\tanh\left(it_2 k_0 n\right)} \tag{2.58}$$

where the subscripts distinguish between the two specimens. Equation (2.58) is solved for refractive index with a numerical root finder such as Newton's method [28], allowing for an explicit solution to Z_m using (2.57). Permittivity and permeability are then calculated directly from the refractive index and the bulk impedance.

An example of the two-thickness inversion in practice is shown in Figure 2.8, which plots both permittivity (a) and permeability (b) for a magnetic absorber made from carbonyl iron powder mixed in a polyurethane matrix. The solid lines are the real and the dashed lines are the imaginary permittivity and permeability. Two sets of curves are shown; the thick lines are from the two-thickness inversion, and the thin lines are a conventional air-backed inversion of this material done with the four-parameter method discussed previously. In practice, the two-thickness inversion is not as robust as the four-parameter inversion. This is partly due to the adjacent conductive backplane, which results in a weak tangential electric field within the material under test. This effect makes the determination of dielectric permittivity less accurate, particularly at lower frequencies where the material is electrically thin. Another potential source of error specific to this inversion stems from the assumption that the intrinsic properties of the two different specimens are identical. Magneto-dielectric materials are often composites, and there can be variation in the properties between specimens, which makes the inversion less accurate as well. Thus, this two-layer inversion should only be used when the materials under test cannot be separated from the conductive substrate to which they are applied.

2.5.3 Model-Based Inversion

While the two-thickness method is useful for determining intrinsic properties of conductor-backed magneto-dielectric materials, it has the disadvantage of requiring two different material samples. A more desirable situation is where the complex permittivity and permeability can be obtained from a single sample of metal-backed material. However, solving for four unknown parameters (real and imaginary permittivity and permeability) based on two measured values (reflection amplitude and phase) is not practical for frequency-by-frequency inversion. Instead, a model-based approach to inversions uses the broadband nature of free-space measurements to fit assumed model parameters to the observed S_{11} data. In other words, rather than trying to fit four intrinsic parameters at each frequency, the model-based algorithm fits six or so model parameters across all the frequencies in the measured bandwidth. To keep the method practical and minimize the number of fitted parameters, this discussion uses Debye models for both the permittivity and the permeability of the material,

$$\varepsilon = \varepsilon_U + \frac{\varepsilon_R - \varepsilon_U}{1 + i\omega\tau} \text{ and } \mu = \mu_U + \frac{\mu_R - \mu_U}{1 + i\omega\tau_m} \tag{2.59}$$

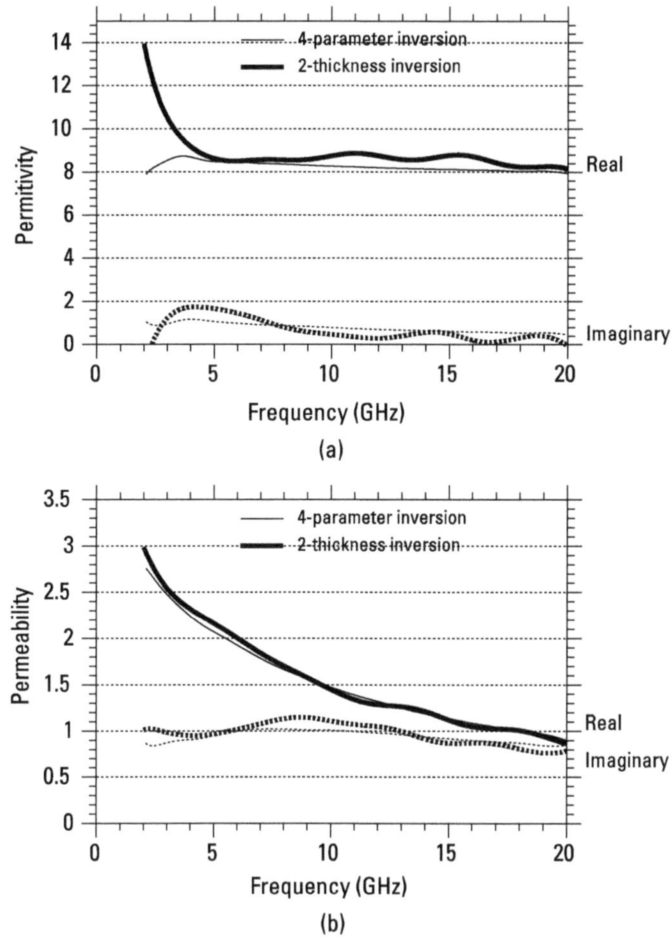

Figure 2.8 Inverted properties of a microwave absorber determined with (a) two-thickness and (b) four-parameter inversions.

where $\omega = 2\pi f$ is the angular frequency. This Debye formulation has three fitted model parameters for permittivity and another three for permeability leading to six fitting parameters (ε_U, ε_R, μ_U, μ_R, τ, τ_m).

An example of the Debye model is presented in Figure 2.9, which shows the real part as a solid line and the imaginary part as a dashed line. Also shown in Figure 2.9 are the meanings of the different fit parameters. Inversion is done by choosing an initial set of fit parameters, calculating trial permittivity and permeabilities and then calculating the resulting conductor-backed S_{11} using

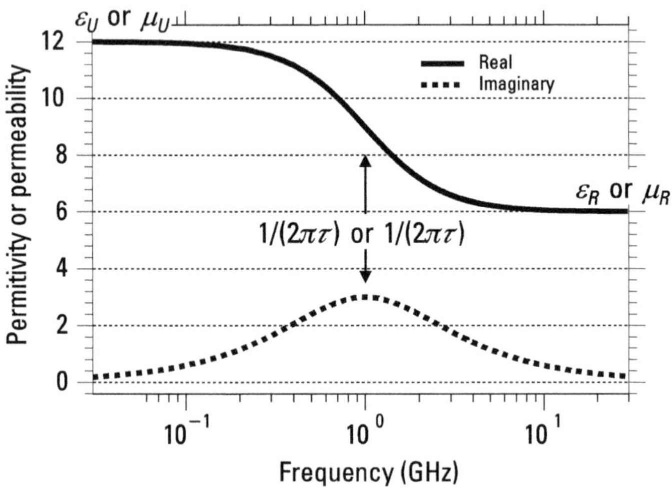

Figure 2.9 Debye model curves showing the characteristic step in the real part and peak behavior in the imaginary part.

(2.37). A least-squares fit over the measurement frequency band can be done to iterate until the best solution is found [28].

This method is approximate, and it is certainly possible to use more complex functions such as Lorentz. Purists will—correctly—point out that the simple Debye model is not a true representation of a real material. However, adding more parameters increases computation time and complexity, so that the inversion is less practical. Furthermore, there will be a certain level of measurement uncertainty as well. Adding further complexity to the fitted dispersion model may not be supported by the limits of the measurement accuracy. That said, it is not uncommon for some absorber materials to have a characteristic conductivity in the dielectric properties. In that case a conductivity term may be added to the permittivity model to account for that conductive material behavior. For magnetic permeability, an additional conductivity term is not required.

Like the two-thickness method, this technique can have limited sensitivity to dielectric permittivity by the presence of a conductive interface next to the specimen. Since the tangential E-field adjacent to a conductive boundary goes to zero, a material that is too thin will only have a weak interaction with the electric field. On the other hand, for a sample that is too thick, the backside reflection of the signal is weak compared to the front and accuracy to the magnetic permeability is reduced [33]. Fortunately, free-space measurements can be relatively wideband so that with an appropriate specimen thickness, the measurement frequencies span regions of different sensitivity to permittivity and permeability.

Figure 2.10 presents an example of using the Debye model to invert measured data, with the permittivity in (a) and the permeability in (b). The solid lines in these plots are real, and the dashed lines are imaginary. Each plot shows two different inversions. The thick lines are the Debye model inversions of a metal-backed specimen, and the thin lines are from the same specimen removed from the metal plate to measure both reflection and transmission for

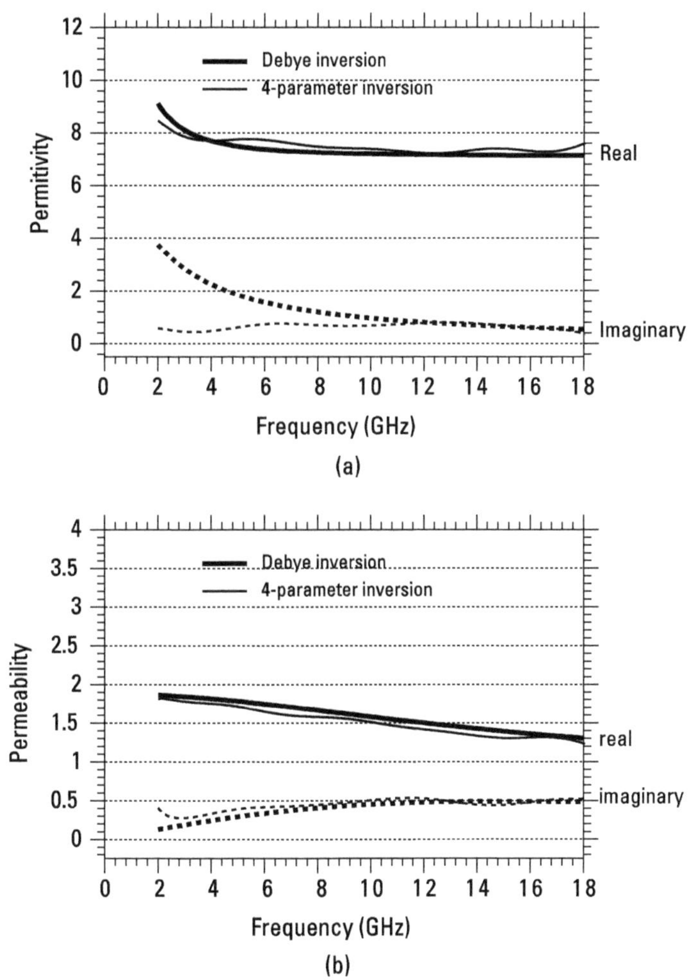

Figure 2.10 Debye and four-parameter inverted permittivity (a) and permeability (b) for a sheet of magnetic absorber.

a four-parameter inversion. The two inversions agree except for the imaginary permittivity at lower frequencies. However, this deviation of the model inversion is expected because of the weaker sensitivity of metal-backed specimens to dielectric properties at lower frequencies as described earlier.

2.6 Absorber Characterization

A common application for RF and microwave materials is absorption. Microwave absorbers are important for a variety of testing applications, for reducing interference between components and antennas, and for reducing the radar cross-section of military vehicles. Absorbers can be constructed from a single material, such as a ferrite operating at frequencies around its ferromagnetic resonance. Such materials have a substantial imaginary permeability that absorbs energy from an incident magnetic field. Other absorber materials consist of composites of magnetic powder mixed into a dielectric matrix. The magnetic properties shown in Figures 2.9 and 2.10 are examples of mixtures of iron powder in a rubber matrix. The role of free-space measurements in characterizing these materials includes the determination of dielectric permittivity and magnetic permeability as discussed previously.

Another role of free-space measurement is to determine absorption or reflection performance. The frequency-dependent reflectivity of an absorber depends on the dielectric and magnetic properties as well as the thickness. A key performance metric used to characterize an absorber is the reflection amplitude in decibels. Sometimes, the performance is described in terms of a "reflection loss," which is the absolute value of the reflection in decibels. Reflection can be measured at normal incidence to the plane of the absorber with a single antenna in a configuration known as a monostatic, which is shown in Figure 2.2(a). Figure 2.11 provides examples of the monostatic reflection performance of a layer of magnetic absorber backed by a metal sheet. The nulls in these curves can be adjusted by tuning either the thickness or the concentration of magnetic powder within the composite. For example, absorber 1 in Figure 2.11 was designed to minimize the reflection at or around 3 GHz. The null position can be raised in frequency by reducing thickness and/or reducing the volume fraction of the magnetic powder.

The frequency-dependent performance of an absorber also depends on angle. The incidence angle can be defined either with respect to the normal or with respect to grazing. Figure 2.12 shows the definition of incidence angle used in this book, which is with respect to the surface normal. At angles other than normal incidence, the polarization of the incident wave also becomes

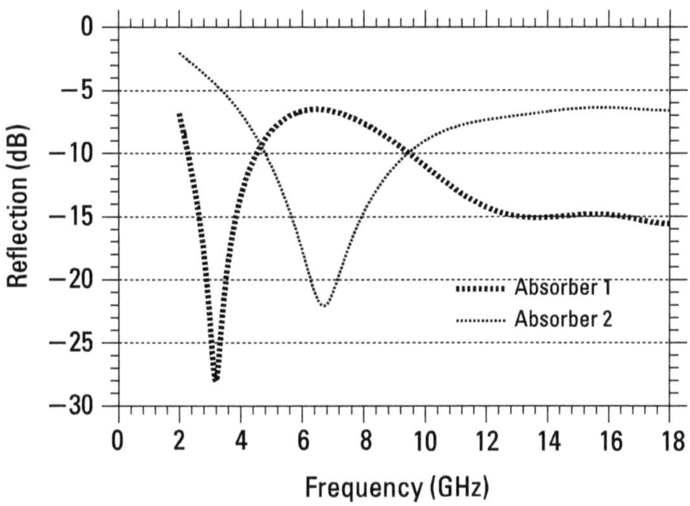

Figure 2.11 Monostatic reflection of two different magneto-dielectric absorbers.

important. A propagating wave traveling through free space has both electric, E, and magnetic, H, field components that are orthogonal to each other. When discussing a wave incident to a planar surface, that wave can be described in terms of two polarization components: transverse-electric (TE) and transverse-magnetic (TM) polarizations. These components are defined with respect to an incidence plane formed by the surface normal and the incident wave vector as shown in Figure 2.12. The TE polarization is the component that has the E-field transverse to this plane, while the TM polarization has the H-field transverse to this plane. For a flat surface, the reflected wave is at an angle that is equal and opposite to the incident angle, following Snell's Law [32]. This reflection is termed specular, and it behaves similar to light being reflected by a flat mirror.

Figure 2.13 shows an example of an absorber's specular reflection performance as a function of several incidence angles. This is a metal-backed layer of carbonyl iron mixed in a polyurethane matrix, and the two plots show what happens for TE and TM polarizations at angles of normal, 45-, 60-, and 75-degrees incidence. At normal incidence, TE and TM reflectivity is the same. But as the incidence angle increases toward grazing, the two polarizations have significantly different behaviors. For single-layer magneto-dielectric absorbers such as this, the TE performance generally degrades at higher incidence angles, and the TM performance shows higher frequency nulls, but an increased reflectivity in the lower frequency null. Some applications are interested in normal

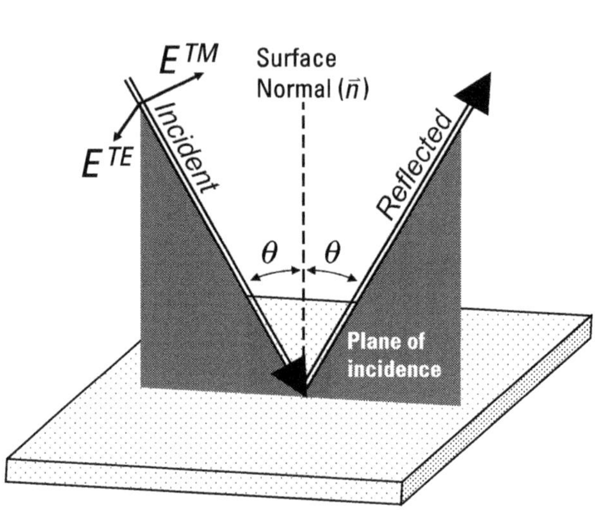

Figure 2.12 Definitions of angle and polarization used in bistatic measurements

incidence performance, while others position the absorber so that it is at an angle with respect to the expected incidence wave, and so are more focused on the oblique angle characteristics. For this reason, bistatic measurement fixtures, which use two different antennas mounted at nonnormal angles, are often used to characterize materials, as shown in Figure 2.2(b).

Typically, an absorber's desired performance is over a wide range of frequencies. To increase the bandwidth of an absorber, it can be constructed from multiple layers, with each layer having different dielectric and/or magnetic properties. For example, a two- or three-layer magneto-dielectric absorber can be constructed from layers with different loadings of magnetic powder. Alternatively, an absorber can be constructed from layers with different loadings of conductive additive, or even from thin sputtered or evaporated metal layers and/or carbon ink layers with nonconductive dielectric layers separating them. This latter absorber paradigm is known as a Jaumann absorber [31].

In many practical applications, absorbers need to be physically thin, as there is not much room to apply the absorber. These relatively thin absorbers depend on interference effects between the different layers to obtain their wideband performance. On the other hand, in testing applications, such as in an anechoic chamber, there is plenty of room to make the absorber electrically thick—many wavelengths thick. In an electrically thick design, the absorber gets its performance by gradually transforming the wave from free space into the absorbing material through a slow gradient. Gradient absorber design may use a slow change in conductive loading from the front to the back of the

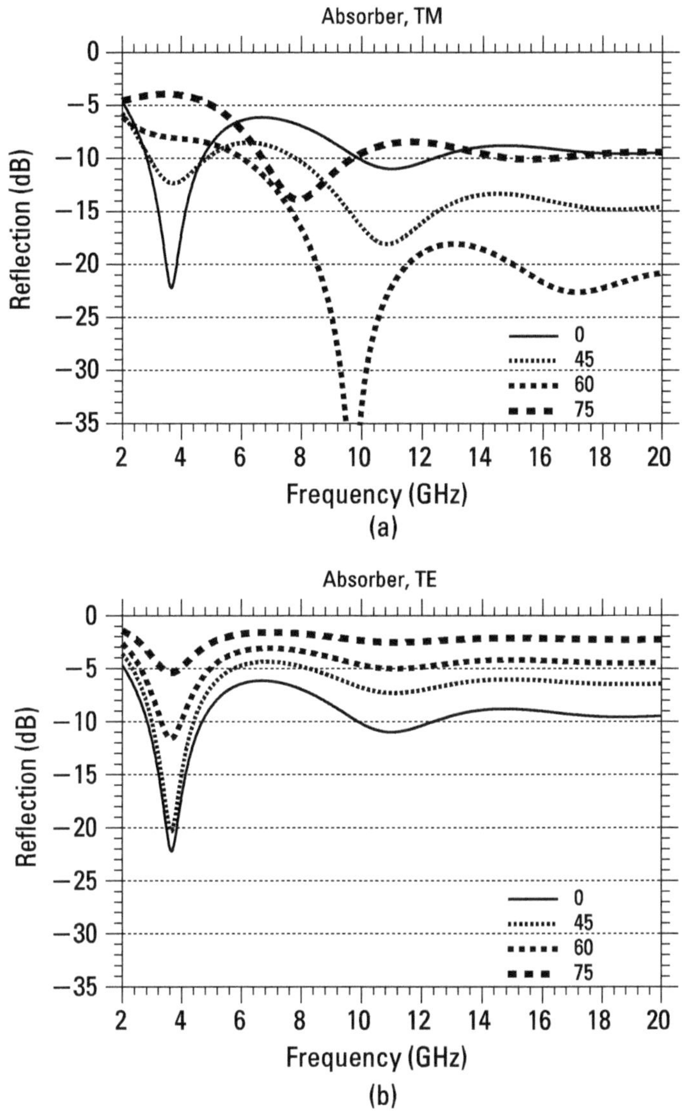

Figure 2.13 Specular performance of a magneto-dielectric absorber as a function of angle.

absorber, or it can use a geometric gradient, such as in the pyramidal absorber shown in Figure 2.14. An incident wave first encounters the tips of a pyramidal absorber, where there is mostly air and just a little bit of lossy foam. As the wave moves toward the base of the absorber, it slowly transitions into a region of less air and more foam. The foam is typically loaded with a carbon ink so that it has a substantial imaginary dielectric permittivity while keeping the real permittivity relatively low. Depending on the thickness of the pyramidal absorber, it can range from several wavelengths to tens of wavelengths thick at the operational frequencies, and the reflection levels can be as low as –60 or –70 dB relative to a conductive surface.

Free-space measurement techniques for pyramidal absorbers must have a good dynamic range to handle these very low reflectivities. A typical measurement setup for an anechoic absorber is shown in Figure 2.15. A monostatic or bistatic configuration can be used; however, a bistatic geometry is often preferred since it usually has better dynamic range. Antennas always have internal reflections contributing to a delayed ring-down within the antenna, so separating the receive antenna from the transmitter reduces the extraneous energy received by the analyzer. Also shown in Figure 2.15 is a moveable table that holds the absorber. As discussed in Section 2.2, response-and-isolation calibration is preferred for free-space measurements; enabling the specimen table to move allows it to be removed for the isolation measurement. The empty table is used as the response standard, and (2.3) is applied. Also critical for these measurements is the use of time-domain gating to eliminate room reflection, antenna ring-down, and the floor reflection.

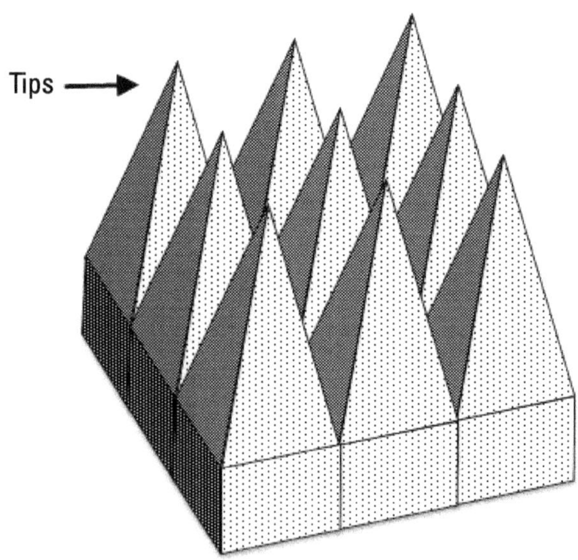

Figure 2.14 Schematic diagram of pyramidal absorber used in anechoic test chambers.

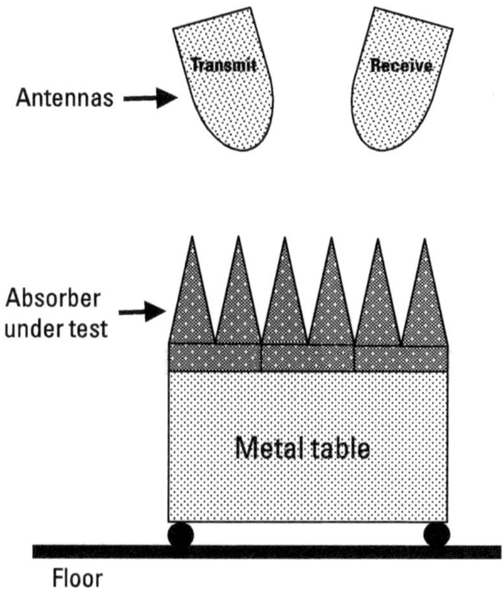

Figure 2.15 Measurement configuration for bistatic testing of an anechoic chamber absorber.

References

[1] Appleyard, R., "Pioneers of Electrical Communication—Heinrich Rudolf Hertz—V," *Electrical Communication*, Vol. 6, No. 2, 1927, pp. 63–77.

[2] Emerson, D. T., "Jagadis Chandra Bose: Millimetre Wave Research in the Nineteenth Century," *IEEE Trans. Microwave Theory and Techniques*, Vol. 45, No. 12, 1997, pp. 2267–2273.

[3] Redheffer, R. M., "The Measurement of Dielectric Constants," in *Technique of Microwave Measurements*, C.G. Montgomery (ed.), New York, NY: McGraw-Hill, 1947, pp. 561–678.

[4] Redheffer, R. M., "Microwave Antennas and Dielectric Surfaces," *J. Appl. Phys.*, Vol. 20, April 1949, pp. 397–411.

[5] Talpey, T. E., "Optical Methods for the Measurement of Complex Dielectric and Magnetic Constants at Centimeter and Millimeter Wavelengths," *L'Onde Electrique*, October 1953.

[6] Hiatt, R. E., E. F. Knott, and T. B. Senior, "A Study of VHF Absorbers and Anechoic Rooms," *University of Michigan Report 5391-1-F for NASA contract NASr-54(L-1)*, Langley Research Center, Hampton, VA, February 1963.

[7] NRL Public Affairs, "75 Years Naval Research Laboratory," press release, 1998.

[8] Stallings, D. C., "Resistive Sheet Measurements," *Report for Project A-2583*, Georgia Institute of Technology, 1980.

[9] Boyd, F. E., "Converging Lens Dielectric Antennas," *NRL Report 3780*, Naval Research Laboratory, Washington DC, DTIC ADB801103, 1950.

[10] Primich, R. I., "Microwave Techniques for Hypersonic Ballistic Ranges," *Planetary and Space Science*, Vol. 6, 1961, pp. 186–195.

[11] Primich, R. I., and F. H. Northover, "Use of Focused Antenna for Ionized Trail Measurements: Part 1. Power Transfer Between Two Focused Antennas," *IEEE Trans. Antennas and Propagation*, March 1963, pp. 112–118.

[12] Braezeale, W. M., "Method and Apparatus for Measuring Moisture Content," U.S. Patent 2,659,860, filed Aug 27, 1949, awarded Nov. 17, 1953.

[13] Culshaw, W., "A Spectrometer for Millimetre Wavelengths," *Proc. IEE—Part IIA: Insulating Materials*, Vol. 100, No. 3, 1953.

[14] Goubau, G., "Beam-Waveguide Antenna," U.S. Patent 2,994,873, 1961.

[15] Datlov, J., J. Musil, and F. Zacek, "Beam Width of Two Antenna Systems for Plasma Diagnostics," *Czechoslovak Journal of Physics*, Vol. 15, No. 10, 1965, pp. 766–768.

[16] Bassett, H. L, "A Free-Space Focused Microwave System to Determine the Complex Permittivity of Materials o Temperatures Exceeding 2000 C," *Ref. Sci. Instr.*, Vol. 42, No. 2, 1971, pp. 200–204

[17] Pentecost, J. L., "Electrical Evaluation of Radome Materials," In *Radome Engineering Handbook, Design and Principles*, J.D. Walton (ed.), New York, NY: Marcel Dekker, 1970.

[18] Walton, J. D., S. H. Bomar, and H. L. Bassett, "Evaluation of Materials in a High Heat Flux Radiant Thermal Energy Environment," *AMMRC CTR 73-16, Final Report for Contract DAAG46-72-C-0189*, 1973.

[19] Musil, J., et al., "New Microwave System to Determine the Complex Permittivity of Small Dielectric and Semiconducting Samples," *Fourth European Microwave Conference*, 1974, pp. 66–70.

[20] Zhang, Z., "Design of the Broadband Admittance Tunnel for High Fidelity Material Characterization," Arizona State University: PhD dissertation, 2005.

[21] Ghodgaonkar, D. K., V. V. Varadan, and V. K. Varadan, "Free-Space Method for Measurement of Dielectric Constants and Loss Tangents at Microwave Frequencies," *IEEE Trans. Instrumentation & Measurement*, Vol. 37, No. 3, 1989, pp. 789–793.

[22] Harris, Fredric J., "On the Use of Windows for Harmonic Analysis with the Discrete Fourier Transform," *Proc. IEEE*, Vol. 66, No. 1 1978, pp. 51–83.

[23] Born, M., and W. E. Wolf, *Principles of Optics* (Sixth Edition), Cambridge, United Kingdom: Cambridge University Press, 1980.

[24] Nicolson, A. M., and G. Ross, "Measurement of Intrinsic Properties of Materials by Time Domain Techniques," *IEEE Trans. Instrumentation & Measurement*, Vol. 19, 1970, pp. 377–82.

[25] Weir, W. B., "Automatic Measurement of Complex Dielectric Constant and Permeability at Microwave Frequencies," *Proc. IEEE*, Vol. 62, 1974, pp. 33–36.

[26] Baker-Jarvis, J., "Transmission/Reflection and Short-Circuit Line Permittivity Measurements," *NIST Technical Note 1341*, 1990.

[27] Boughriet, A.-H., C. Legrand, and A. Chapoton, "Noniterative Stable Transmission/Reflection Method for Low-Loss Material Complex Permittivity Determination," *IEEE Trans. Microwave Theory and Techniques*, Vol. 45, No. 1, 1997, pp. 52–57.

[28] Press, W. H., et al., *Numerical Recipes* (Third Edition): *The Art of Scientific Computing*, Cambridge, United Kingdom: Cambridge University Press, 2007.

[29] Baker-Jarvis, J., "Transmission/Reflection and Short-Circuit Line Methods for Measuring Permittivity and Permeability," *NIST Technical Note 1355*, 1992.

[30] Roberts, S., and A. Von Hippel, "A New Method for Measuring Dielectric Constant and Loss in the Range of Centimeter Waves," *J. Appl. Phys.*, Vol. 17, 1946, pp. 610–616.

[31] Knott, E. F., J. F. Shaeffer, and M. T. Tuley, *Radar Cross Section* (Second Edition), Norwood, MA: Artech House, 1993.

[32] Ramo, S., J. R. Whinnery, and T. Van Duzer, *Fields and Waves in Communication Electronics* (Third Edition), Hoboken, NJ: John Wiley & Sons, 1994.

[33] Geryak, R. D., and J. W. Schultz, "Extraction of Magneto-Dielectric Properties from Metal-Backed Free-Space Reflectivity," *Antenna Measurement Techniques Association (AMTA) Symposium Proceedings*, San Diego, CA, Oct. 6–11, 2019.

3

Microwave Nondestructive Evaluation

3.1 Sensors/Antennas

In manufacturing and other industrial environments, electromagnetic materials or structural components may need inspection to ensure that they are within desired specifications. This is known as quality assurance (QA). However, in many of these applications the components under test may be too large for mounting in traditional laboratory fixtures. Cutting specimens from a larger component is then the only option for laboratory measurement, whether it is free-space or another wideband system such as a coaxial airline or waveguide. This is costly since it requires the manufacture of extra sacrificial components that are destroyed during testing. QA testing with these devices also requires additional time and resources to prepare the specimens for measurement. Sometimes a manufacturing process uses so-called witness coupons that are smaller and less expensive than a full component and that fit in a laboratory fixture. However, measurement of witness coupons is not the same as measuring the deliverable material or component. A more direct method to verify components is with a measurement sensor small enough to be brought to the part under test rather than the other way around. The idea is to conduct NDE or nondestructive inspection (NDI) of a component in situ, which can even happen in a production line as a component is being manufactured. This section describes some of the RF sensors that may be used for this purpose.

A relatively simple NDE sensor is an open-ended waveguide that is placed adjacent to the surface under test [1]. Waveguides are hollow metal tubes that act as transmission lines to propagate RF energy [2]. An open-ended waveguide can be held against a surface and emit RF energy to probe the response of that surface. Under ideal conditions, an analytical formulation can be derived to relate the probe response to the dielectric properties and thickness of a simple material under test. However, this requires a flat, single-layer specimen so that the probe can be fully flush against the surface. Many components are curved, which makes the use of a contact probe problematic and a noncontact sensor with some finite standoff distance more practical.

An open-ended waveguide can also function as a noncontact probe as shown schematically in Figure 3.1. One end of the waveguide probe includes a coax-to-waveguide transition to attach an RF cable for connection to a microwave source and receiver. For noncontact operation, the probe is maintained at some known standoff distance from the surface under test. This standoff allows for the measurement of surfaces that aren't perfectly flat. Noncontact measurements such as these are also useful when touching the surface would cause damage, such as right after sprayed-on coatings are applied. For either circular or rectangular waveguides, there will be some limited-frequency

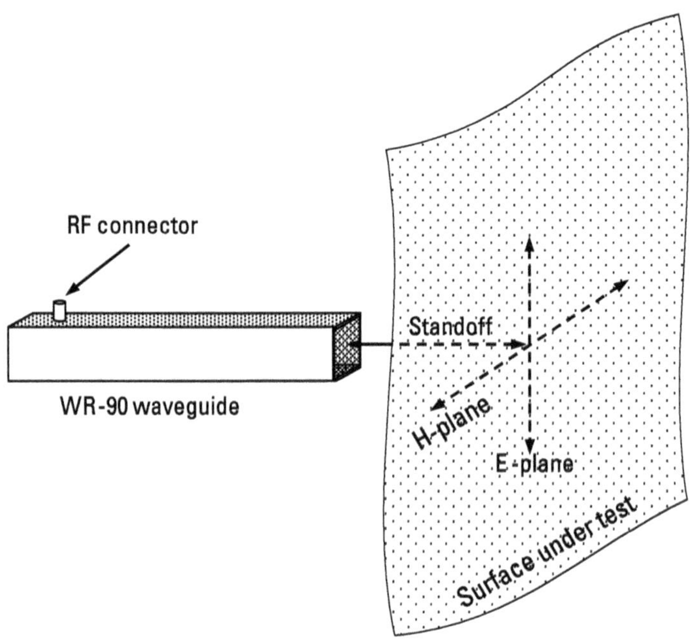

Figure 3.1 Schematic drawing of an open-ended X-band waveguide probe with an approximately 5-cm standoff from a surface under test.

bandwidth over which they operate, which depends on the internal dimensions of the waveguides. For example, a common waveguide band centered around 10 GHz is known as X-band, which is used as the example in Figure 3.1, and WR-90 is a standard that specifies the internal dimensions of X-band waveguides to be 0.9 by 0.4 inches. The fundamental propagation mode of RF energy in a rectangular waveguide has the E-field oriented parallel to the short dimension of the waveguide. The energy that radiates forward from an open-ended waveguide has an E-field polarization in that same direction, and the corresponding magnetic or H-field is orthogonal. Because of that, two principal planes are defined, parallel to either the E-field or H-field of the radiated energy.

The idea of a noncontact probe is to emit a beam of RF energy that illuminates a small area of the surface under test. The illuminated spot then reflects some of the incident energy back to the probe. A useful way to parameterize the performance of such a spot-probe is to determine the amplitude and phase taper of the illuminated area, and the tool usually employed for this type of calculation is a computational electromagnetic (CEM) code. CEM software divides up a defined volume such as the region around a probe and surface into facets or cells and applies Maxwell's equations to each of those divisions [3]. In this way, the waves generated by the probe and interacting surface are rigorously simulated with the only significant approximation being the coarseness with which the simulation space is divided up. The example probe calculations shown in this section used a CEM code based on the finite-difference time-domain (FDTD) method [4].

Figure 3.2 shows the calculated amplitude and phase on a surface under test that is spaced 5 cm from the end of the open-ended waveguide probe in Figure 3.1. The data in Figure 3.2 is along the E-plane direction, but the illumination spot has similar characteristics in the H-plane direction. The solid line is the illumination amplitude in decibels, and thin vertical dotted lines are drawn where the amplitude is down 3 dB from the peak. The larger dashed lines are the phase of the illuminated energy. For an ideal plane wave, the phase would be approximately constant. However, the energy emitted by the open-ended waveguide is divergent, and the phase varies rapidly from the center of the spot to the edge. The effect of this fast-phase taper is that the illumination spot size rapidly increases as the standoff distance increases. With a divergent beam, the reflected energy continues to spread out so that the probe only receives a portion of the energy reflected by the illuminated spot. Small errors in the standoff distance will then change the amount of energy received by the probe, increasing the error in the measured reflection coefficient. This effect is sometimes called space loss.

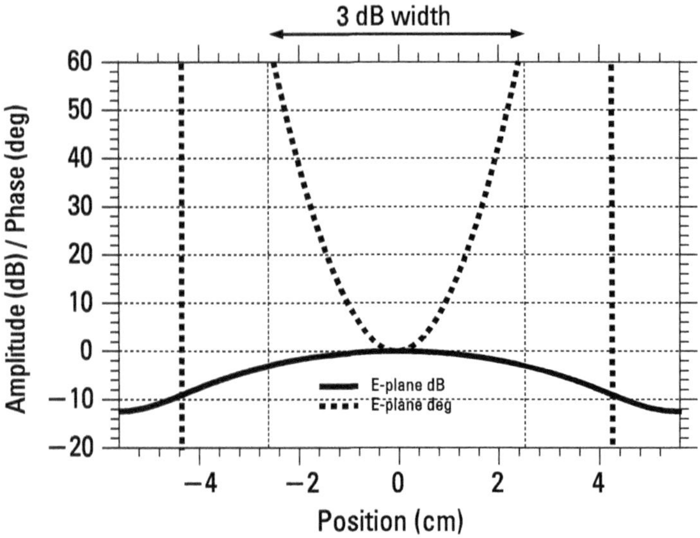

Figure 3.2 Amplitude and phase of the radiated energy parallel to the E-plane with an approximately 5-cm standoff distance from the end of the simple waveguide probe.

The rapidly diverging beam from an open-ended waveguide makes it a less-than-ideal spot probe. Performance can be significantly improved by inserting a dielectric rod into the waveguide as shown in Figure 3.3. Dielectric rod antennas for far-field radiation and antenna arrays were explored as far back as the 1940s [5]. The dielectric rod acts as a transition from the waveguide to free space, and it guides the RF energy so that the sensor is more directive. The dielectric rod of the example design in Figure 3.3 is assumed to have the same permittivity as polystyrene, a low-loss plastic commonly used in microwave applications with a real permittivity of $\varepsilon \sim 2.54$. The design is machined from a single piece of plastic and includes a tapered section that is inserted into the waveguide (outlined with a dotted line). The probe also includes a transition from the end of the waveguide to a constant cross-section rod and a short taper at the very end designed to minimize internal reflections within the probe.

Assuming the same 5-cm standoff distance between the end of the dielectric rod probe and the surface under test, the illumination spot has the amplitude and phase shown in Figure 3.4. While the 3-dB width of the illumination area is very similar to the open-ended waveguide of Figure 3.2, the phase taper is significantly less—about 35 degrees at the 3-dB point versus the over 60-degree phase taper for the open-ended waveguide without the dielectric rod. This less-severe phase taper within the 3-dB width is consistent with improved localization of the probe illumination. The more directive beam is

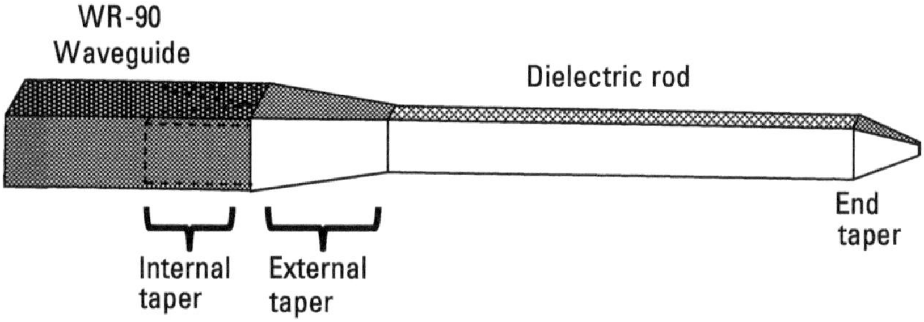

Figure 3.3 A dielectric rod probe constructed by inserting an optimized dielectric rod into an open-ended waveguide.

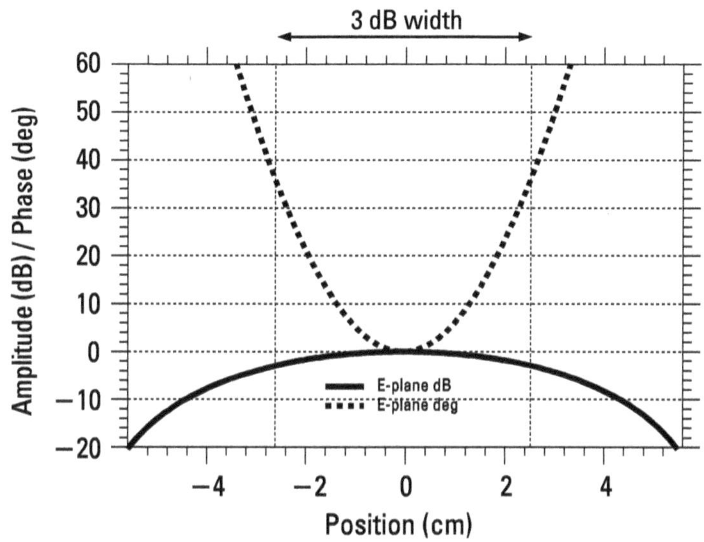

Figure 3.4 Amplitude and phase of the radiated energy parallel to the E-plane with an approximately 5-cm standoff distance from the end of the dielectric rod probe.

also evident by comparing the amplitude tapers for the two probes at positions farther away from the center of the illumination. Antenna gain, a measure of the power radiated by an antenna in a specified direction [6], is another way to characterize the performance of an NDE sensor. The increased directivity of the dielectric rod antenna is easily seen in the gain patterns for these probes, shown in Figure 3.5. The solid line is the pattern in the E-plane for the dielectric rod probe, while the dotted line shows the E-plane pattern for the open-ended waveguide. Confining more energy to the desired illumination spot improves signal to noise while also reducing interference from probe

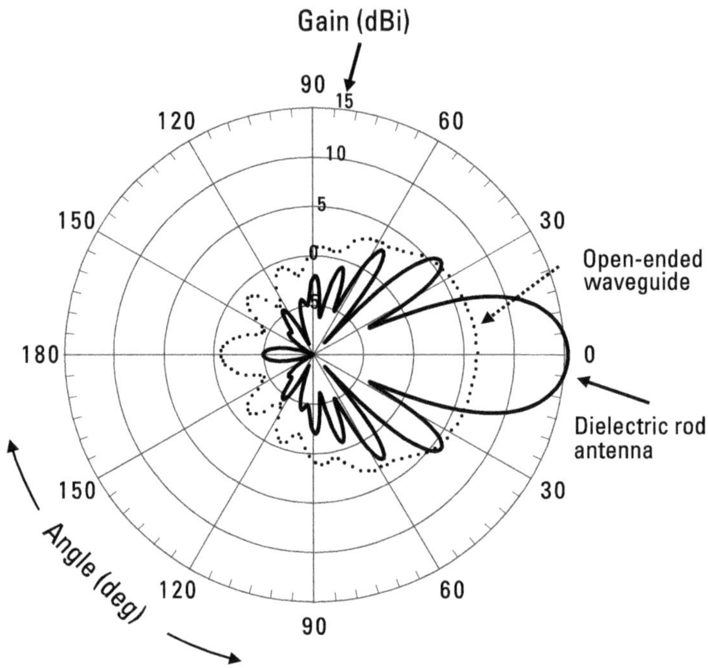

Figure 3.5 Polar plot of far-field gain for dielectric rod probe (solid line) and open-ended waveguide (dotted line).

energy interacting with other nearby structures. For this reason, the dielectric rod probe has found frequent use as a microwave NDE sensor.

One of the primary drawbacks for the dielectric rod antenna probe is its limited bandwidth. Recall the example WR-90 probe, which operates from 8 GHz to about 12 or 13 GHz. Below that frequency range the waveguide's inner dimensions are too small to allow propagation of the RF energy [7]. This is known as the frequency cutoff for the fundamental propagation mode within the waveguide. Above 12 or 13 GHz, higher-order modes can be excited in the X-band probe, which complicate its radiation characteristics. Measurements over a broader range of frequencies can provide additional information about a surface under test, so increased bandwidth probes are desired for enhancing the utility of microwave NDE. Increasing probe bandwidth is realized by moving beyond a simple waveguide-based design. This can be accomplished by inserting dielectric rods into broadband ridged-horn antennas instead [8]. The antenna acts as a transition from a standard coaxial RF cable to the dielectric waveguide that forms the end of the probe. Ridged-horn antennas have bandwidths on the order of 10:1, much greater than simple waveguide feeds.

Perhaps the most effective way to achieve a wideband microwave NDE probe, however, is by designing the transition and dielectric components together [9, 10]. Combining transition and dielectric components into a unified optimization further improves microwave performance and bandwidth. Other considerations for microwave NDE sensors are also important. Specifically, microwave NDE is usually done in a manufacturing or industrial environment, so a probe that is rugged enough to withstand harsh environments or rough handling is important. Electrostatic discharge is also a concern with conventional antennas, so having a probe that is encapsulated with insulating dielectric minimizes the chance of damaging the sensitive microwave receiver components in the measurement system.

3.2 Dealing with RF Cables

RF cables or transmission lines are almost always part of a microwave or millimeter wave measurement system. These RF cables connect the microwave source/receiver to the test fixture and are, therefore, subject to environmental variations such as temperature or pressure. Environmental variations cause changes in the overall phase and amplitude of signals that travel through the cables. In some cases, testing also requires physical motion of the cable(s), which creates another source of phase and amplitude error. In laboratory settings, great care is taken to design a test apparatus or methodology to minimize the movement of the test cables so that these position-induced phase errors are also minimized. However, microwave NDE applications in nonlaboratory settings often use scanning sensors. Mounting sensors on translation stages or industrial robots enables reach over a surface of interest, but payload limitations may require that the microwave analyzer is mounted elsewhere. So, an RF cable connecting a sensor to the microwave analyzer will necessarily flex, inducing measurement errors.

In other cases, measurements may be required over periods of time where ambient temperatures can change enough to cause errors. This can happen in settings where ovens are used in proximity to the measurement system. It can also be an issue when measurements need to be done over many hours of time during a manufacturing process. Even in a laboratory setting where heated or cooled measurements are desired, measurements necessarily take hours to conduct, and the oven or furnace is a source of significant temperature drift within even a temperature-controlled laboratory environment.

A number of methods have been introduced for compensation of cable-induced errors. For example, Hahn and Halama [11] accounted for phase variation in a long RF cable by terminating it with additional microwave circuitry to provide a controlled reflection. They then measured phase variations with an additional microwave circuit at the source and employed motorized cable stretching to compensate. In another example, Roos [12] used local mixing of intermediate frequency (IF) signals at the test fixture to reduce phase variations from the RF cables. They then combined that with a separate phase-stable reflection reference at the test fixture to monitor and compensate for phase errors induced by the RF cables. While these methods provide phase corrections, they also increase measurement-system complexity by adding microwave circuitry. Alternatively, a simpler method can be used, one that does not require specialized circuitry at the measurement fixture: In-situ reflections that already exist in the measurement fixture can be monitored to obtain a reference signal that directly quantifies the cable-induced measurement errors [13]. This simpler method leverages the collection of wide bandwidth signals so that time-domain gating can be employed to isolate the signals of interest.

Figure 3.6 shows the signals measured from a wideband microwave probe [14]. In this example, a microwave network analyzer excites the probe at a series of frequencies stepped from 2 to 20 GHz. The analyzer is connected to the probe by an approximately 4.5-m RF cable, and the probe illuminates a material specimen at some distance in front of it. The specimen reflection is received by the probe and transmitted to the microwave analyzer through

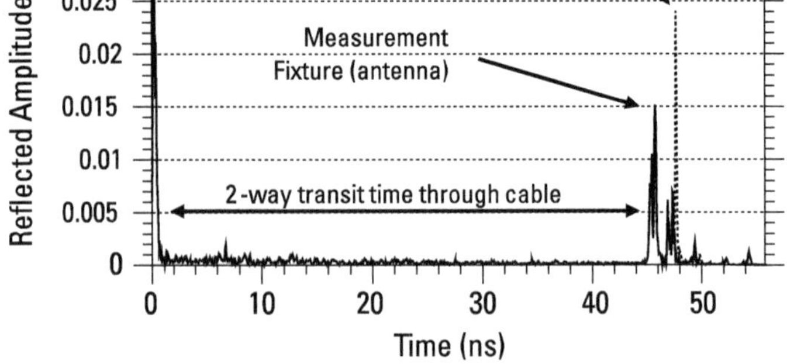

Figure 3.6 Measured time-domain signal from a wideband probe system showing just the probe (solid curve) and response with a metal plate in front of the probe (dashed curve).

the RF cable. The data in Figure 3.6 shows what occurs after the reflected signal has been mathematically transformed from frequency domain to time domain with a Fourier transform, and several reflections are evident as peaks in these signals. As the data indicates, the round-trip time for the signal to travel from the microwave analyzer to the probe is approximately 45 ns. A second major peak is also evident approximately 2 ns after the initial probe reflection, and this second peak represents the primary reflection from a test or calibration specimen, depending on what is being measured. In other words, the key to separating the RF cable phase and amplitude changes is the ability to separate the fixture or probe reflection from the specimen under test. The probe reflection should not change, and if it does, then we use that to quantify the cable effects.

As discussed in Chapter 2, the calibration procedure for measurement data includes vector subtraction of foreground and background signals from the signal of interest. Foreground signals are unwanted reflections that occur before the signal of interest, and background signals occur after the signal of interest. If these unwanted signals are not properly subtracted, they then impact the desired signal in the frequency domain. In this example the reflection of the probe antenna is an unwanted foreground signal. As Figure 3.6 indicates, some of the reflections from the probe may be immediately adjacent to or even overlap the specimen under test. This vector subtraction is done for both the calibration standards and the specimen under test. However, if the ambient temperature changes, then thermal expansion can cause the length of a cable to change between calibration and specimen measurements, which imposes an erroneous phase shift that is different for the specimen measurement than it is for the calibration measurement(s). Similarly, if the cable is physically moved or disturbed during the measurement, that cable displacement can also impose a phase or amplitude change that degrades the quality of the background/foreground subtraction.

Figure 3.7 shows a flow chart illustrating one version of a time domain correction method for addressing the cable phase errors. The first two steps of this flow chart are partially illustrated in Figure 3.6, which shows the received signals from a reflective metal calibration plate as well as no specimen (free space). Comparing the time-domain signals with and without a calibration or unknown specimen under test makes it possible to discern the reflections caused by just the measurement fixture. For calibration, the measurement fixture reflections are subtracted from the known reference measurement as well as from subsequent measurements of specimens. However, when environmental or physical changes in the cable cause the measurement fixture reflections to occur at a slightly different time, the unwanted parts of the signal are not

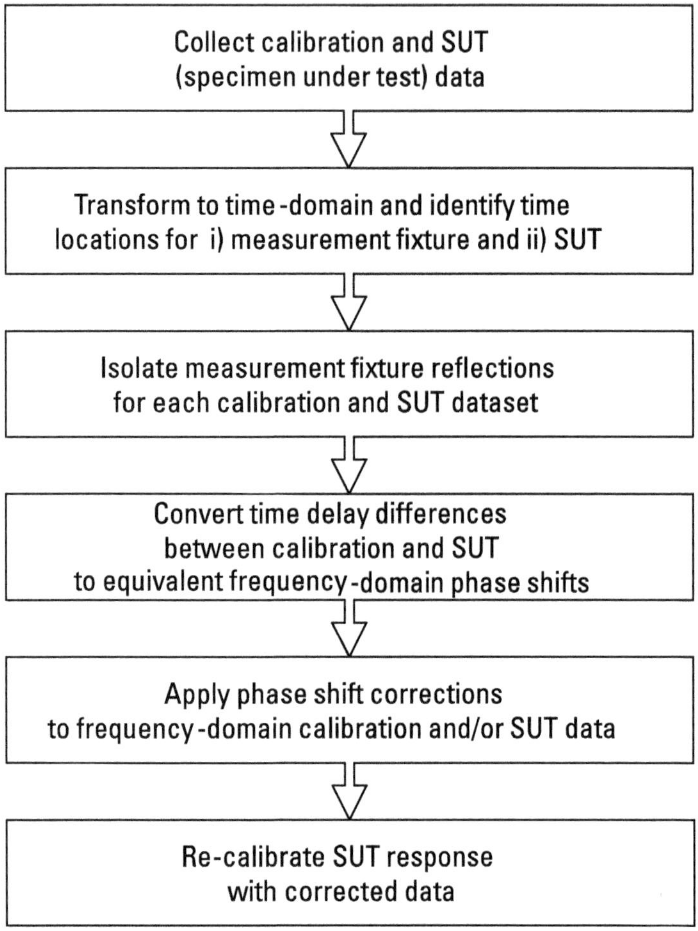

Figure 3.7 Process flow for RF cable phase correction due to thermal drift or flexing.

properly subtracted. In the frequency domain, this time shift is equivalent to a frequency-dependent phase error that degrades the vector subtraction of the foreground or background (i.e., the probe reflections in this example). Thus, the second and third steps of the method in Figure 3.7 use these time-domain signals to determine the exact location of the measurement fixture reflections in time so that they can be monitored during subsequent data collections.

The fourth step in Figure 3.7 calculates the time delay/phase error imposed by environmental effects or cable motion, and Figure 3.8 shows two measurements of a reflected signal from a probe connected to an RF cable. One curve is from before the cable was moved, and the other shows the cable

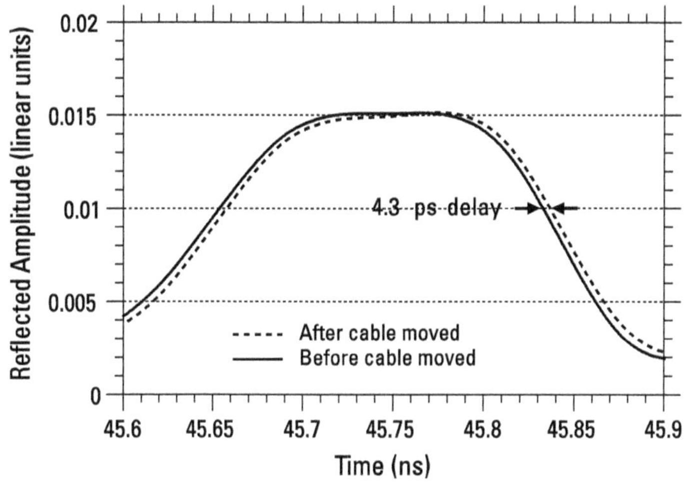

Figure 3.8 The measured time-domain effect of cable flex on the reflection from a sensor probe connected to the cable.

after the move, demonstrating that the cable motion imposed a 4.3-ps time delay. At 2 GHz, this time delay corresponds to an approximately 3-degree phase error, while at 18 GHz, the same time delay is equivalent to a 28-degree phase error. An automated algorithm can be used to determine the delay by subtracting the two signals while iteratively shifting one of them in time relative to the other. The minimum subtracted value occurs when the shifted signal best overlaps the other signal in time, and this corresponds to the delay. The fifth step of the phase-correction method uses the quantified phase error or time delay to apply a phase correction. The phase correction can be applied by multiplying the signal to be corrected, S, with the exponential function of radial frequency, ω, times the time delay, t, multiplied by the square root of negative one, $S^{\text{corrected}} = S^{\text{uncorrected}} e^{-i\omega t}$.

Another variation of the time-domain correction procedure is illustrated in the flow graph of Figure 3.9. This example differs from Figure 3.7 by determining the phase offsets *after* the signal from the measurement fixture is transformed back into frequency domain instead of in time domain. In frequency domain the two measurement fixture signals are ratioed to determine the phase shift. That phase shift, $\theta(f)$, can be fitted to a line to determine a frequency-domain correction. Cable motion may additionally result in a small amplitude error, and the mean of the amplitude ratio between the two signals, α, can also be calculated. The signal of interest is corrected by multiplying it in frequency domain by the amplitude and phase correction factors combined, $S^{\text{corrected}} = S^{\text{uncorrected}} \alpha e^{i\theta}$.

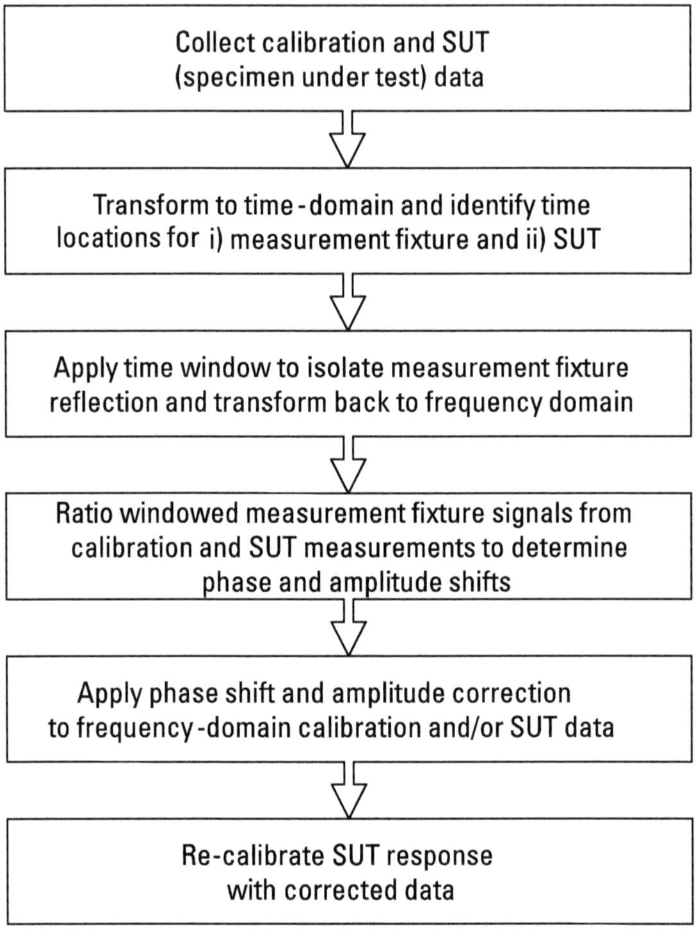

Figure 3.9 An alternate correction process that includes both phase and amplitude correction.

Figure 3.10 shows the time-domain reflection after vector subtractions of the foreground and background from a measurement of a material specimen. The dashed line shows the vector subtraction done without correcting for the extra time delay imposed by the cable motion. The solid curve shows the same subtraction done after the material specimen measurement was corrected for the 4.3-ps phase delay shown in Figure 3.8. Without the proper phase correction, the measurement fixture reflections that are immediately adjacent or that overlap the specimen reflections may not be fully subtracted, thereby increasing the measurement error.

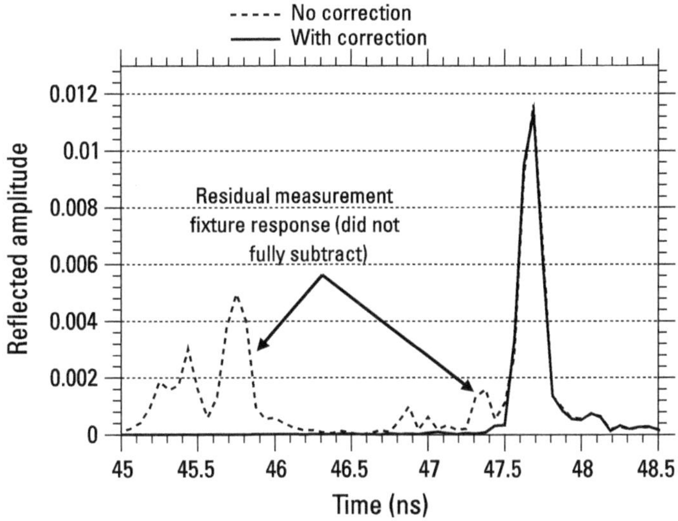

Figure 3.10 Vector subtraction of empty clear site from specimen measurement, with and without correction applied, and plotted in the time domain.

The final step of these cable-correction methods calibrate the measured data after including the appropriate phase and amplitude corrections. Figure 3.11 shows an example of a fully calibrated reflection measurement of an absorber material backed by a conductive sheet. Three different curves are shown in Figure 3.11, all of which were calibrated using a response-and-isolation methodology described in Chapter 2. The response measurement was of an ideal reflector—a flat metal plate—while the isolation measurement was of no specimen—free space. When the time-domain cable-correction method is used, the phase and amplitude correction is applied at each vector subtraction step. In other words, the isolation measurement is vector-subtracted from the corrected specimen-under-test data, and the isolation measurement is also vector-subtracted from the corrected response data. The final calibrated reflectivity, $S^{\text{calibrated}}$, of the specimen is then the ratio of the subtracted specimen data to the subtracted response data.

$$S^{\text{calibrated}} = \frac{S^{\text{corrected}}_{\text{specimen}} - S_{\text{isolation}}}{S^{\text{corrected}}_{\text{response}} - S_{\text{isolation}}} \tag{3.1}$$

The thin solid line of Figure 3.11 shows the initial calibrated result, where care was taken to avoid moving the RF cable between the calibration and specimen measurements. In this example, a network analyzer was

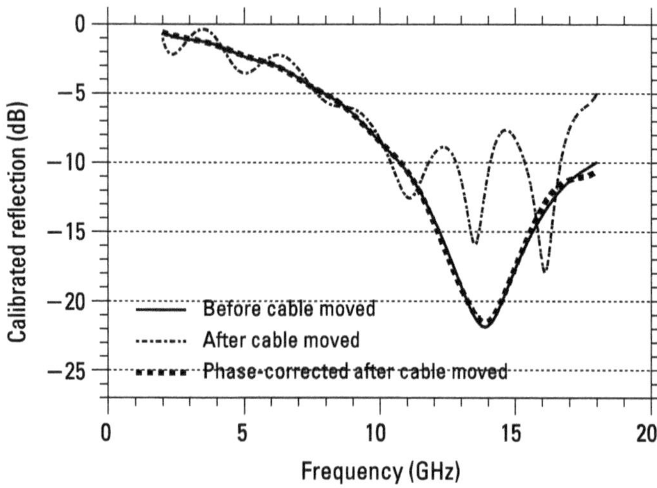

Figure 3.11 Effect of correction method on measured reflection coefficient of a metal-backed absorber.

used along with a microwave spot probe that was held in close proximity to the specimen under test. The spot probe illuminated a small area of the specimen, providing a localized measurement of the material reflectivity. The thinner dash-dot line shows the same specimen measured after the RF cable was moved, but without any phase correction. The thicker dashed line shows the specimen measurement after the cable was moved but when a phase and amplitude correction is made by applying the above-described method. As this data shows, cable movement can significantly degrade the accuracy of an RF measurement, but the abovementioned correction method can almost fully account for these errors.

While the use of algorithmic cable corrections is effective, there is still the matter of wear. Robotic or other repeated motion during measurements can cause fatigue in RF cables after so many cycles. Under these conditions, RF cables tend to degrade incrementally rather than catastrophically so that the degradation is not immediately obvious. In a manufacturing situation, this slow decline of performance is problematic since it requires additional measurements or checks to verify the quality-assurance measurements. A better solution is to eliminate or dramatically reduce the use of RF cables in the first place. Transmission measurements that use both amplitude and phase require two microwave sensors to be connected to a single microwave analyzer so that they can be phase-locked together, making elimination of RF cables impractical. However, reflection-only measurements operate with just a single sensor.

In this case, the microwave analyzer can be directly connected to the sensor with a short barrel connector instead of a longer cable [15]. Figure 3.12 shows a notional microwave NDE reflectometer, with just a single probe integrated with a miniaturized analyzer that both transmits to illuminate the surface under test and then receives what is reflected from that surface. In the past, microwave analyzers were too large and heavy to do this. However, this newer paradigm of an integrated probe and analyzer significantly enhances the ability to construct robotic and handheld microwave NDE systems for a wider range of industrial applications.

3.3 Thickness Inversions

Structural composites are often used as radomes to protect antennas on vehicles, towers, or the ground. These radomes are tuned to have specific RF performance and thickness so that they maximize transmission of electromagnetic energy at specific frequencies. The composite layers as well as any coatings must be within certain specifications for radomes to function properly. Similarly, some air-, ground-, and sea-borne applications require absorber coatings to isolate antennas, reduce radar scatter, or reduce electromagnetic interference.

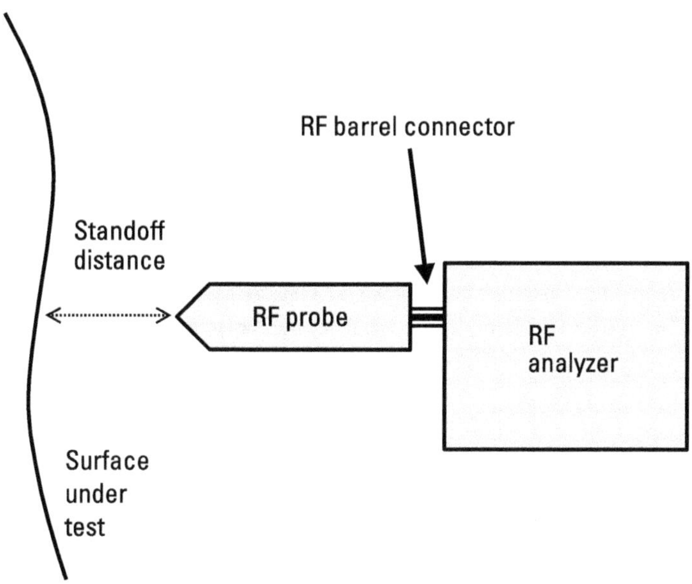

Figure 3.12 RF NDE probe directly connected to an RF transmitter/receiver and held a known distance away from the surface being measured.

These coatings can be spray-applied with industrial robots or by hand. However, variations in the deposition hardware, physical and environmental conditions during coating application, and changes in the coating viscosity can cause thickness variations that degrade electromagnetic performance. Nondestructive measurement of layer thickness(es) is therefore an ongoing need for many modern engineering applications.

Microwave probes can be used to measure thickness in manufacturing or field applications. One such method for determining thickness is to use electromagnetic energy reflected from a coated substrate and to analyze that electromagnetic energy in time domain for waveforms that indicate interfacial reflections [16]. In particular, electromagnetic energy is reflected from interfaces between materials with two different dielectric or magnetic properties, such as air-to-coating and coating-to-substrate. However, if the wavelength of the electromagnetic energy is large relative to the coating thickness, these interfacial reflections will overlap, making them difficult to distinguish from each other. In other words, this method requires that the electromagnetic wavelength be small relative to the thickness of the coating. With coating thicknesses typically ranging from 10s of microns and up, this method requires electromagnetic energy at terahertz frequencies, which have wavelengths less than or in the range of these coating thicknesses. At these high frequencies, however, underlying substrate properties and surface roughness can have a strong influence on the reflected waveforms and bias the data so that measurement accuracy is reduced.

An alternative microwave method that does not require the operable wavelength to be so small is also possible. Instead, an aggregate electromagnetic reflection of all the interfaces together is compared to a theoretical or empirical model to determine coating thickness [17]. With this method, a microwave NDE probe can interrogate a surface under test as shown in Figure 3.12. Standoff distance, a function of the probe antenna design and the desired interrogation area, enables coatings and structures to be measured either wet or dry. The probe is connected to a microwave source or receiver, such as a network analyzer, and radiates energy to interact with the surface under test. The probe then receives a reflection reduced by some amplitude, depending on the properties of the coating and structure underneath.

Key to this method is the use of multiple frequencies to capture the dispersive nature of reflection from a surface. This is because measurement at just a single frequency may not provide a unique solution for many coatings or structures. For example, an absorber coating will typically have at least one null where the reflection amplitude is a minimum at some frequency. If the thickness of the same coating material is varied, then that null will occur at different frequencies. This is illustrated in Figure 3.13, which shows a notional frequency-dependent reflectivity of an absorber coating on a conductive surface. Figure 3.13 shows two curves, each corresponding to a different thickness of the same coating material. If the reflection measurement is made at a single frequency such as frequency *A* in Figure 3.13, then it does not identify the thickness uniquely. However, if both frequency *A* and frequency *A′* are measured, then the coating thickness is uniquely identified. Some coatings may have multiple nulls, and even if reflection phase is also included, solution uniqueness is improved by scanning a range of frequencies.

Once measurements are performed with a probe apparatus, the measured data is calibrated and processed to remove measurement errors as described in Chapter 2. If phase is used in the inversion calculation, the probe-to-specimen distance must be well-known or accurately match the distance that was used to calibrate the probe against a conductive plate. With a handheld fixture, this can be accomplished by integrating a low-dielectric spacer with the probe so that placing it in contact with the surface maintains a constant offset distance. In a robotic system, additional devices such as ultrasonic distance sensors or laser range finders can be used to exactly locate the probe to specimen distance.

Figure 3.14 shows example measurements with a wideband microwave probe integrated with a miniaturized vector network analyzer operating at

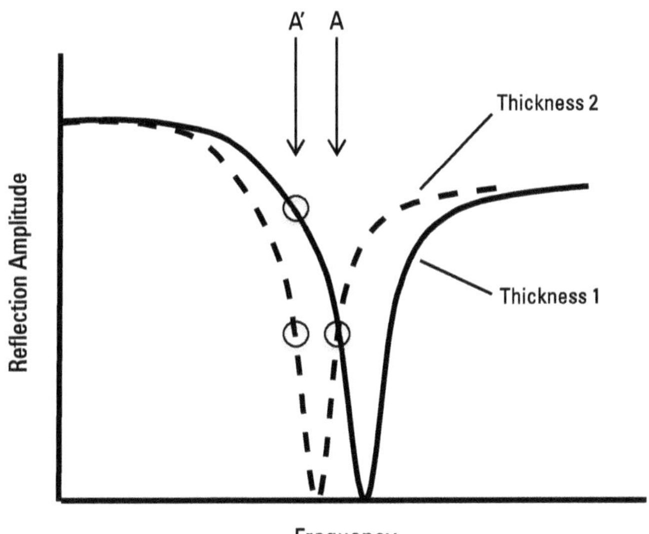

Figure 3.13 Characteristic reflection behavior of a notional absorber coating for two different thicknesses, where measurement only at frequency *A* does not uniquely distinguish them but measurement at both frequency *A* and *A′* does.

4–18 GHz. The specimen under test was a commercially available coating that was applied to a metal surface. The measured reflectivities were obtained successively after each thin coat of the material was deposited. This data was calibrated as described previously and shows that the spectral reflection characteristic of the coated surface is a strong function of the number of coats applied. In this case, each coat was deposited by a robotic spraying system. The measurements were made immediately after the coating application so that while most of the solvent had flashed away, the coating was still tacky. The spectral characteristics measured depended on the number of coats applied, and the data makes it clear that there is a correlation between reflection spectra and thickness.

Chapter 2 described methods for calculating the intrinsic properties of materials. Similar analyses can be used to derive a calculation of the thickness (or number of coats) as a function of frequency-dependent reflection. Figure 3.15 shows the application of an analytical model to fit the measured reflection data. In this case, a previously measured sample of the coating material was characterized using standard laboratory methods to obtain the coating's intrinsic properties: dielectric permittivity and magnetic permeability in the appropriate frequency range. Based on these intrinsic properties, an analytical model is constructed to calculate the expected reflection for a given coating thickness. Examples of such models can be also found in standard textbooks such as [18]. Equation (3.2) relates the measured reflection coefficient, S_{11}, to the thickness, t, of a single layer coating on a conductive surface.

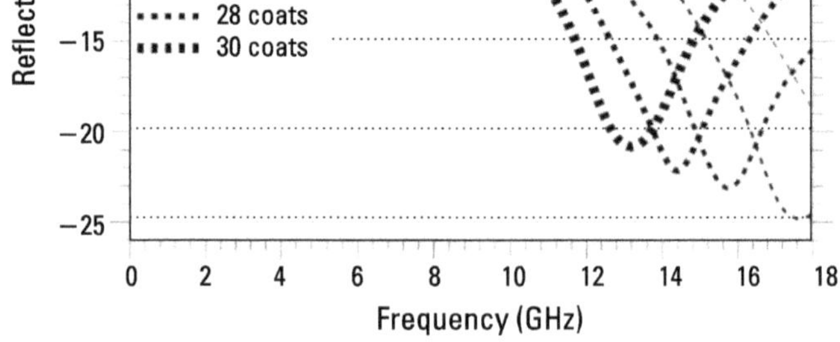

Figure 3.14 Reflection data from a microwave probe measurement of a commercial absorber coating.

$$S_{11} = \frac{Z - Z_0}{Z + Z_0} \text{ and } Z = Z_0 \sqrt{\frac{\mu_r}{\varepsilon_r}} \tanh\left(-i2\pi \frac{t}{\lambda_0} \sqrt{\mu_r \varepsilon_r}\right) \tag{3.2}$$

where λ_0 is the free-space wavelength, Z_0 is the impedance of free space, and μ_r and ε_r are the relative magnetic permeability and relative dielectric permittivity of the coating.

In the Figure 3.15 results, (3.2) was iteratively compared to the measured reflection data while varying thickness. Standard iterative solvers or root-finding methods may be used. In this case a Nelder-Mead solver [19] was applied to minimize the square of the difference between the model prediction and measured data. For this minimization an error function can take the form of

$$\text{Error} = \sum_f \left\| S_{11}^{\text{model}}(t_1, f) \right| - \left| S_{11}^{\text{measurement}}(f) \right\|^2 \tag{3.3}$$

where f is frequency, and t_1 is the thickness. This error function is computed iteratively for different thicknesses until the error is minimized. The extracted thickness may then be used to provide feedback to the coating-application process or for QA purposes.

This method also can be used for multilayer coatings where each layer of the coatings is a different material (has a different dielectric permittivity and/or magnetic permeability), as shown in Figure 3.16. When multiple layers

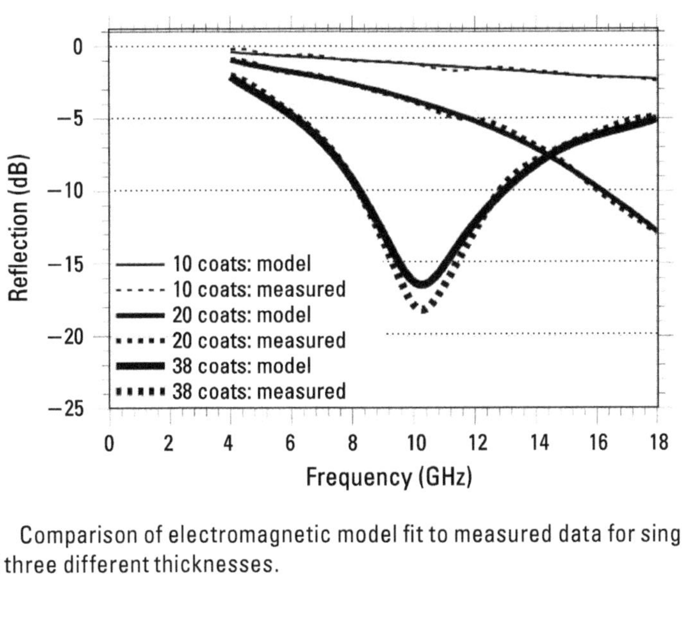

Figure 3.15 Comparison of electromagnetic model fit to measured data for single-layer coatings of three different thicknesses.

with different composition are applied in sequence, measurements are conducted as the layers are applied to monitor their respective thicknesses. For layers beyond the first layer, a slightly modified error function can be used to prevent propagation of residual errors. This modified error function uses the following ratios of the reflection data for both the measurement and the model of a two-layer coating:

$$\Delta S_{11}^{\text{model}} = \frac{S_{11}^{\text{model layer 1}}\left(t_1, f\right)}{S_{11}^{\text{model layers 1+2}}\left(t_2, f\right)} \tag{3.4}$$

and

$$\Delta S_{11}^{\text{measurement}} = \frac{S_{11}^{\text{measurement layer 1}}\left(f\right)}{S_{11}^{\text{measurement layers 1+2}}\left(f\right)} \tag{3.5}$$

Equations (3.4) and (3.5) are for a two-layer coating where t_2 is the thickness of the top layer as shown in the middle of Figure 3.16, and t_1 is implicit in the model and is fixed based on the last model/measurement fit before the second material was deposited. The error function that is then minimized is given by

$$\text{Error} = \sum_f \left\| \Delta S_{11}^{\text{model}}\left(t_2, f\right) \right| - \left| \Delta S_{11}^{\text{measurement}}\left(f\right) \right\|^2 \tag{3.6}$$

For subsequent layers, variations of these equations are used, where the thicknesses of the underlayers are fixed, and only the thickness of the unknown top layer is varied. Once the best-fit curves such as those shown in Figure 3.15 are obtained, the fit parameter is the calculated coating thickness.

In some cases, it may be necessary to determine the thickness of multiple layers in a structure without having first determined the underlying layer thicknesses. For example, advanced radomes may have three or five layers of composite shell and low-dielectric honeycomb. These multiple layer thicknesses can be determined simultaneously; however, this is a more difficult inversion than just a single thickness. Multivariate optimization can be applied much in the same way that a single variable is solved, but there must be enough measurement data to ensure that the thickness inversions do not have multiple solutions. This is possible when sufficient frequency bandwidth is available from a wide-bandwidth microwave NDE probe.

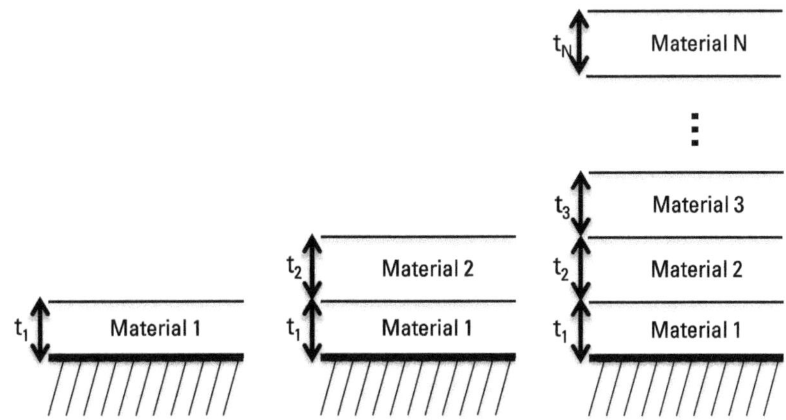

Figure 3.16 The profile of coatings with one, two, or many layers of different coating materials.

3.4 Thickness and Property Inversion

Chapter 2 discussed the inversion of dielectric and magnetic properties from free-space measurements which requires a priori knowledge of the specimen thickness. Section 3.3 described a situation where the intrinsic properties are known but the thickness is not. Sometimes in a manufacturing situation, both the thickness and the electromagnetic properties are unknown and must be measured. If the part is large or curved or cannot be touched because of sensitivity to damage, conventional thickness measurement, such as that obtained with a micrometer, is not an option. However, with some other assumptions about the component under test, it is possible to use microwave NDE methods to determine both intrinsic properties and thickness at the same time. This can be done by (1) taking data over a wide enough bandwidth so that there is some feature such as a null in the measured data and (2) making additional assumptions including nondispersive intrinsic properties and a limited number of layers.

To illustrate this idea, the following walks through an example that uses a pair of spot probes to map the electromagnetic properties of a dielectric window with a conductive coating applied on it for electromagnetic interference (EMI) mitigation. In other words, the conductive coating is designed to minimize transmission of microwave energy while still enabling optical transmission. A drawing of a test panel is shown in Figure 3.17. This panel is

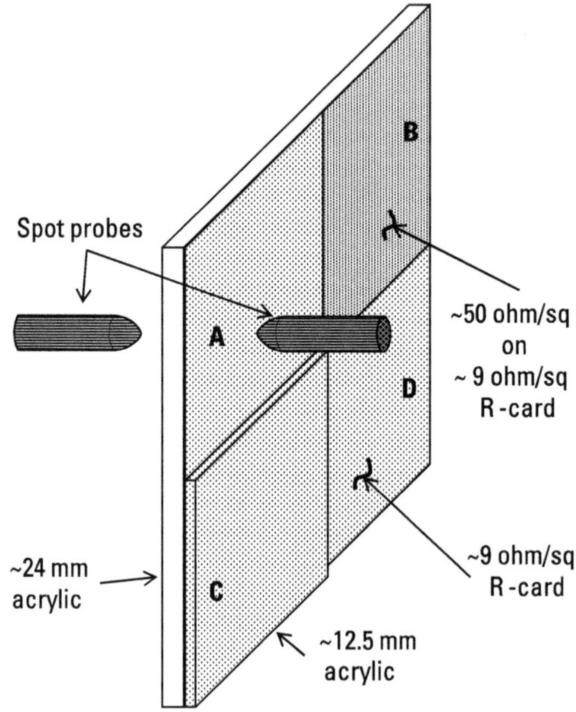

Figure 3.17 Drawing of an acrylic test panel to demonstrate microwave mapping with a pair of robotically mounted probes measuring transmission.

an approximately 24-mm-thick acrylic-coated with a resistive sheet of about 9-Ω/square sheet impedance. One quadrant (bottom left) of the test panel also has an additional 12.5-mm layer of acrylic while another quadrant (top right) has an additional 50-Ω/square sheet of restive material on it. The panel was scanned with a pair of wide-bandwidth probes on either side to measure transmission through it as a function of position. While this example scenario is somewhat contrived, it is representative of a real-life scenario: using robotic probes to map the insertion loss through an aircraft transparency in production or after repair.

This example scenario is based on transmission, so a two-port microwave network analyzer collected data from 4–18 GHz by stepping through 1,601 different frequency measurements. A simple response calibration was applied, consisting of a single measurement of nothing or clear site in between the microwave-mapping probes. Calibrated insertion loss is calculated as a ratio of the insertion loss with the specimen to that of the clear site. For both the calibration and the specimen measurement, the probes were separated by a distance of 14 cm. Additionally, the data was processed with time-domain windowing methods to reduce errors from multipath reflections. The data set shown here has a 0.5-ns-wide window applied. The insertion loss was measured every 1 cm across the panel to create a map of panel performance. An example of the measured S_{21} at one position is shown in Figure 3.18. This data was from the probes centered at position A as designated in Figure 3.17. The transmission amplitude shows an oscillatory behavior as a function of frequency, which is caused by reflections at the front and back of the acrylic slab constructively and destructively interfering with each other.

Once calibrated, the insertion loss data is used to invert desired properties of the material under test. In this example, the substrate thickness and the sheet impedance of the conductive layer were calculated. Using the transmission line theory described in Chapter 2, an expression relating the transmission (S_{21}) to the thickness (t) and sheet impedance (Z_S) of this two-layer material can be derived,

$$S_{21} = \frac{2Z_s\left(\Gamma^2 - 1\right)T}{\left(2Z_s + Z_0\right)\left(\Gamma^2 T^2 - 1\right) + Z_0\left(T^2 - 1\right)} \tag{3.7}$$

where Z_0 is the impedance of free space, $\Gamma = (\mu - \sqrt{\mu\varepsilon})/\mu + \sqrt{\mu\varepsilon})$, and $T = e^{-ik_0 t\sqrt{\mu\varepsilon}}$. Since the substrate is acrylic, its relative permittivity is already known and assumed to be a fixed value ($\varepsilon = 2.6$) in the inversion calculation.

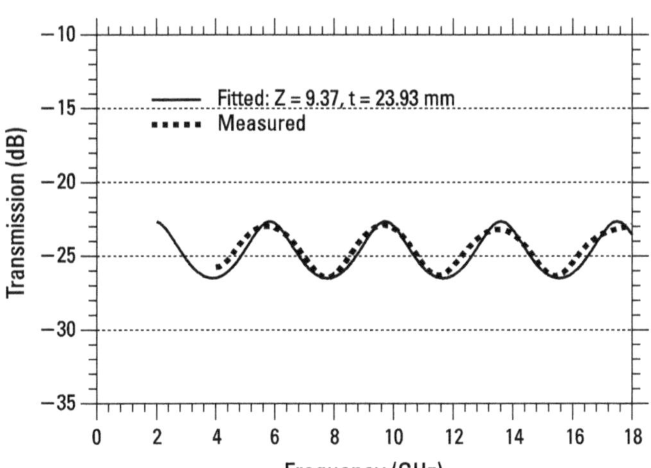

Figure 3.18 Insertion loss measured and modeled of a panel of acrylic covered with a resistive window tint layer.

A calculated substrate thickness and sheet impedance is then obtained by minimizing the difference between (3.7) and the measured S_{21} with a standard multivariate minimization algorithm [19]. Figure 3.18 shows the comparison of the idealized two-layer model compared to the measured transmission, as well as the corresponding values of Z_S and t for this model.

This inversion procedure can be repeated at each measurement location to map the inverted properties. Figure 3.19 shows the inverted thickness and sheet impedances for two lines across the test panel, corresponding to line scans from position A to B and from position C to D in Figure 3.17. In the top plot, segment AB shows no significant change in thickness, while segment CD shows the step thickness change from the 24-mm-thick panel to the quadrant that has an additional 12.5 mm covering on it. The inverted impedance in the bottom plot shows that segment AB includes an area with approximately 9.5-Ω/square resistance as well as the quadrant that also has an additional 50-Ω/square resistance on top. When a 50-Ω sheet is placed on top of a 9.5-Ω area, they add in parallel so that the net effect is to lower the total sheet impedance by a few ohms. Line CD has a consistent impedance going across the whole segment, except that there is an edge effect from the thickness change that appears to create additional uncertainty in the inverted impedance. This is because the sudden change in acrylic thickness causes diffraction of the incident wave.

3.5 Defect Detection

Another potential application for microwave NDE is defect detection in structural composites. One common type of defect of concern for structural rigidity is delamination. In composite laminates, delamination can occur because of weak bonding between the fibers and the matrix, or it can happen between two dissimilar layers where the adhesive bond has degraded. This section illustrates how microwave NDE methods can be used to detect delamination within a composite.

Figure 3.20 shows a calculation of the effect of delamination on the transmission amplitude of a fiberglass composite with a relative permittivity of 4.5–0.1i. The dielectric composite was assumed to be 12.7-mm-thick, and this data was calculated with transmission line theory. The top plot shows the transmission amplitude that occurs when there is a small air gap or delamination layer in the middle of the composite. The different curves in this plot correspond to gap widths from 0 to 0.2 mm. The data shows a small but detectable perturbation of the transmission coefficient compared to a solid composite

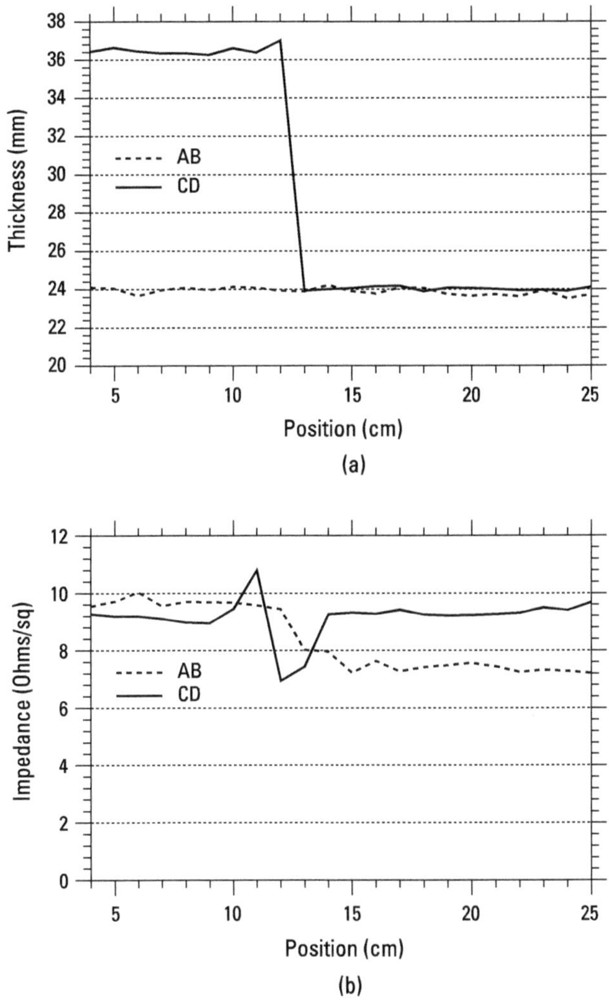

Figure 3.19 Inverted sheet impedance and thickness versus position for two linear scans of an acrylic test panel with a resistive coating.

with no delamination gap. The periodic characteristic of these curves is due to constructive and destructive interference between the front and rear faces of the composite panel, and the periodicity is driven by the overall thickness of the fiberglass slab relative to microwave wavelength. While this calculation shows the ideal case where the thickness of the slab is well-known, an important question is: Can we distinguish between the effects of a delamination gap versus a variation in the overall thickness of the composite when thickness is not well-known or consistent? To answer this, Figure 3.20(b) shows the effect on the transmission coefficient as the overall thickness is varied in a composite with no delamination. In this case, the transmission coefficient is also perturbed, but close examination shows that the character of that perturbation is different than it is with the delamination gap.

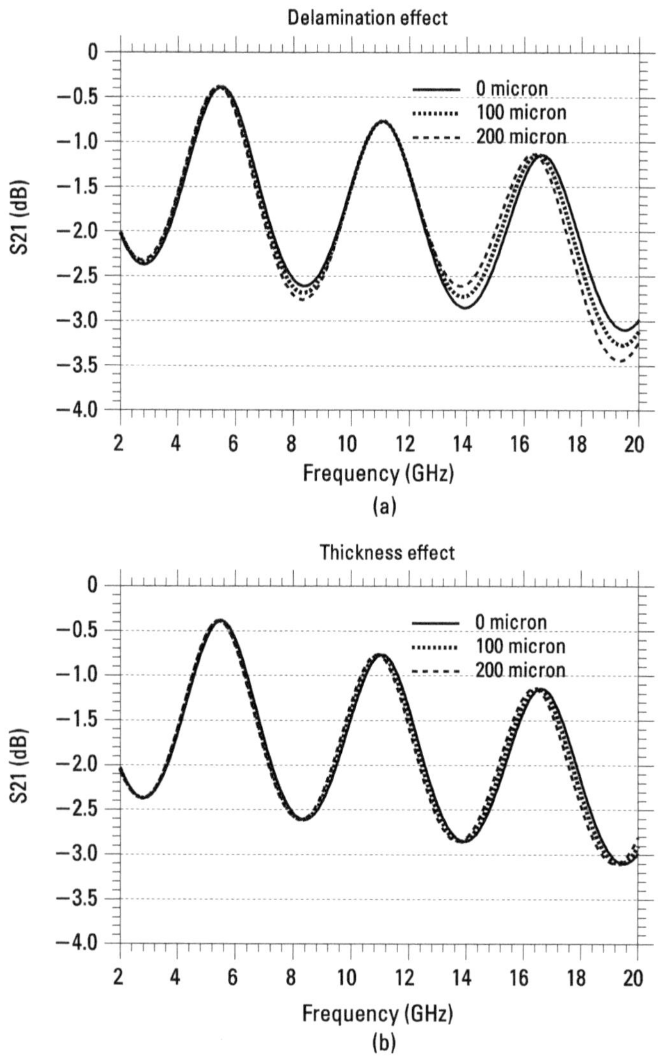

Figure 3.20 Calculated transmission amplitude for a 12.7-mm-thick fiberglass panel with (a) a delamination in the middle or (b) an equivalent increase in thickness.

A clearer picture of the difference caused by a delamination gap versus thickness variation can be seen in Figure 3.21. The curves in Figure 3.21 use the same data as in Figure 3.20 but replotted after first dividing them by the S_{21} of the ideal or reference 12.7-mm composite. Figure 3.21(a) shows the effect of varying the delamination gap, while Figure 3.21(b) shows the effect

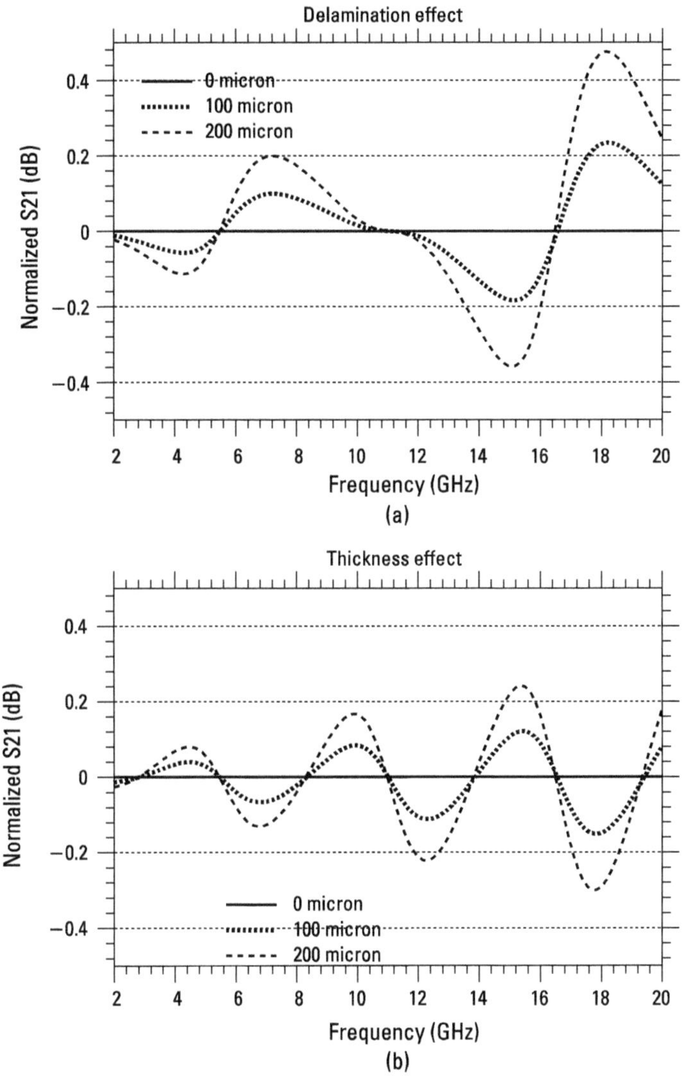

Figure 3.21 Calculated transmission amplitude normalized to an ideal 12.7-mm-thick composite showing the effect of a delamination gap in the panel (a) or a simple thickness increase of the panel (b).

of varying the overall thickness on a nondelaminated composite. Without a delamination, the change in thickness imposes a simple oscillation with a monotonically growing envelope versus frequency. In contrast, the effect of a delamination gap in the middle of the slab is to impose a much more complicated frequency dependence. While different, this behavior is a bit too subtle to use as an obvious identification of the presence or severity of delamination.

In a similar vein, Figure 3.22 shows the effect of delamination (top) and simple thickness increase (bottom) on the reflection coefficient (S_{11}), after it has been normalized by the reflection from the ideal composite panel. In this case, the thickness change of the panel adds a feature in the middle of the measured spectra that is not present for the delamination defect. These differences show that reflection is also a good candidate for differentiating between delaminations and thickness changes. Thus, in either reflection or transmission, there is a difference in the microwave behavior of an internal gap that can be differentiated from simple thickness variations. A key feature, no matter whether reflection or transmission is used, is that the detection of these behaviors requires measurement over a sufficiently broad frequency range to capture the oscillatory nature of the effect.

In practical application, the frequency-dependent transmission or reflection data is somewhat complex to interpret and require an algorithm to convert microwave data to a simple parameter that indicates the presence and severity of a delamination. One way to do this is to apply an idealized model, much like in Section 3.4. For this composite example, a simple one-layer slab model can be used, and measured data compared to this model to see how well it fits. Specifically, a multivariate fit of thickness and dielectric permittivity were made to all the measured data, whether or not there was a delamination gap. The reason for including permittivity as a variable fit parameter is that in many situations the dielectric permittivity may not be exactly known or if there is inhomogeneity the permittivity can vary somewhat with position.

Without a delamination, the data should fit the ideal theoretical model with minimal error. When a delamination is present, the fit will not be exact, and a finite difference between the measured microwave parameters and theoretical model will exist. This finite difference is related primarily to the presence of the delamination gap. Summing residual error over the measured frequencies should then give an indication of the size of an air gap associated with the delamination,

$$S_{xx}^{\text{residual}} = \sqrt{\sum_{\text{frequency}} \left(\left| S_{xx}^{\text{theory}} \right| - \left| S_{xx}^{\text{measured}} \right| \right)^2} \tag{3.8}$$

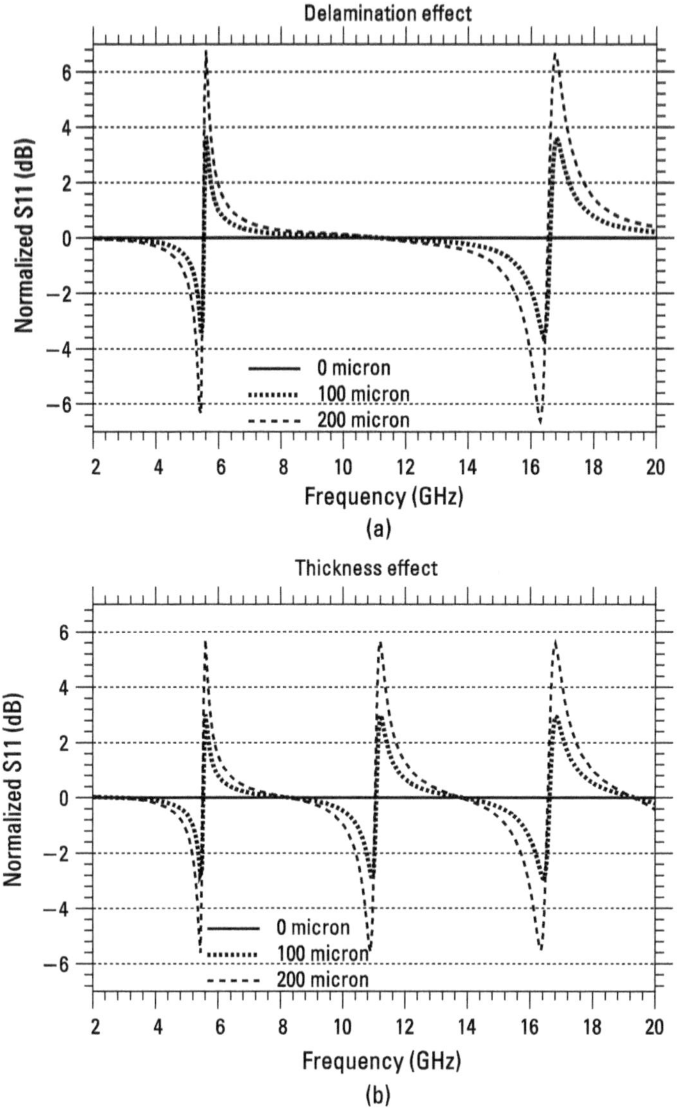

Figure 3.22 Calculated reflection amplitude normalized to an ideal 12.7-mm-thick composite showing the effect of a delamination gap in the panel (a) or a simple thickness increase of the panel (b).

Equation (3.8) only uses the amplitude and does not require the phase of the measured S-parameters. This can be advantageous for measurements in a factory or field environment, since it reduces the importance of exact positioning of the microwave mapping probes. In some situations, phase can be accurately captured as well and provides another measured parameter for characterizing the health of the material under test.

To demonstrate the effectiveness of this algorithm, Figure 3.23 shows its application to measured data. In this case, wide-bandwidth microwave spot probes were used with a fiberglass composite panel that had a controlled delamination. The internal delamination was created by sandwiching two 6.35-mm-thick fiberglass panels together. For the variable delamination case, the panels were clamped tightly together at one end and the other end had a 0.45-mm spacer inserted between the panels to create a variable air gap from left to right. For the no-delamination case, the panels were clamped at both ends. The spot probes were scanned from one end to the other to measure the transmission and reflection characteristics and the multivariate fit algorithm was applied to these data. The S-parameters from each measurement location were fit to equations similar to those found in Section 2.4.3 to obtain a best thickness and permittivity, and the summed residual fit-error as a function of probe position is plotted in Figure 3.23. Figure 3.23(a) shows the residual for transmission (S_{21}) data while Figure 3.23(b) is the residual for reflection (S_{11}) data. Both parts of Figure 3.23 have two curves: (1) one corresponding to the fiberglass panels with a continuously varying gap in-between and (2) the other corresponding to the panels tightly clamped on both ends. The residual for the fully clamped specimen is unchanging with position, indicating no significant gap between them. The residual for the specimen with an air gap in the middle is monotonically increasing, corresponding to the varying gap going from the clamped end to the spacer.

Ideally, the fully clamped specimen should have a residual signal of 0. Instead Figure 3.23 shows that there is a nonzero baseline residual error. One potential source of uncertainty is that the fit algorithm assumes that the permittivity is single-valued and not a function of frequency. In reality, permittivity may vary slightly with frequency. Additional error sources may include microwave measurement uncertainty, material inhomogeneity (i.e., if voids are present or if there are variations in the fiber weave), and the fact that even when clamped, surface roughness of the fiberglass induces a very small air gap layer. Also, the residual signal is both stronger and less noisy for reflection than for transmission, even though the transmission and reflection residuals were summed over the same number of frequency points. In practical situations only one side of a structure or component may be accessible, and the use of just a single probe with a one-port microwave analyzer can be cheaper and easier than a two-sided microwave transmission system.

The preceding example of Figure 3.23 is just a simple, single-material layer, and multilayer structures such as radomes on aircraft nosecones also

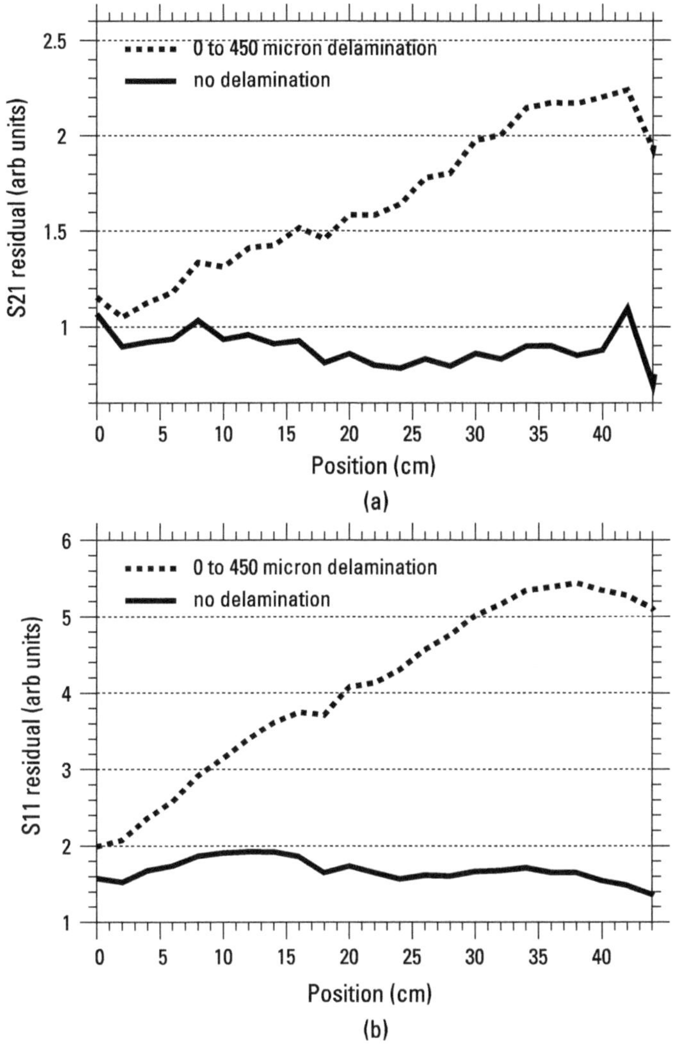

Figure 3.23 Calculated residual fit error for (a) transmission or (b) reflection for fiberglass panels with a controlled delamination gap.

exist. These are often constructed from three layers—inner and outer fiberglass shells that are separated by a honeycomb spacer layer. In this case, microwave measurement data can be fitted to a multilayer model constructed by cascading layers together with the network analysis formalism described in Chapter 2. Assumptions about the materials can be made, and a multivariate optimization can compare the measured data to an idealized model for calculating parameters such as layer thicknesses and fit error. Like the example above these fitted parameters can detect the presence of defects such as water ingress or delamination.

The complexity of multilayer systems does mean that a wider variety of defects may be present. So, it is helpful to have additional analysis algorithms for improving the reliability of defect detection. An alternative method that can be used either in place of or as a supplement to the network analysis theory method is spectral fitting. Radomes are optimized to have maximum transmission at the frequencies of the radar or antenna system that they are designed to protect. This is a characteristic frequency dependence based on the dielectric properties and thicknesses of the radome layers. The spectral amplitude dependence of a radome-reflection response can be empirically fitted to a Lorentzian null superimposed on top of a linear background [20],

$$S_{11,\mathrm{amp}}(f) = (a_1 + bf)\left(\left(\frac{a_2}{\pi}\right)\frac{\frac{w}{2}}{(f - f_0)^2 + \left(\frac{w}{2}\right)^2} + 1\right) \tag{3.9}$$

The reflection phase spectra exhibits a phase shift at the radome's resonant frequency, and can be fitted to the following:

$$S_{11,\mathrm{phase}}(f) = a_1 + bf + a_2 \tanh\left(\frac{(f - f_0)}{w}\right) \tag{3.10}$$

where in both expressions f denotes frequency, a_1 is a frequency-independent offset, b is the overall slope, f_0 is the frequency of the radome null, a_2 is the amplitude of the null, and w is the linewidth of the null. In this approach the amplitude and phase spectra are measured and then fit to the above expressions. The resulting five fit parameters from each of the spectra can then be analyzed to show whether the radome is healthy or defective and provide indications of what defects may be present.

References

[1] Bakhtiari, S., S. I. Ganchev, and R. Zoughi, "Open-Ended Rectangular Waveguide for Nondestructive Thickness Measurement and Variation Detection of Lossy Dielectric Slabs Backed by a Conducting Plate," *IEEE Transactions on Instrumentation & Measurement*, Vol. 42, No. 1, Feb 1993, pp. 19–24.

[2] Pozar, D. M., *Microwave Engineering* (Fourth Edition), Hoboken, NJ: Wiley, 2012.

[3] Sadiku, M. N. O., *Computational Electromagnetics with Matlab*, Boca Raton, FL: CRC Press, 2019.

[4] Maloney, J. G. et al., "Antennas," Chapter 14 in (Third Edition), A. Taflove, S. C. Hagness (eds.), Norwood MA: Artech House, 2005, pp. 607–676.

[5] Mueller, G. E., and W. A. Tyrrell, "Polyrod Antennas," *The Bell System Technical Journal*, Vol. 26, No. 4, 1947, pp. 837–851.

[6] Balanis, C. A., *Antenna Theory Analysis and Design* (Fourth Edition), Hoboken NJ: Wiley, 2016.

[7] Harrington, R.F., *Time-Harmonic Electromagnetic Fields*, New York, NY: Wiley-Interscience, 2001.

[8] Diaz, R., et al, "Compact Broad-Band Admittance Tunnel Incorporating Gaussian Beam Antennas," U.S. Patent 7889148B2, Feb 15, 2011.

[9] Chen, C. C., K. R. Rao, and R. Lee, "A New Ultrawide-Bandwidth Dielectric-Rod Antenna for Ground-Penetrating Radar Applications," *IEEE Trans. Antennas and Propagation*, Vol. 51, No. 3, March 2003, pp. 371–377.

[10] Schultz, J. W., et al., "A Comparison of Material Measurement Accuracy of RF Spot Probes to a Lens-Based Focused Beam System," *Antenna Measurement Techniques Association (AMTA) Symposium Proceedings*, Tucson, AZ, Oct. 12–17, 2014, pp. 421–427.

[11] Hahn, H., and H. J. Halama, "Compensation of Phase Drift on Long Cables" U.S. Patent 3434061, March 1969.

[12] Roos, M. D., "Method for Removing Phase Instabilities Caused by Flexure of Cables in Microwave Network Analyzer Measurements," U.S. Patent 4839578, June 1989.

[13] Schultz, J. W., et al., "Correction of Transmission Line Induced Phase and Amplitude Errors in Reflectivity Measurements," U.S. Patent 20160103197A1, Oct. 2014.

[14] Schultz, J. W., and J. G. Maloney, "Correction of Transmission Line Induced Phase and Amplitude Errors in Reflection and Transmission Measurements," *Antenna Measurement Techniques Association (AMTA) Symposium Proceedings*, Austin, TX, Oct. 30–Nov. 4, 2016.

[15] Zaostrovnykh, S. A., et al. "Measurement module of virtual vector network analyzer," US Patent 10094864B2, Aug. 2012.

[16] White, J. W., et al., "System and method to measure the transit time position(s) of pulses in time domain data," U.S. Patent 8457915B2, July 2007.

[17] Schultz, J. W., et al., "Non-Contact Determination of Coating Thickness," U.S. Patent US10203202B2, April 2014.

[18] Born, M., and E. Wolf, *Principles of Optics* (Seventh Edition), Cambridge, United Kingdom: Cambridge University Press, 2002.

[19] Press, W. H., et al., *Numerical Recipes: The Art of Scientific Computing* (Third Edition), Cambridge, United Kingdom: Cambridge University Press, 2007.

[20] Freeman, R., et al., "A Microwave Spot Probe Method for Scanning Aircraft Radomes," *Proc. 12th Int. Symposium on NDT in Aerospace 2020*, Williamsburg, VA, Oct. 6–8, 2020.

4

Focused-Beam Methods

4.1 Focused-Beam System Design

A free-space focused-beam system is used to determine material properties such as dielectric permittivity, magnetic permeability, or sheet impedance. These parameters are called intrinsic properties because they do not depend on the size or shape of the specimen. Also, they cannot be directly measured by a focused beam, and instead an extrinsic property is measured. For free-space techniques such as the focused beam, the intrinsic properties are then derived from the extrinsic ones using the theory outlined in Chapter 2. What a focused-beam system does measure is the electromagnetic scatter from a material specimen. Focused-beam systems, as well as the coaxial airline and waveguide methods discussed in Chapter 5, measure the scattering parameter matrix or S-parameters of a specimen, which depend on both intrinsic properties and thickness. These scattering parameters represent the signals that are transmitted or reflected by a material specimen when illuminated with an incident wave.

A focused-beam system is designed to be a convenient instrument for obtaining these scattering parameters with reasonable accuracy. A typical lens-based focused-beam system is shown schematically in Figure 4.1. Electromagnetic radiation from a feed antenna is directed to a dielectric lens, which focuses that radiation to a minimized radius, where a planar specimen is positioned. A reciprocal lens and antenna are on the other side to measure the transmission through the specimen. Additionally, the feed antenna and lens can collect the reflected energy from the specimen. By placing a flat specimen at the focus, a reasonable approximation to plane-wave illumination is achieved. As an alternative, mirrors can be used as focusing elements instead of lenses [1]. It is even possible to create a focused beam with a phased-array antenna [2]. This chapter assumes a lens-based design; however, many of the same principles may be used for the other focusing methods.

At optical frequencies, implementation of focusing systems is well described by geometrical optics, an approximation that is effective when the optical components are hundreds of wavelengths or more in dimension. However, lens systems at microwave frequencies manipulate wavelengths from millimeters to a large fraction of a meter. Practical considerations then drive the size of microwave optical components to operate at less than 10 wavelengths in dimension. Under these conditions, the theory of geometrical optics is only approximate, and a more general theory is needed to describe microwave phenomena. In this quasi-optical regime, diffraction effects become important, and Gaussian optical theory provides a more accurate model of a microwave focused-beam system. Specifically, the geometrical optical theory of a lens' focus concentrates light into a single point, while Gaussian optics accounts for a minimum spot size due to diffraction. Gaussian beam theory as applied to microwave- and millimeter-wave systems is reviewed extensively elsewhere [3]. So, this book gives only a brief outline of some of the pertinent aspects of the theory important for materials measurement.

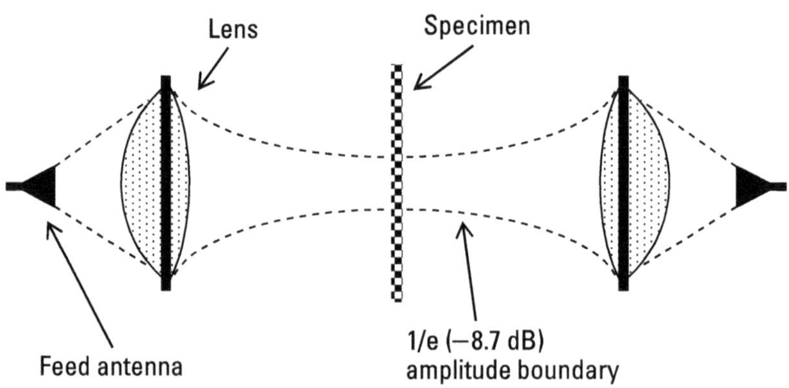

Figure 4.1 Notional geometry of a focused-beam measurement system.

4.1.1 Gaussian Beam Basics

In Gaussian optics theory, the radiating beam is assumed to have an amplitude taper described by a Gaussian function. From Maxwell's equations a general wave equation can be derived [4]. With assumptions of a linear, time-invariant medium and a slowly varying envelope of the waves or paraxial approximation, the wave equation can be simplified. For an axially symmetric, z-directed beam in Cartesian coordinates, the paraxial wave equation is expressed by

$$\frac{\partial^2}{\partial x^2}u + \frac{\partial^2}{\partial y^2}u - 2ik\frac{\partial}{\partial z}u = 0 \tag{4.1}$$

The fundamental Gaussian mode is a solution to this wave equation given by

$$u = a(z)e^{\frac{-ik\left(x^2+y^2\right)}{2q(z)}} \tag{4.2}$$

where $k = 2\pi/\lambda$ is the wavenumber, and a and q are two complex functions of z that depend on the configuration of the beam. This solution assumes that the wave numbers follow $k_x, k_y << k$ and that the fields are time-harmonic with a time dependency of $\exp(i\omega t)$, where t is time and ω is angular frequency. Such a beam is shown schematically in Figure 4.2, along with a few defined parameters. Plugging this solution into the paraxial wave equation (4.1) enables further evaluation of the unknown functions a and q. In particular, q is called the complex beam parameter and can be written in terms of the parameters shown in Figure 4.2 [3],

$$q = z + iZ_R \tag{4.3}$$

where the origin of z is defined to be at the minimum radius of the beam, and

$$Z_R = \frac{k}{2}w_0^2 \tag{4.4}$$

is called the Rayleigh range (also sometimes called the confocal parameter). w_0 is called the beam waist and is the minimum radius, which occurs at the beam focus. The beam radius at other locations along the axis of propagation is given by

$$w(z) = w_0\sqrt{1+\left(\frac{z}{Z_R}\right)^2} \tag{4.5}$$

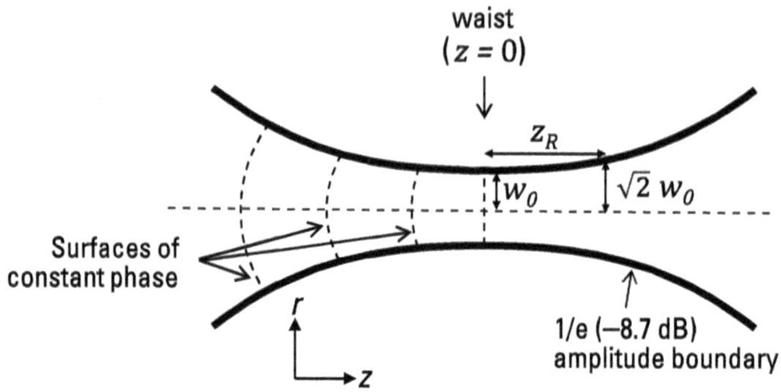

Figure 4.2 Features of a focused Gaussian beam.

At this radius, the amplitude of the beam is down by a factor of $1/e$ relative to the middle.

A convenient definition for specifying depth of focus is the Rayleigh range, Z_R. Rearrangement of (4.5) shows that this Rayleigh range is the distance from the beam waist location to where the spot size has increased to $w(Z_R) = \sqrt{2}w_0$. In a material-measurement apparatus, the inversion of material properties such as those derived in Chapter 2 assumes an ideal plane wave. If a thick specimen extends outside of this Rayleigh range, then it will experience additional phase curvature that reduces the accuracy of the plane-wave assumption. Additionally, when a material specimen is placed in the focused beam, this Rayleigh range is reduced by the permittivity and permeability of the material since $k = k_0\sqrt{\varepsilon_r\mu_r}$ in (4.4). Thus, while thick specimens usually work well in a wideband focused-beam system, high-index materials specimens shouldn't be so thick that they extend outside of the Rayleigh range.

Going back to the wave equation solution, (4.2), and evaluating the function, $a(z)$, in terms of the above-defined parameters while assuming that the on-axis amplitude at the beam waist is unity gives a more specific expression for the wave equation solution,

$$u(x,y,z) = \frac{w_0}{w(z)} e^{\left(\frac{-r^2}{w^2(z)}\right)} e^{\left(\frac{-ikr^2}{2R(z)}\right)} e^{i\tan^{-1}\left(\frac{z}{Z_R}\right)} \tag{4.6}$$

where $r^2 = x^2 + y^2$ is the distance from the axis. In this expression, we have also introduced another parameter,

$$R(z) = z\left(1 + \left(\frac{Z_R}{z}\right)^2\right) \tag{4.7}$$

which is the radius of curvature of the phase front as a function of position along the z-axis. At the beam focus $z=0$, $w_0=w(z=0)$, and (4.6) simplifies to

$$u(z=0)=e^{\left(\frac{-r^2}{w_0^2}\right)} \tag{4.8}$$

As (4.7) indicates, the phase front at the focus has an infinite radius of curvature equivalent to a flat-phase front. This flat phase helps ensure that the approximation of plane-wave illumination is valid at the specimen location. The inversion algorithms usually used in free-space methods assume plane-wave interactions, so this flat phase front improves the accuracy of intrinsic property determination.

In geometrical optics, the radius of curvature is an important parameter used to evaluate the transformation of optical rays from a point source through shaped reflectors or lenses. In the fundamental-mode Gaussian beam equations above, there is an alternate expression for the complex beam parameter that is analogous to this geometric radius of curvature,

$$\frac{1}{q}=\frac{1}{R}-i\frac{2}{kw^2} \tag{4.9}$$

From (4.9), we see that q is effectively a complex radius of curvature that provides a convenient mechanism for transforming Gaussian beams through optical elements such as lenses or mirrors using a similar formalism to that applied in geometrical optical design.

The Gaussian assumption is also useful because it provides straightforward expressions for the performance of microwave lens systems. For example, integrating the power density, $|u^2|$, in cylindrical coordinates gives the fraction of power outside a radius, r_1 at an arbitrary z,

$$\Delta P(r>r_1)=e^{-2r_1^2/w^2} \tag{4.10}$$

This quantity is also the same as the relative power density at that same radius. Equation (4.10) is useful for the design of a measurement fixture since it provides guidance on the minimum specimen size. For example, a common rule of thumb for minimum lateral specimen dimensions (in the x and y directions) is that they should be no smaller than the −20-dB radius of the beam, which is just over one and a half times the beam waist. In this case, the fractional power going around the specimen will be no more than 1%. However, if a specimen has a very high insertion loss, then a larger size is required to reduce the fractional power outside the specimen from dominating the transmitted signal. Similarly, (4.10) provides an estimate for the required radius of the lens, since the lens must be large enough to (1) encompass the energy radiated from the feed antenna and (2) achieve the desired focused spot for the specimen illumination.

As a framework for describing the interaction of microwave energy with lenses and planar sheets, Gaussian beam theory provides a reasonable approximation for the design of a focus beam measurement system. However, the paraxial approximation assumed by this theory is still an approximation and can ultimately limit the accuracy of the plane-wave assumption. Furthermore, when measurement hardware is built, deviations from this approximation may require experimental adjustment of the system, often by repositioning the relative positions of the feed antennas, lenses, and specimen, to account for deviations from ideal assumptions. With that in mind, the Gaussian beam and geometric optics approximations are sufficient to determine an optimum lens shape for a set of design constraints, and measurement examples shown later in Section 4.2 are based on a system designed with these approximations. Section 4.1.2 provides a description of the theory to design lenses in a focused-beam material-measurement system.

4.1.2 Lens Design

The first step in designing a focusing lens is to define the feed antenna in terms of the Gaussian parameters described in Section 4.1.1. The 3-dB beamwidth is a commonly measured parameter for directive antennas, and we can translate the 3-dB beamwidth into an equivalent Gaussian beam representation, which is the width where the beam amplitude is down by $1/e$ from the center. From this, we can express (4.10) in terms of angle. With the paraxial approximation we estimate $r_1^2/w^2 \approx \theta_{3\text{dB}}^2/\theta_{1/e}^2$, where $\theta_{1/e}$ is the angle from the origin of the beam axis to where the normalized field amplitude is down to $1/e$, and $\theta_{3\text{dB}}$ is the half-angle that defines the 3-dB beamwidth. We then calculate the $1/e$ angle from (4.10) by knowing that the power at the 3-dB angle is half the power at the beam center, $e^{-2\theta_{3\text{dB}}^2/\theta_{1/e}^2} = \frac{1}{2}$, which results in $\theta_{3\text{dB}} = 0.589\theta_{1/e}$. Assuming a Gaussian beam emanating from the antenna, we calculate the equivalent waist of that beam from the $1/e$ far-field beamwidth. The growth of the $1/e$ radius of the beam can be defined in terms of an angle from the beam origin (i.e., the phase center of the antenna), $\theta = \tan^{-1}(w/z)$. In the far field ($z >> Z_R$), (4.5) can be rearranged to show that $w/z \to w_0/Z_R$. Then using the paraxial approximation along with (4.4) we express the far-field divergence angle in terms of the beam waist,

$$\theta_{1/e} = \tan^{-1}\left(\frac{\lambda}{\pi w_0}\right) \approx \frac{\lambda}{\pi w_0} \quad (4.11)$$

Once we have these characteristic parameters of the feed, we then determine a design for the lens that is optimized for that feed. The design of a typical lens-based focused-beam measurement system uses a bi-convex lens to transform the source radiation from the feed antenna into a focused beam, and geometrical optics design approximations determine the lens shape. In microwave lenses [3, 5, 6], the usual design concept is that each half of the bi-convex lens transforms rays between the divergent radiation from the focus to a collimated wave in the middle of the lens. One side of a lens is designed based on the feed antenna, and the other side of a lens is designed based on the desired focused beam size at the specimen under test. This concept of converting a diverging beam to a collimated beam is shown in Figure 4.3 and is applied in the lens design equations that follow in the rest of Section 4.1.2 and Section 4.1.3. For illustration, the following discussion applies specifically to the "input" half-lens, which transforms the antenna radiation into

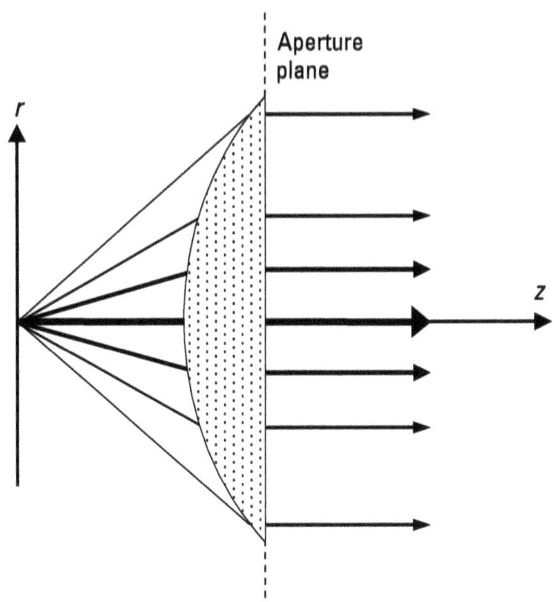

Figure 4.3 Sketch of half-lens showing the principal of transforming from a point source to a collimated beam in terms of optical rays.

an approximately collimated beam at the aperture plane of Figure 4.3. An analogous set of equations can be used design the "output" half of the lens.

The shape of the lens is derived by applying Fermat's principal, which states that the total transit time for a ray emanating from a point source and traveling through the lens must be the same no matter what path that ray takes through the lens. Assuming rotational symmetry and the geometry in Figure 4.4, we derive a formula to describe the thickness as a function of radius. For this calculation, the radiating antenna is assumed to be a point source located at the origin, *O*, and at a distance from the lens apex defined as the focal length, f_i. X is then the distance from the origin to any point on the lens surface. The lens has an index of refraction, n, and is assumed to have negligible loss.

To collimate the diverging rays from the origin, the electrical distance from *O* to any point along the aperture plane must be a constant. When the angle, θ is zero, the electrical length between the origin and the aperture plane is given by

$$\text{Electrical length}\left(\theta = 0\right) = f_i + nt_i \tag{4.12}$$

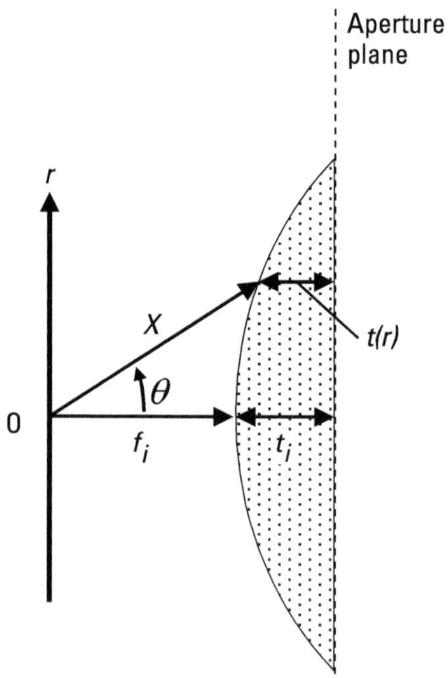

Figure 4.4 The parameters that define the lens shape based on Fermat's principle.

where t_i is the maximum thickness of the lens. Straightforward geometry then enables us to calculate the electrical length at other angles,

$$\text{Electrical length}(\theta) = X + nt = \sqrt{r^2 + \left(f_i + \left(t_i - t\right)\right)^2} + nt \quad (4.13)$$

where t is the thickness and is a function of r. Comparing (4.12) and (4.13) determines the shape of the lens,

$$t(r) = t_i + \frac{f_i - \sqrt{f_i^2 + \left(\frac{n+1}{n-1}\right) r^2}}{n+1} \quad (4.14)$$

The shape of the output lens that focuses at a given focal length is also provided by a similar equation (4.14). An alternate expression for the shape of the lens in terms of X is given by similarly evaluating the geometry shown in Figure 4.4 along with Fermat's principal stated another way: $X + nt = f_i + nt_i$. Noting that $\sin\theta = r/X$ and that $\tan\theta = r/(f_i + t_i - t)$, then

$$X = \frac{(n-1) f_i}{n\cos\theta - 1} \quad (4.15)$$

For $n > 1$, this defines a hyperbola of revolution, where the focal point of the lens is at the focus of the hyperbola.

In Cartesian coordinate systems, the standard expression for a hyperbola is given by

$$\frac{x^2}{a^2} - \frac{y^2}{b^2} = 1 \quad (4.16)$$

In the coordinates defined in Figure 4.3, $x \to z$ and $y \to r$. Recognizing from Figure 4.4 that $X^2 = z^2 + r^2$, and after some algebra, the standard hyperbolic form for the lens defined by (4.16) is described by the following equation

$$\frac{z^2}{a^2} - \frac{r^2}{b^2} = \frac{z^2}{\frac{f_i^2}{(n+1)^2}} - \frac{r^2}{\frac{f_i^2 (n-1)}{(n+1)}} = 1 \quad (4.17)$$

With this expression for the lens shape based on geometrical optics (4.17), we then propagate a Gaussian beam through that lens and iterate as needed for effects due to the Gaussian beam phenomena. The position and diameter of the lens must account for the radiated pattern of the feed antenna as well as the desired beam waist at the focal point where the specimen is placed. More practically the lens diameter is usually limited by cost and space constraints, and this ultimately sets the lower-frequency limit of the designed lens.

For a fixed-lens diameter, the antenna position is set so that the antenna illuminates the lens area efficiently but without over-illuminating. Over-illumination will create unwanted diffraction from the structure holding the lens. A first estimate of the antenna-lens spacing can be determined by setting a design criterion for the power at the lens edge, and a good rule of thumb for material measurement systems is that power at the edge is −20 dB or lower compared to the center. Measured antenna patterns or the Gaussian beam characterization of the antenna described in (4.11) can be used to determine the illumination pattern as a function of distance.

There is an additional complicating factor for setting this separation between the feed antenna and the lens. Size scales typically desired at microwave frequencies result in a nontrivial lens thickness, which is likely to be a significant fraction of the total lens-antenna separation. This thick-lens effect causes the beamwidth exiting the lens to narrow due to refraction by the lens. Thus, we need an expression for propagating the Gaussian beam though the lens that accounts for this thickness effect.

4.1.3 ABCD Matrix Design

Gaussian beam propagation through a lens can be characterized in terms of an ABCD matrix formalism (also known as a ray-transfer matrix) that is widely used in geometrical and Gaussian beam analysis [7, 8]. The ABCD matrix allows the output of an optical element to be written in terms of its input. An input ray is characterized by a vector with components r and θ as defined in Figure 4.4. The ABCD matrix is than applied to determine the vector that represents the output ray,

$$\begin{pmatrix} r' \\ \theta' \end{pmatrix} = \begin{pmatrix} A & B \\ C & D \end{pmatrix} \begin{pmatrix} r \\ \theta \end{pmatrix} \text{ or } \begin{matrix} r' = Ar + B\theta \\ \theta' = Cr + D\theta \end{matrix} \tag{4.18}$$

where the primed components are for the output beam. The coefficients A, B, C, D, are determined by the specific properties of a given optical element, such as a lens, mirror, or interface between different media.

Because the paraxial approximation is assumed, $\tan\theta \approx \theta$, which is the slope of the ray under consideration. From geometric optics, the radius of curvature of the ray at a given distance from the origin is then $R \approx r/\theta$, and the above transformation can be restated in terms of R,

$$R' = \frac{AR + B}{CR + D} \tag{4.19}$$

Equation (4.19) is generalized for a Gaussian beam by replacing the geometrical radius of curvature with the corresponding Gaussian radius of curvature (i.e., the complex beam parameter) defined in (4.9),

$$q' = \frac{Aq + B}{Cq + D} \tag{4.20}$$

To use this formulism for a specific lens design, we must have a library of different ABCD matrixes to apply. For the simplest case of a length, l, of free space, it is straightforward to show that this ABCD matrix is

$$M_1 = \begin{pmatrix} 1 & l \\ 0 & 1 \end{pmatrix} \tag{4.21}$$

This same matrix also applies in a homogeneous medium of arbitrary index. For a planar interface between two regions with difference indices of refraction, the ABCD matrix is

$$M_2 = \begin{pmatrix} 1 & 0 \\ 0 & \frac{n_1}{n_2} \end{pmatrix} \tag{4.22}$$

where the refractive index of the medium the ray is coming from is n_1, and the index of the medium the ray is going toward is n_2. This second matrix can be verified by applying it to (4.18) and showing that it satisfies Snell's law for refraction, when the paraxial approximation is applied.

An often-used transformation is for the thin lens, which is a focusing element consisting of one or two curved interfaces and where the physical separation and thickness is neglected. The equation that describes the thin lens transformation is given by [8]

$$M_{\text{thin lens}} = \begin{pmatrix} 1 & 0 \\ \frac{-1}{f} & 1 \end{pmatrix} \tag{4.23}$$

where f is the focal length of the lens. Unfortunately, the typical design for a focused-beam material-measurement system requires a lens with a thickness that is a significant fraction of the separation between the antenna and the lens output. This thin lens formula ignores the effect of that thickness and thus is too approximate to be useful in microwave measurement systems. In the lens design described here, we have a hyperbolic-shaped surface that can be described by the following ABCD matrix [9, 10]:

$$M_3 = \begin{pmatrix} 1 & 0 \\ \frac{-1}{an(n+1)} & \frac{1}{n} \end{pmatrix} \tag{4.24}$$

where a is the semimajor axis of a hyperbola. For the hyperbolic shape we developed above, we have already obtained a relationship between this parameter and the input parameters of the lens shape. From (4.17) we can write, $a = f_i/(n+1)$.

We are evaluating a half-lens, which includes a hyperbolic surface, a distance within the lens medium, and a flat transition from the lens back to air. To construct a matrix that includes these three parts we simply cascade together the appropriate matrixes for each of these parts of the system. Because the output vector of the beam parameter is on the right of this cascade, the matrixes are placed in order from right to left rather than left to right,

$$M_2 M_1 M_3 M_1 = \begin{pmatrix} 1 & 0 \\ 0 & n \end{pmatrix} \begin{pmatrix} 1 & t_i \\ 0 & 1 \end{pmatrix} \begin{pmatrix} 1 & 0 \\ \frac{-1}{an(n+1)} & \frac{1}{n} \end{pmatrix} \begin{pmatrix} 1 & f_i \\ 0 & 1 \end{pmatrix} = \begin{pmatrix} 1 - \frac{t_i}{f_i n} & f_i \\ \frac{-1}{f_i} & 0 \end{pmatrix} \tag{4.25}$$

where the ABCD matrixes multiplied above correspond to the distance from the feed antenna to the lens (M_1), the hyperbolic lens face (M_3), the thickness of the lens (M_1), and the flat face of the lens (M_2). These regions are shown in Figure 4.5, along with the antenna-to-lens-face distance, f_i, the lens thickness t_i, and the refractive index, n. This matrix is then applied to the beam emanating from the antenna to obtain the beam parameter at the output plane of the lens,

$$q_{\text{output}} = \frac{\left(1 - \frac{t_i}{f_i n}\right) q_{\text{antenna}} + f_i}{\left(\frac{-1}{f_i}\right) q_{\text{antenna}}} \tag{4.26}$$

Another way to derive the effect of the lens on the beam emanating from the antenna is to directly map the incident field amplitude from one side of the lens to the other side while accounting for the nonzero z-depth of the convex side. Referring to Figure 4.6, the incident field amplitude on one side of the

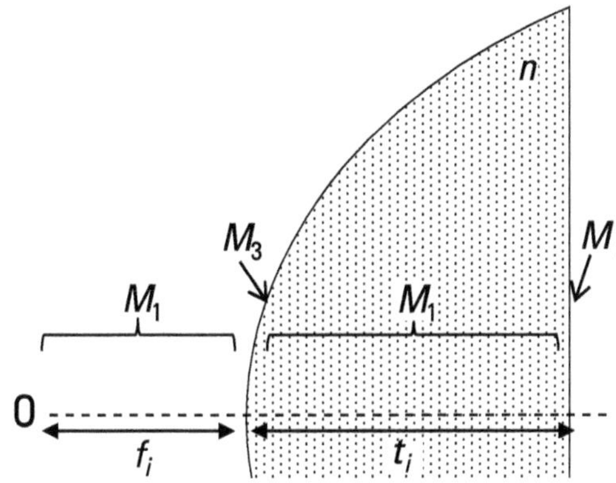

Figure 4.5 Geometry of lens showing the regions corresponding to different cascaded ABCD matrixes ($M_{total} = M_2M_1M_3M_1$) of (4.25).

lens, multiplied by the differential area on that side, must be mapped to the equivalent field amplitude and differential area on the planar side of the lens,

$$u(\theta)\delta\theta = u'(r)\delta r \tag{4.27}$$

where u represents the field amplitude on the convex side, and u' is the field amplitude on the planar side of the lens. This is shown schematically in Figure 4.6. Assuming rotational symmetry, we can rewrite $\delta\theta$ and δr as infinitesimal line segments in spherical coordinates ($\delta\theta = Xd\theta$ and $\delta r = dr$)

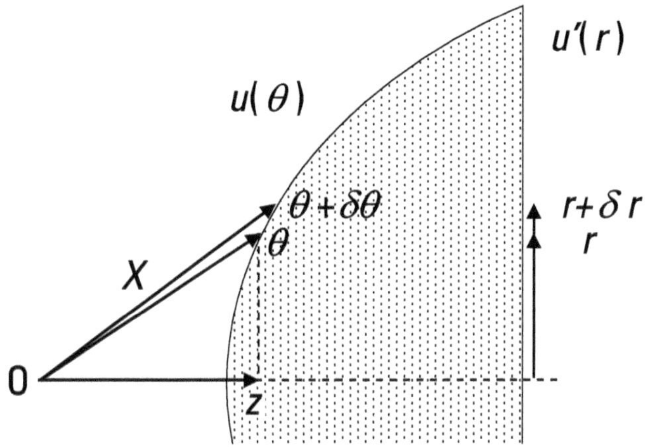

Figure 4.6 Schematic showing relevant variables for the translation of a beam profile from one side to the other of a half-lens.

and reformulate (4.27) to obtain the unknown field amplitude profile on the planar side of the lens in terms of the field on the convex side,

$$u'(r) = X\frac{u(\theta)}{dr/d\theta} \tag{4.28}$$

From Figure 4.4, we know that $r = X\sin\theta$, and from (4.15) we know the angular dependence of X, thus we can evaluate the derivative,

$$\frac{dr}{d\theta} = \frac{d}{d\theta}\left(\frac{\sin\theta(n-1)f_i}{n\cos\theta - 1}\right) = \frac{X(n-\cos\theta)}{n\cos\theta - 1} \tag{4.29}$$

Because $u(\theta)$ is the field amplitude on the hyperbolic surface, it varies with z. Using the geometry of Figure 4.4, we can rewrite the expression that defines the hyperbolic surface so that it is a function of z,

$$X = \frac{(n-1)f_i}{n\dfrac{nz}{X} - 1} = f_i(1-n) + nz \tag{4.30}$$

After some algebra, this can be rearranged to obtain an expression for z that follows the hyperbolic surface as a function of r,

$$z = \frac{nf_i}{n+1} + \sqrt{\frac{r^2}{n^2-1} + \frac{f_i^2}{(n+1)^2}} \tag{4.31}$$

Using the above equations, we can also rewrite the expression for the transformed field amplitude (4.28) in terms of the z that defines the hyperbolic surface,

$$u'(r) = \frac{-f_i}{nf_i - (n+1)z}u(\theta) \tag{4.32}$$

The expression for z can then be inserted into (4.6) to determine $u(\theta)$ on the surface of the hyperbolic lens, and this in turn is inserted into (4.32) to calculate the new field distribution at the output aperture of the half-lens.

To illustrate the use of these equations for a lens design, Figure 4.7 shows a plano-convex lens, optimized to provide a collimated beam from a feed antenna with a 50-degree 3-dB beamwidth at 8 GHz. In this design, the permittivity of the lens is assumed to be 2, and the maximum radius of the lens is 70 cm. The design of this lens began with the feed antenna, which was characterized in terms of an equivalent Gaussian beam waist. The distance of the antenna relative to the lens was optimized so that the power level of energy exiting the lens at the maximum radius was approximately −20 dB from the center of the beam. Once this focal distance was established, then the above derived methods can be used to calculate the convex shape. Figure 4.7 also shows the $1/e$ input beam radius as a function of z, since this corresponds to the Gaussian beam radius defined in (4.5). Notice that applying either Fermat's principal or the ABCD method results in a reduction of both the $1/e$ and 3-dB beam radii after the energy is refracted through the plano-convex lens as compared to what geometrical optics would determine.

The calculated field profile at the output side of the lens is shown in Figure 4.8. The profiles shown in this plot were calculated by three different methods: (1) ABCD matrix method, (2) Fermat's principal, and (3) numerical calculation with a full-wave CEM code based on the FDTD method. The ABCD matrix method and Fermat's principal curves were calculated with equations described above ((4.26) and (4.32), respectively) and are approximate. The FDTD curve was calculated by modeling a two-dimensional representation of the lens. A distribution of plane waves was injected with appropriate weights to provide the Gaussian beam emanating from the feed antenna. The fields adjacent to the output plane of the lens were then sampled and plotted. Because the ABCD matrix and Fermat's calculations are approximate, they

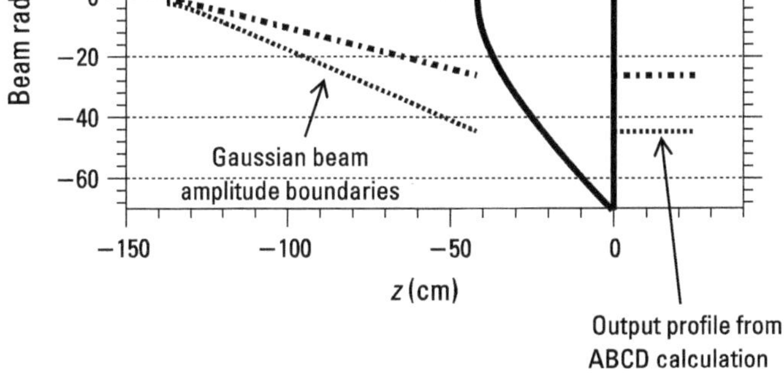

Figure 4.7 Example collimating lens design showing shape and calculated beamwidths when optimized at 8 GHz for a 50-degree beam width feed antenna.

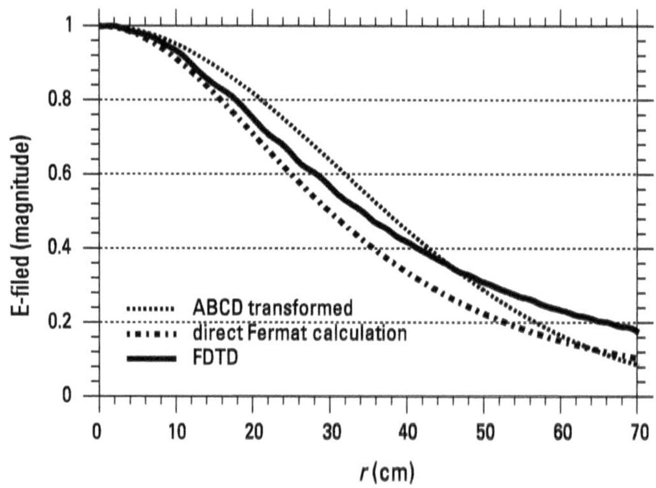

Figure 4.8 Output beam profile for an example-collimating lens calculated by three methods: ABCD matrix method, Fermat's principal, and FDTD full-wave simulation code.

disagree somewhat with each other and with the more rigorous FDTD calculation. There is also a slight amount of ripple in the FDTD-simulated profile due to multiple reflection effects from the lens/air interfaces, which are also not accounted for by the approximate theories.

Full-wave electromagnetic simulations such as the FDTD simulations shown here provide the most accurate representation of the lens behavior. However, they are time-consuming and may be practical only for verifying the design performance. Fortunately, the approximate methods described above provide sufficient accuracy to achieve most design goals. In addition, the actual feed antennas typically deviate somewhat from the idealized Gaussian beam representation across frequency. In other words, the phase centers of many wideband feed antennas vary with frequency, so that experimentally refined antenna-lens distance adjustments may be necessary once a focused-beam system is built.

4.1.4 Lens System Construction

Lenses are typically manufactured from simple dielectric materials such as a polystyrene, polytetrafluoroethylene, or tooling foam, which have low-to-moderate dielectric permittivities. The dielectric constant must be high enough to provide sufficient phase delay for manipulating the electromagnetic energy, but low enough to keep reflections at the lens/air interface to a manageable level. If temperature is a concern, lenses can be manufactured from ceramic foam.

Feed antennas for focused-beam measurement systems are usually horn antennas with linear polarization. These feed antennas can be standard-gain waveguide horns, which have good performance but are band-limited. Alternatively, they can be wideband ridged horns that have increased bandwidth, but at a cost of a somewhat higher voltage standing-wave ratio. With the use of time-domain gating as described in Chapter 2, the decreased match of a wideband horn is often outweighed by this bandwidth increase. Dual-polarization antennas can also be used and are especially convenient for characterizing anisotropic specimens. But dual-polarization horns may increase uncertainties when measuring strongly anisotropic materials. In principle, the best accuracy is obtained with antennas that have good polarization purity, a Gaussian-like beam pattern, and moderate gain in the 10–15-dBi range so that they illuminate a lens placed a reasonable distance in front of the antenna aperture. Selection of an appropriate feed also depends on the wavelength band of interest.

With a fixed physical aperture, the gain characteristics of a typical feed antenna are proportional to A/λ^2, where A is the area of the antenna aperture. When transformed into a focused beam with a bi-convex lens, this results in a beam waist that is proportional to wavelength (or inversely proportional to frequency). Thus, the minimum size of the specimen under test is driven by the lowest frequency of interest. Because of this frequency dependence, a convenient way to specify the beam of a focused-beam system is in terms of a frequency-scaled parameter, $k_0 w_0$, where $k_0 = 2\pi/\lambda_0$ is the wave number in free space.

A common beam configuration for microwave frequencies used in a number of measurement laboratories is $k_0 w_0 \approx 8$. In this case, the beam waist varies from 19 cm at 2 GHz to 2 cm at 18 GHz. Following the –20-dB rule-of-thumb for minimum lateral specimen size and using (4.5), a minimum specimen for this system should be no smaller than 58 cm across at 2 GHz, or 28 cm across at 4 GHz. It is possible to have smaller specimens by designing a system with reduced $k_0 w_0$. However, this also increases measurement errors due to deviation from the plane-wave assumption, as will be shown later in this chapter. At lower frequencies, this error is tolerated because of the practicalities of keeping specimen size reasonable. For example, I previously oversaw the design and construction of a 1.8-m-diameter lens system operating down to 500 MHz. To keep the specimen size reasonable, this system was designed with $k_0 w_0 = 5.5$. The lenses were constructed from cast polypropylene that was machined on a large vertical bore mill and they weighed approximately 1,300 kg each. On the other hand, at millimeter frequencies, specimen size constraints are not as restrictive, and lens systems with larger k_0w_0 values can be more practically constructed. Increasing k_0w_0 improves measurement accuracy by more effectively simulating plane-wave illumination with the specimen under test.

For bi-convex lenses, a convenient way to construct the bi-convex lens is from two half-lenses held together by a structural ring. This flexible design also has the advantage of allowing different lens faces to be switched out for different waist sizes. For example, measurement of inhomogeneous materials can require a larger illumination area to encompass the characteristic length scales in the material. Replacing one of the lens halves with a longer focal length face allows reconfiguration of the waist size to accommodate this inhomogeneity. In some lens systems, the lens and horn are mounted together with a fixed spacing. In this case, the horn/lens pair must be changed out together to switch to another frequency band. Alternatively, the lens and horn can be mounted separately on a common rail system so that the same lens can be used with different feed horns.

4.2 Focused-Beam Measurement Examples

The focused-beam technique can measure a wide range of specimens including agricultural materials [11], magnetodielectrics [12], and novel patterned materials such as metamaterials or frequency selective surfaces [13]. As a noncontact method, it can also measure at low or high temperatures [14]. To provide insight into the characteristic data obtained with focused-beam measurements, Sections 4.2.1 and 4.2.2 present representative examples of some experimental measurements. These examples illustrate the use of focusing lenses to obtain intrinsic material properties. With the relative complexity of these fixtures combined with the wide variety of possible specimen characteristics, it is easy to acquire data but sometimes difficult to obtain *accurate* data. Some of the examples presented here represent challenging measurements that can stress the accuracy of these measurement systems and thus demonstrate some of the measurement issues that can occur, as well as strategies for overcoming them.

4.2.1 Dielectric Measurements

A homogeneous, nonmagnetic specimen provides a first illustration of focused-beam measurement data. A widely known and well-characterized dielectric material used in microwave applications is cross-linked polystyrene, known by the trade name of Rexolite®. It has a real dielectric permittivity that is approximately 2.53 at microwave frequencies. It also has very little frequency dispersion so the imaginary part of the permittivity is small—smaller than can be typically measured using transmission and reflection techniques. Figure 4.9 shows the inverted real and imaginary permittivity determined for a 3.3-mm-thick polystyrene specimen measured in a focused-beam system. For comparison, real and imaginary permittivity determined with different

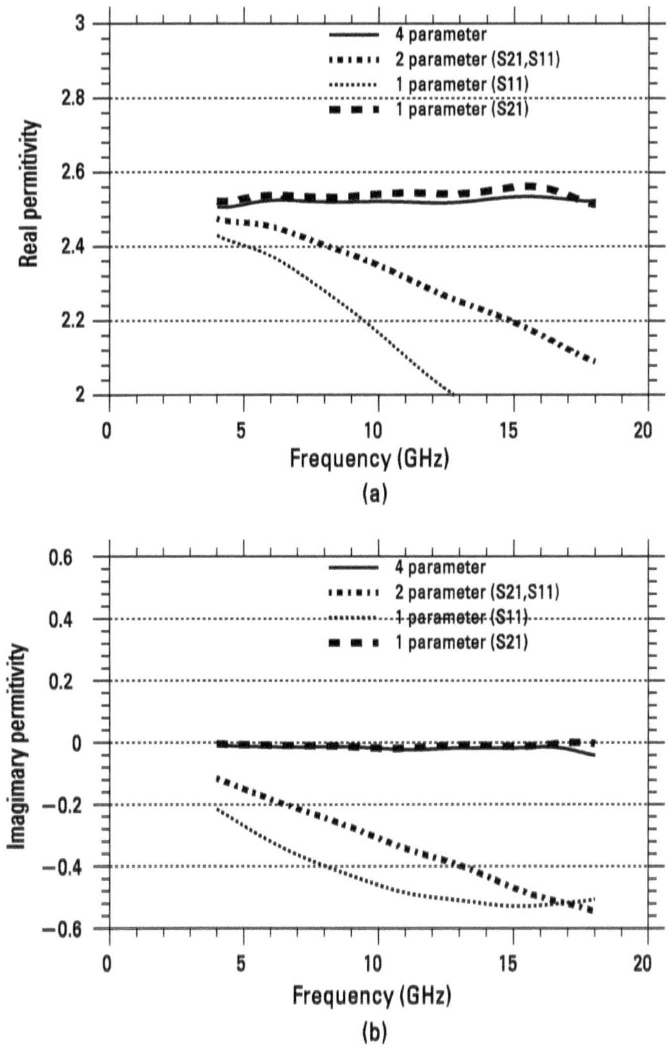

Figure 4.9 (a) Real and (b) imaginary permittivity of crosslinked polystyrene (Rexolite®) measured with a focused beam and inverted from various network scattering parameters.

inversion methods on the same measured data are plotted. These inversions were described in Chapter 2 (Sections 2.4.3, 2.4.4, and 2.4.7). In all four cases, a fixed time-domain window of 0.75 ns was applied after calibration. The four-parameter method shows excellent agreement with the known permittivity and dielectric loss of polystyrene. The two-parameter and one-parameter, reflection (S_{11}) methods, however, show significant deviations in the form of unexpected frequency dispersion in the real and imaginary permittivity. The sign convention used for these data is $\varepsilon^* = \varepsilon' - i\varepsilon''$, so a negative imaginary permittivity indicates gain. Material gain is not physically possible, so these dispersive data (two-parameter and S_{11} inversions) are erroneous. On the other hand, the four-parameter and S_{21} imaginary data are very close to zero, and similarly, any excursion into negative imaginary permittivity is a result of measurement uncertainty.

The common factor between the two-parameter and S_{11}-only inversions causing erroneous results is the phase of the S_{11} data. This data is inaccurate, because the reference plane defined by the metal calibration plate and the front face of the specimen are not in the same position. This can happen when the specimen is not flat and is instead slightly bowed. Even with a small amount of bowing—just a fraction of a millimeter displacement between specimen and reference—significant reflection phase errors can occur. These errors increase as frequency increases because the fixed physical displacement increases in electrical size. One method used in the past to correct this error involves manually fitting the data with a reflection phase correction that eliminates nonphysical imaginary permittivities and permeabilities. However, this is an unsatisfactory solution that is at best semi-empirical.

On the other hand, the four-parameter inversion accounts for specimen/reference plane differences by measuring reflection in both directions. This limits the phase error to uncertainty in the calibration plate thickness. Confirmation of this phase error phenomena can be seen in the one-parameter, transmission (S_{21}) data shown in Figure 4.9. This data shows good agreement with the four-parameter results. The S_{21} inversion does not include reflection phase, so the errors associated with reflection reference plane displacement are nonexistent. Therefore, S_{21} inversion is typically preferred for nonmagnetic specimens. Unfortunately, it is not an option when magnetic specimens are measured, and that is the primary advantage of the four-parameter method.

The data in Figure 4.9 shows that the four-parameter inversion minimized errors from small, submillimeter specimen displacements. Figure 4.10 shows the measured real and imaginary permittivity of the same specimen with the four-parameter inversion when displaced by large distances. These plots each show five curves where the specimen was displaced in 6.35-mm increments.

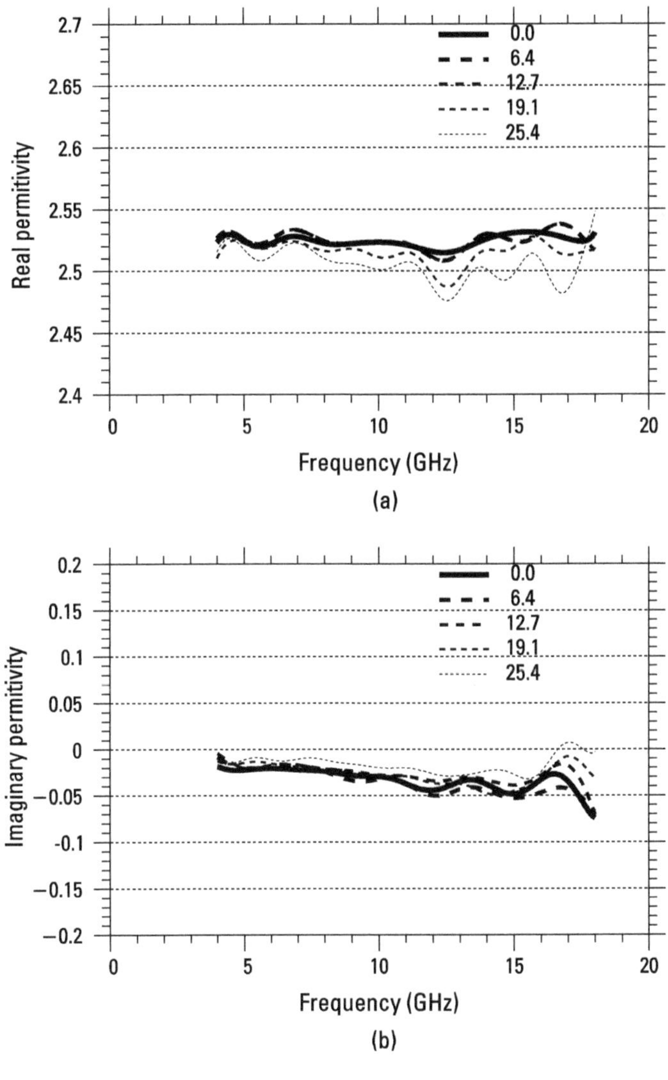

Figure 4.10 (a) Real and (b) imaginary permittivity of cross-linked polystyrene measured with four-parameter inversion as a function of specimen displacement.

Even at the largest displacement of 25.4 mm from the calibration plate reference plane, the measurement error has only increased slightly—still less than a few percent for the real permittivity. This slight error increase is likely due to the finite focal depth of the Gaussian beam.

As an alternative method for minimizing the reflection phase error, a mechanical or noncontact laser micrometer fixture can be incorporated into the focused-beam system and used to physically measure the difference between the specimen and the calibration plate location. The additional apparatus required for this position measurement, however, may not be as convenient as the four-parameter method, since it adds more steps to the measurement procedure, and measuring scattering parameters in both directions is easily automated. Nevertheless, the four-parameter method assumes single-layer, reciprocal specimens. So, if multilayer specimens are measured, and reflection phase is important, then micrometer measurements could be necessary for accurately obtaining reflection phase.

Another example measurement of a homogeneous dielectric material is shown in Figure 4.11. The data in Figure 4.11(a) shows the measured real permittivity for a thin polyimide sheet with a thickness of only 75 microns (0.003 inches). At 10 GHz, this specimen thickness is approximately 1/400th of a wavelength, so it exhibits a very small perturbation on the transmitted wave in a focused-beam system. Two curves are shown in the upper plot of Figure 4.11, each corresponding to a different calibration. In this case, only the transmission coefficient is used to invert permittivity, so the calibration depends primarily on a clear site measurement (S_{21} with no specimen). Both curves are based on the same specimen measurement; however, the first calibration was conducted a few minutes before the specimen measurement, and the second calibration occurred a few minutes after the specimen measurement. The total time between the two calibrations was nine minutes, and the specimen measurement occurred in the time between the two.

Figure 4.11(b) shows the phase difference between the two calibration measurements. In low-loss dielectric materials, the phase delay that is measured is dominated by the real part of the dielectric permittivity and its effect on the speed of light through the specimen. So, the small difference between the phases of the two calibrations is responsible for the difference in measured permittivity shown in Figure 4.11(a). These differences indicate a phase drift of the measurement apparatus occurring over the course of the measurement. If that phase drift is approximately linear, and the measurement was made equidistant in time from the two calibrations, then a simple average of these two curves will correct for that phase drift. The expected permittivity for this polyimide measurement is approximately $\varepsilon = 3.5$, which roughly corresponds to the average value between the two measured curves. Because this specimen is so thin, and the measured phase is so small, it is susceptible to phase drifts of even a tenth of a degree or less.

Laboratory-grade modern network analyzers have highly stable sources, so phase drift from the internal electronics is generally small. For that reason, the dominant source of phase drift in a focused beam apparatus is usually

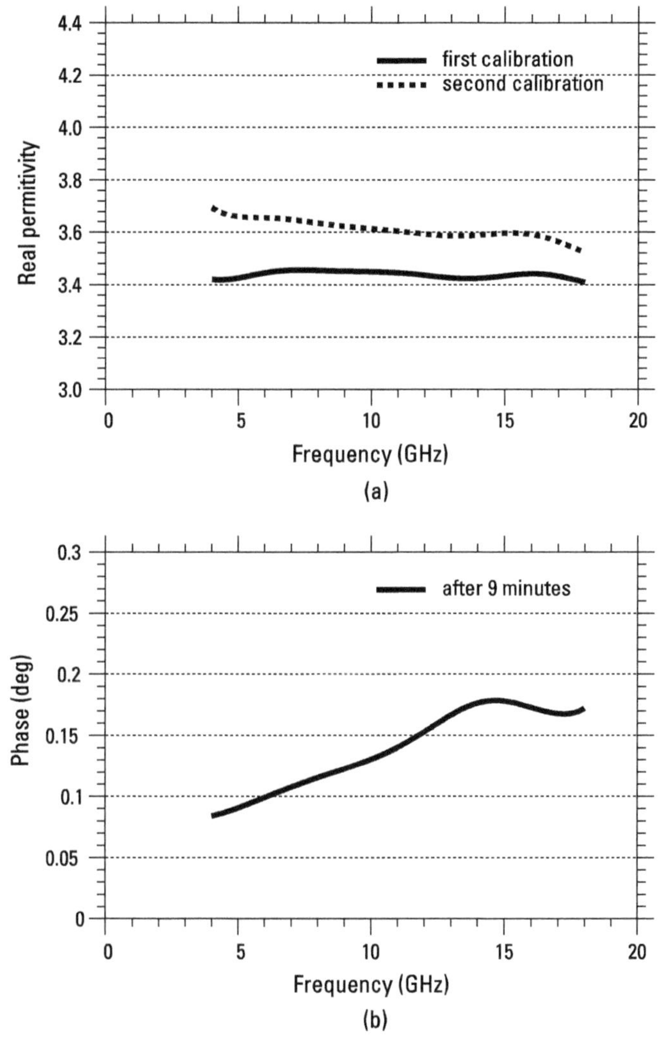

Figure 4.11 (a) Measured permittivity for a thin (0.003″/76 microns) polyimide sheet for two different calibrations, before and after the specimen measurement. (b) Relative phase drift during the nine-minute period between the two calibration measurements.

ambient temperature changes that affect the RF cables. In the measurement of Figure 4.11, temperature changes occurred due to the cycling of the laboratory air conditioning. In particular, the microwave cables that connected the network analyzer to the feed antennas were coaxial transmission lines with the center conductor separated from the outer conductor by a dielectric spacer. Teflon is often used as this spacer, and its dielectric permittivity is temperature-dependent [15]. Teflon also undergoes a crystalline phase transition near room temperature reflected in the thermal expansion coefficient, which shows a rapid change right around room temperature [16]. From these effects, the electrical length of a Teflon cable experiences a relatively rapid temperature-dependent change near room temperature, sometimes referred to as the Teflon knee.

Because of this Teflon knee, ambient temperature changes of even a fraction of a degree can induce nontrivial phase errors in the cables, such as the phase drift observed in Figure 4.11. In thicker specimens, the phase delay induced by the specimen is much larger, and this small phase drift is not a significant factor. However, in the 76-microns thick specimens of Figure 4.11, the material induced phase delay is small so that the cable-caused phase drift is more noticeable. Consequently, it is important to monitor for phase-drift error and correct it when it occurs. For example, maintaining a steady temperature in the laboratory is important, as is calibrating often enough so that phase drift between calibrations is minimized. In cases such as Figure 4.11, where even a very small phase drift dominates, some laboratories have resorted to actively cooling their microwave cables. This can be as simple as wrapping the microwave cable with copper tubing that circulates chilled water. Cooling the cables to temperatures away from the Teflon-knee temperature reduces variations in phase to an almost negligible level, even over longer (one- to three-hour) measurement intervals. A more convenient solution, however, is to use the cable phase correction method described in Chapter 3.

4.2.2 Magneto-Dielectric Measurements

Moving beyond simple dielectric specimens, example inversions for a magnetic material measured in a focused-beam system are shown in Figure 4.12 for dielectric permittivity and in Figure 4.13 for magnetic permeability. This data is the product of a commercial magnetic absorber material made from carbonyl iron powder mixed with an elastomer, and four different inversions are compared. Looking first at the NRW and iterative two-parameter inversions, both methods overlay, because they are based on the same scattering parameters. Because the magnetic specimen was flexible, both of these inversions also have significant errors due to the displacement or bowing of the specimen relative to the reference plane defined by the metal calibration plate.

Also on these plots are curves from a "corrected" NRW method and the four-parameter inversion method. The corrected NRW method applied a phase offset corresponding to a net 0.32-mm displacement of the specimen from the calibration plate. This displacement was measured with a mechanical

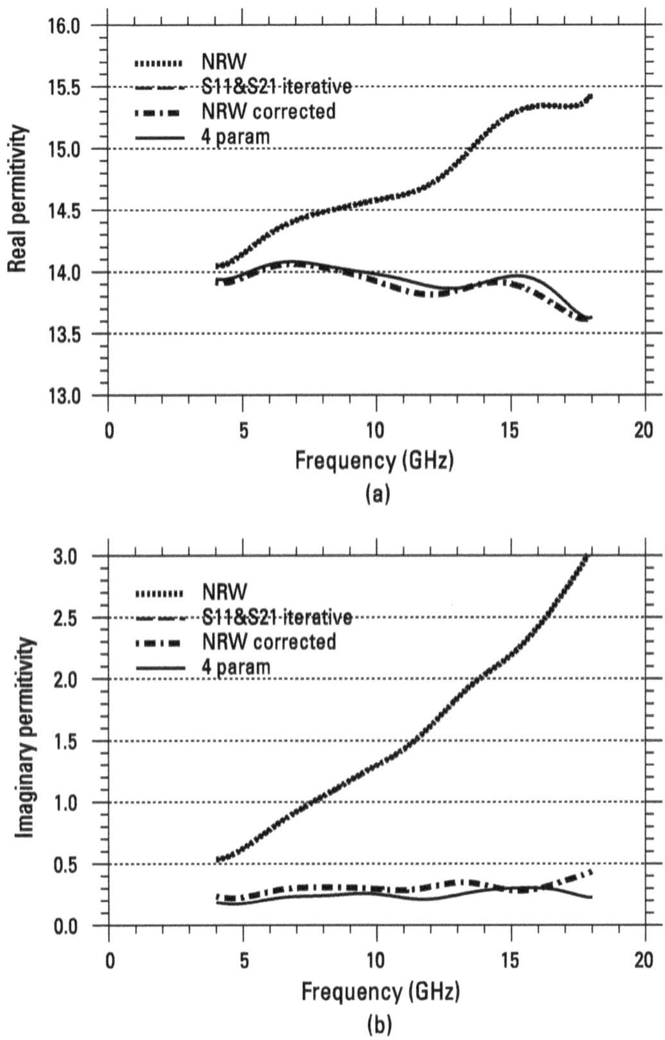

Figure 4.12 (a) Real and (b) imaginary permittivity for a commercial magnetic absorber using various inversion algorithms.

micrometer fixture positioned in the fixture after the metal calibration plate was measured. It obtained a reference location for the metal plate. The metal plate was then removed, the specimen inserted, and the micrometer adjusted to determine the mechanical offset at a position corresponding to the center of the illuminating beam. The micrometer was then removed so that it would not interfere with the microwave beam during specimen measurement.

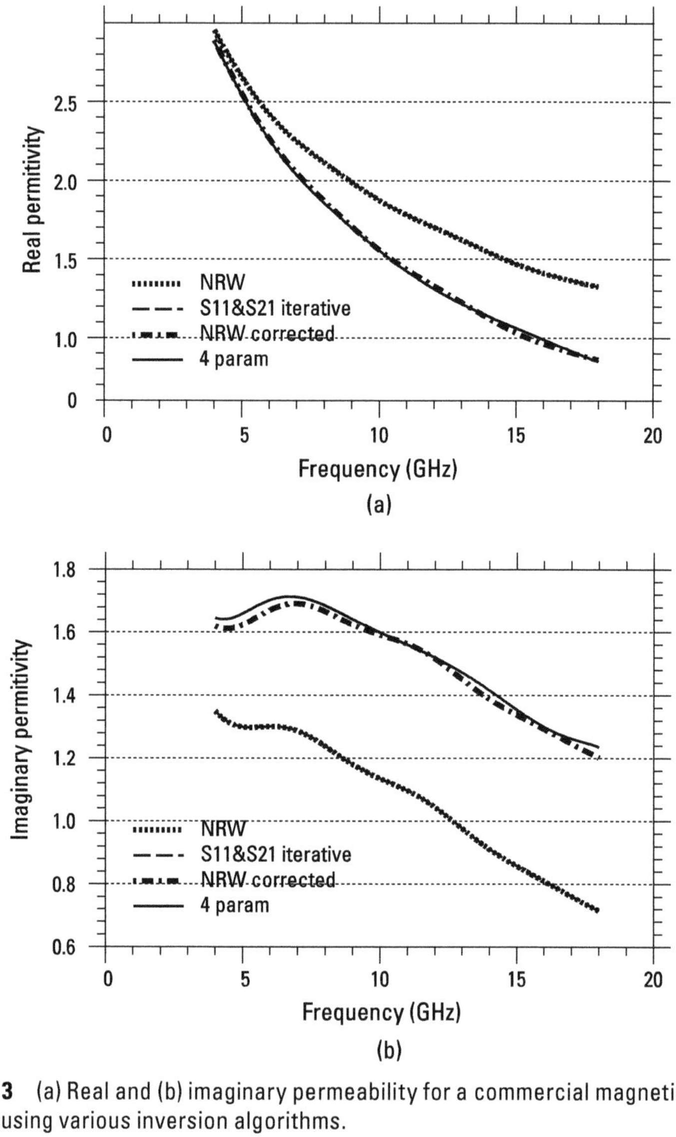

Figure 4.13 (a) Real and (b) imaginary permeability for a commercial magnetic absorber using various inversion algorithms.

The agreement between this corrected NRW and the four-parameter method further demonstrates the importance of accounting for the reflection phase due to specimen displacement. The net effect of the phase error in the uncorrected inversions was a frequency dependent offset of both the permittivity and permeability. The amount of offset increased with frequency, because the electrical size of the 0.32-mm specimen displacement also increased with frequency. One final observation: unlike the other two inversion methods, the corrected NRW and the four-parameter curves do not exactly overlay. This is because the NRW method is based on two of the four scattering parameters, while the four-parameter also included the other two scattering parameters. There may be some inhomogeneity in the iron distribution through the thickness of the material, and measuring reflections in both directions averages out this inhomogeneity.

A second set of inverted properties for a much more challenging magneto-dielectric material specimen is shown in Figure 4.14. This material is a highly anisotropic composite of magnetic inclusions in a polymer matrix, designed to work as an EMI absorber in consumer electronics devices. In contrast to the more spherical shape of the iron in the example covered in Figures 4.12 and 4.13, the magnetic inclusions in Figure 4.14 are flattened in shape so that they have a high magnetic permeability and loss at VHF and UHF frequencies. The high aspect ratio of the inclusions also gives them a large dipole moment resulting in very high dielectric permittivity. Because the magnetic relaxation is primarily at frequencies in the VHF and UHF bands (30 MHz–3 GHz), the focused beam data of Figure 4.14 shows just the high-frequency tail of the permeability relaxation. The real part of the permittivity dips below zero in this part of the relaxation, consistent with behavior predicted by the Lorentz relaxation function discussed in Chapter 1. To capture lower-frequency permeability of magnetic relaxations such as this, methods such as coaxial airlines discussed in Chapter 5, or magnetic probes or permeameters such as those described in Chapters 7 and 8 can be used without requiring electrically large specimens.

The inverted data of Figure 4.14 was obtained with the four-parameter inversion method, which eliminates phase errors in the reflection coefficients. While using the more traditional NRW or S_{11} and S_{21} iteration methods results in significant errors in the magnetic specimens of Figures 4.12 and 4.13, the more extreme permittivity of the EMI specimen dramatically increases sensitivity to phase errors. Thus, attempts to use the NRW or two-parameter iteration methods do not even converge, and the four-parameter method is the only feasible option for this hard-to-measure specimen at these frequencies.

Calculating material inversions from scattering parameter data is sometimes made more difficult because the relationships between the scattering parameters and the intrinsic properties are not monotonic, meaning that there are frequently multiple solutions. The occurrence of multiple solutions becomes worse when material specimens are electrically thick. When iterative methods are used, the initial guess values must be reasonably close, or the solver may converge onto an incorrect solution. Obtaining reasonable

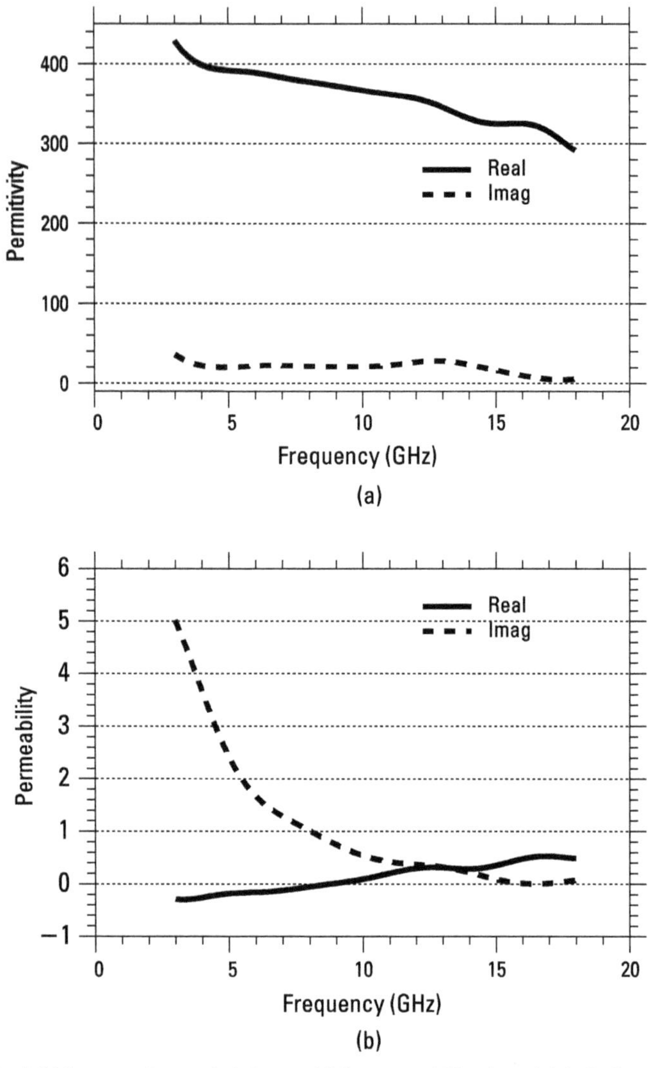

Figure 4.14 (a) Measured permittivity and (b) permeability for a high-index, 0.25-mm (0.010″) thick commercial EMI absorber material.

results usually requires some foreknowledge of the expected range for the intrinsic properties. In difficult-to-measure specimens such as that of Figure 4.14, these difficulties are exacerbated. Thus, it is sometimes advantageous to supplement focused-beam measurements with other techniques, such as cavity or transmission line methods in order to have better bounds on the expected properties for the inversion.

4.3 Measurement Uncertainties

Uncertainty analysis for transmission/reflection measurements is challenging and usually requires more work than the measurement itself. An inversion algorithm to calculate permittivity and permeability often includes numerical root finding, and the relationship between permittivity and permeability and the scattering parameters is not a simple functional relationship. Primary factors that contribute to uncertainty in focused beam measurements include the following:

- Inherent network analyzer accuracy limits;
- Specimen position/flatness uncertainty;
- Specimen dimension uncertainty;
- Focusing error (deviation from true plane wave);
- Multipath noise.

There are several methods for adding up the total uncertainty from these contributions, ranging from Monte-Carlo calculations to analytic uncertainty estimates. Monte-Carlo methods have the advantage of more rigorously representing the functional dependencies by explicitly incorporating the governing equations that relate measured variables to the calculated parameters. However, the Monte-Carlo method requires more computational resources since it must evaluate the governing equations repeatedly to obtain a statistically significant sampling. Analytical methods instead calculate the uncertainties with a function that reduces the need for multiple evaluations of the governing equations. While faster, these analytical functions are approximate.

The Taylor series method is a commonly used analytical method for uncertainty calculation. However, it may include truncation errors depending on the linearity of the relationship between the measured variables and the calculated parameters. It does have the advantage of a simple implementation relative to other methods so is reviewed in the rest of this section. Other more advanced uncertainty propagation methods based on numerical methods are reviewed elsewhere [17].

An example of a Taylor series uncertainty estimate, also known as a root mean square (RMS) error analysis, is as follows [18]

$$\delta\varepsilon = \sqrt{\sum\nolimits_{i,j}\left(\frac{\partial\varepsilon}{\partial\left|S_{ij}\right|}\delta\left|S_{ij}\right|\right)^2 + \sum\nolimits_{i,j}\left(\frac{\partial\varepsilon}{\partial\Phi_{ij}}\delta\Phi_{ij}\right)^2 + \left(\frac{\partial\varepsilon}{\partial t}\delta t\right)^2} \tag{4.33}$$

$$\delta\mu = \sqrt{\sum\nolimits_{i,j}\left(\frac{\partial\mu}{\partial\left|S_{ij}\right|}\delta\left|S_{ij}\right|\right)^2 + \sum\nolimits_{i,j}\left(\frac{\partial\mu}{\partial\Phi_{ij}}\delta\Phi_{ij}\right)^2 + \left(\frac{\partial\mu}{\partial t}\delta t\right)^2} \tag{4.34}$$

where the uncertainties in ε and μ are calculated by a sum of squares evaluation. These equations apply to any of the inversion methods described above depending on which S-parameters are included. The δ operator designates the standard uncertainty. The measured variables included in these sums are amplitude of the scattering coefficient, $|S_{ij}|$, the phase of the scattering coefficient, Φ_{ij}, and specimen thickness, t. The specific form of these equations depends on which method is used to invert permittivity and/or permeability. In this error propagation method, the standard uncertainties are multiplied by a weighting factor, which is the derivative of the calculated quantity with respect to the measured variable.

The various weighting factors can be precomputed by taking the relevant derivatives of the constitutive equations for the applicable inversion algorithm. An example of using this Taylor uncertainty estimate method is given for the case of waveguide and coaxial transmission lines in [19, 20]. These analyses are easily adapted to the free-space measurement method. Sections 4.3.1–4.3.5 outline the dominant uncertainty sources in focused-beam measurements. These sources can be used to build an uncertainty model that depends on the inversion algorithm being used, as well as the measurement hardware. Conversely, improving measurement accuracy is principally addressed by devising new strategies for reducing these uncertainty sources.

4.3.1 Transmission Line Errors

Because of mismatches at the various junctions in the system (e.g., network analyzer ports, horns, lenses, and cables), undesired reflections will corrupt the measured scattering parameters. Time-domain measurements such as those in Figure 2.4 have shown that the two largest mismatches are usually the discontinuities at the network analyzer port and at the input of the antenna. Interference from these mismatch reflections must be removed with calibration and data processing to allow more accurate measurement of the specimen signals. Even with calibration, however, there remains residual error in the measured signal. This can be illustrated through a simplified transmission line model, as shown in Figure 4.15. This model describes the characteristics of a reflection (S_{11}) measurement, where for illustration only one side of a focused-beam system (i.e., one horn and lens) are modeled. The focused-beam model includes shunt impedances for the mismatches of the network analyzer port (Z_1), the antenna (Z_2), and the reflection from the specimen (Z_3). Z_0 is the intrinsic impedance of the transmission line, which is free-space in this case, and is therefore approximately 377Ω.

This transmission line model can evaluate the effectiveness of the calibration. For example, we assume the response and isolation calibration method described in Chapter 2 by (2.3). For reflection, this calibration technique uses two standards: an electrical short as the response ($S_{11} = 1$) and a matched load as the isolation ($S_{11} = 0$). For focused-beam measurements, the short is a conductive plate with an area much greater than the incident beam, oriented normal to the incident beam. The matched load is created by removing the metal plate. Referring to Figure 4.15, the reflection coefficient at the beginning of the transmission line (e.g., as measured by the network analyzer) due to the combined effects of the network analyzer port, antenna, and specimen can be computed by a cascaded matrix formulation of the various reflections, where each reflection is represented by a matrix similar to (2.50). Assuming $Z_3 = 0$ for the short calibration standard and $Z_3 = Z_0$ for the load calibration standard, the reflection S-parameters for those cases can be applied to obtain

$$S_{11}^{\mathrm{cal}} = \frac{Z_3 - Z_0}{Z_3 + Z_0 + \zeta Z_3} \tag{4.35}$$

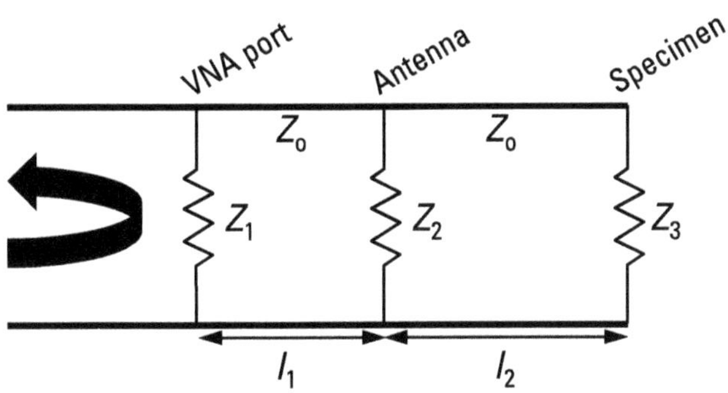

Figure 4.15 Transmission line model for estimating effects of horn and lens reflections.

where

$$\zeta = \frac{Z_0\left(e^{-ia} - e^{-ib}\right) - 2\left(Z_2 e^{-ia} + Z_1 e^{-ib}\right)}{Z_0\left(\cos b - \cos a\right) + Z_1\left(e^{-ib} - e^{-ia}\right) - 2iZ_2 \sin a - 2\frac{Z_1 Z_2}{Z_0}e^{-ia}} \tag{4.36}$$

and where $a = \beta(l_3 + l_2)$, $b = \beta(l_3 - l_2)$, β is the propagation constant $(2\pi/\lambda)$, l_2 and l_3 are the network analyzer—antenna and antenna—specimen separations, and Z_0 is the line impedance.

The voltage-reflection coefficient of a load impedance, Z_3, in an ideal transmission line with line impedance Z_0, is given by

$$S_{11}^{\text{ideal}} = \frac{Z_3 - Z_0}{Z_3 + Z_0} \tag{4.37}$$

Comparison of (4.35) and (4.37) shows that application of the response-and-isolation calibration does not completely eliminate the effects of the antenna and lens mismatches. There is a residual term, ζZ_3, that is proportional to the reflection properties of the specimen under test. Note that when minimizing the other mismatches, Z_1 and $Z_2 \to \infty$, ζ goes to zero, and this term disappears. Thus, reducing the impedance mismatches at the network analyzer port and antenna can improve measurement accuracy.

The amplitude of ζ can be estimated by assuming that the antenna has a voltage standing wave ratio (VSWR) of approximately 2, so that the ratio of line impedance to the antenna impedance is, $Z_0/Z_2 \sim 0.5$. Assuming that the contributions from other mismatches can be neglected compared to the antenna, then $|\zeta|$ can be as much as 0.25. With this value of $|\zeta|$, the ζZ_3 term can be significant, and additional data processing must be used to reduce this systematic error and increase the accuracy of the calculated reflection coefficient. In other words, response-and-isolation calibration reduces the effects of the primary mismatch reflections but does not eliminate the residual multipath reflections that interact with the specimen. Because of this, time-domain gating is used to further refine the measured data, and a more sophisticated error model that directly simulates the time-domain gate effects may be necessary. This can be achieved by applying time-domain processing to this error model for a desired frequency range and calculating the remaining error. However, this is no longer a purely analytical approach.

4.3.2 Focusing Error

Determining the transmission and reflection coefficients from a material specimen assumes that the incident energy is equivalent to an ideal, far-field, plane wave. In reality, the focused beam underilluminates a specimen with a tapered amplitude profile. This Gaussian-like amplitude taper means the focused illumination is not a true plane wave but rather an approximation. To determine the effect of the finite-beam width on transmission and reflection from a target, Petersson and Smith compared the transmitted and reflected power to that from an infinite illumination case via plane-wave spectrum analysis [21]. Figure 4.16 illustrates the impact of a tapered beam in system performance in terms of a plane-wave spectrum representation. The incident field is expressed as a superposition of plane waves. This series of incident plane waves propagates through a planar specimen, and the propagated plane waves are summed to determine transmitted and reflected power. Specifically, this scenario is analyzed in terms of the time-average power passing through input and output reference planes defined on either side of the specimen.

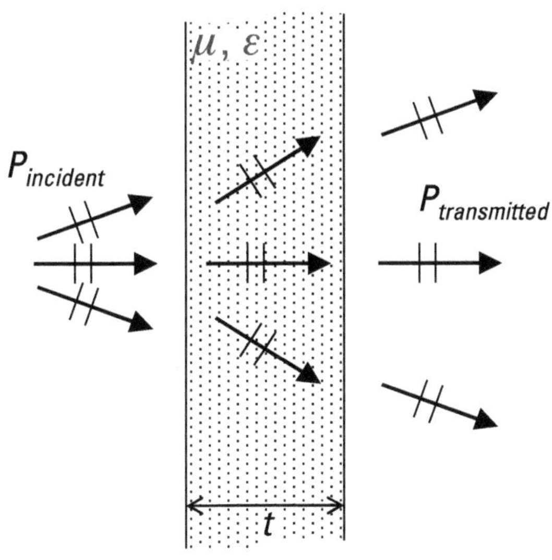

Figure 4.16 Sketch of problem geometry showing plane-wave propagation through specimen.

The plane waves are split into transverse-electric and transverse-magnetic (parallel and perpendicular polarization) components to express the transmission and reflection coefficients (Fresnel coefficients). The following expressions for incident, transmitted, and reflected power, complete without any approximations, [(4.38) to (4.45)] are then derived [22]:

$$P_i = \frac{\pi}{4\eta_0} w_0^2 \left|E_0\right|^2 \left(k_0 w_0\right)^2 \int_{\tilde{k}_r=0}^{1} \left(1-\frac{\tilde{k}_r^2}{2}\right) \frac{\tilde{k}_r}{\tilde{k}_z} e^{-\frac{\tilde{k}_r^2 (k_0 w_0)^2}{2}} d\tilde{k}_r \tag{4.38}$$

$$P_t = \frac{\pi}{8\eta_0} w_0^2 \left|E_0\right|^2 \left(k_0 w_0\right)^2 \int_{\tilde{k}_r=0}^{1} \frac{\tilde{k}_r}{\tilde{k}_z} \left[\tilde{k}_z^2 \left|T_\perp\left(\tilde{k}_r\right)\right|^2 + \left|T_{||}\left(\tilde{k}_r\right)\right|^2 \right] e^{-\frac{\tilde{k}_r^2 (k_0 w_0)^2}{2}} d\tilde{k}_r \tag{4.39}$$

$$P_r = \frac{\pi}{8\eta_0} w_0^2 \left|E_0\right|^2 \left(k_0 w_0\right)^2 \int_{\tilde{k}_r=0}^{1} \frac{\tilde{k}_r}{\tilde{k}_z} \left[\tilde{k}_z^2 \left|R_\perp\left(\tilde{k}_r\right)\right|^2 + \left|R_{||}\left(\tilde{k}_r\right)\right|^2 \right] e^{-\frac{\tilde{k}_r^2 (k_0 w_0)^2}{2}} d\tilde{k}_r \tag{4.40}$$

where

$$T_\perp\left(\tilde{k}_r\right) = \frac{4\tilde{k}_z \tilde{k}_{z\mu} e^{-i\tilde{k}_{zs} k_0 t}}{2\tilde{k}_z \tilde{k}_{z\mu} + \tilde{k}_z^2 + \tilde{k}_{z\mu}^2 + \left[2\tilde{k}_z \tilde{k}_{z\mu} - \tilde{k}_z^2 - \tilde{k}_{z\mu}^2\right] e^{-i2\tilde{k}_{zs} k_0 t}} \tag{4.41}$$

$$R_\perp\left(\tilde{k}_r\right) = \frac{\tilde{k}_z^2 - \tilde{k}_{z\mu}^2 - \left[\tilde{k}_z^2 - \tilde{k}_{z\mu}^2\right] e^{-i\tilde{k}_{zs} k_0 t}}{2\tilde{k}_z \tilde{k}_{z\mu} + \tilde{k}_z^2 + \tilde{k}_{z\mu}^2 + \left[2\tilde{k}_z \tilde{k}_{z\mu} - \tilde{k}_z^2 - \tilde{k}_{z\mu}^2\right] e^{-i2\tilde{k}_{zs} k_0 t}} \tag{4.42}$$

$$\mathrm{T}_{||}\left(\tilde{k}_r\right) = \frac{4\tilde{k}_z \tilde{k}_{z\varepsilon} e^{-i\tilde{k}_{zs} k_0 t}}{2\tilde{k}_z \tilde{k}_{z\varepsilon} + \tilde{k}_z^2 + \tilde{k}_{z\varepsilon}^2 + \left[2\tilde{k}_z \tilde{k}_{z\varepsilon} - \tilde{k}_z^2 - \tilde{k}_{z\varepsilon}^2\right] e^{-i2\tilde{k}_{zs} k_0 t}} \tag{4.43}$$

$$R_{||}\left(\tilde{k}_r\right) = \frac{\tilde{k}_z^2 - \tilde{k}_{z\varepsilon}^2 - \left[\tilde{k}_z^2 - \tilde{\mathrm{k}}_{z\varepsilon}^2\right] e^{-i\tilde{k}_{zs} k_0 t}}{2\tilde{k}_z \tilde{k}_{z\varepsilon} + \tilde{k}_z^2 + \tilde{k}_{z\varepsilon}^2 + \left[2\tilde{k}_z \tilde{k}_{z\varepsilon} - \tilde{k}_z^2 - \tilde{k}_{z\varepsilon}^2\right] e^{-i2\tilde{k}_{zs} k_0 t}} \tag{4.44}$$

$$\tilde{k}_z = \sqrt{1-\tilde{k}_r^2},\ \tilde{k}_{zs} = \sqrt{\mu\varepsilon - \tilde{k}_r^2},\ \tilde{k}_{z\mu} = \frac{\tilde{k}_{zs}}{\mu},\ \tilde{k}_{z\varepsilon} = \frac{\tilde{k}_{zs}}{\varepsilon} \tag{4.45}$$

and where t is the specimen thickness, $k_0 = 2\pi f/c$ is the free-space wavenumber, and μ and ε are the relative permeability and permittivity of the specimen.

The tilde symbol denotes when the k-vector is already normalized to k_0. The power reflection and transmission coefficients for the finite beam are then computed as

$$R_{\text{beam}} = \frac{P_r}{P_i} \quad \text{and} \quad T_{\text{beam}} = \frac{P_t}{P_i} \tag{4.46}$$

which can then be integrated numerically. The relative error due to the finite beam is determined by comparing these reflection and transmission coefficients to that from an ideal plane wave. The plane wave reflection and transmission power coefficients, R_{pw} and T_{pw}, are computed from the above equations when $\tilde{k}_r = 0$. The relative error in the scattering parameters due to the finite beam size is then

$$\delta S_{11} = \frac{\sqrt{R_{\text{beam}}} - \sqrt{R_{pw}}}{\sqrt{R_{pw}}} \quad \text{and} \quad \delta S_{21} = \frac{\sqrt{T_{\text{beam}}} - \sqrt{T_{pw}}}{\sqrt{T_{pw}}} \tag{4.47}$$

The relative error in the transmission coefficient is plotted for representative cases in Figure 4.17 as a function of wavelength-normalized specimen thickness. Note that the wavelength normalization includes the effect of the permittivity and permeability on wavelength within the material. Three different curves are shown on each plot, each corresponding to a different dielectric permittivity of the specimen. This data shows that for a beam radius of $k_0 w_0 = 8$, the systematic errors are a fraction of a percent, with the general trend that error gradually increases with increasing specimen thickness. Also comparing the two plots shows that error levels are lower when the specimen has a magnetic permeability greater than 1.

Figure 4.18 compares the relative transmission (a) and reflection (b) error as a function of electrical thickness for an $\varepsilon = 10$ dielectric slab with several different beam diameters ($k_0 w_0 = 6$, 8, and 12). This data shows that increasing the beam diameter reduces the systematic error from focusing. Figure 4.18 also shows that the relative error becomes exceedingly large in S_{11} at specimen thicknesses corresponding to integral multiples of $\lambda/2$. These errors are due to the inherent resonance caused by interference between the front and back surface of the specimen. Under this condition, S_{11} becomes increasingly small. Excepting for these $\lambda/2$ cases in S_{11}, the plane-wave approximation results in systematic error less than 0.1% for a $k_0 w_0 = 8$ focused-beam system.

With expressions relating the focusing error to permittivity and thickness, it is also possible to apply a focusing correction to measured data in a

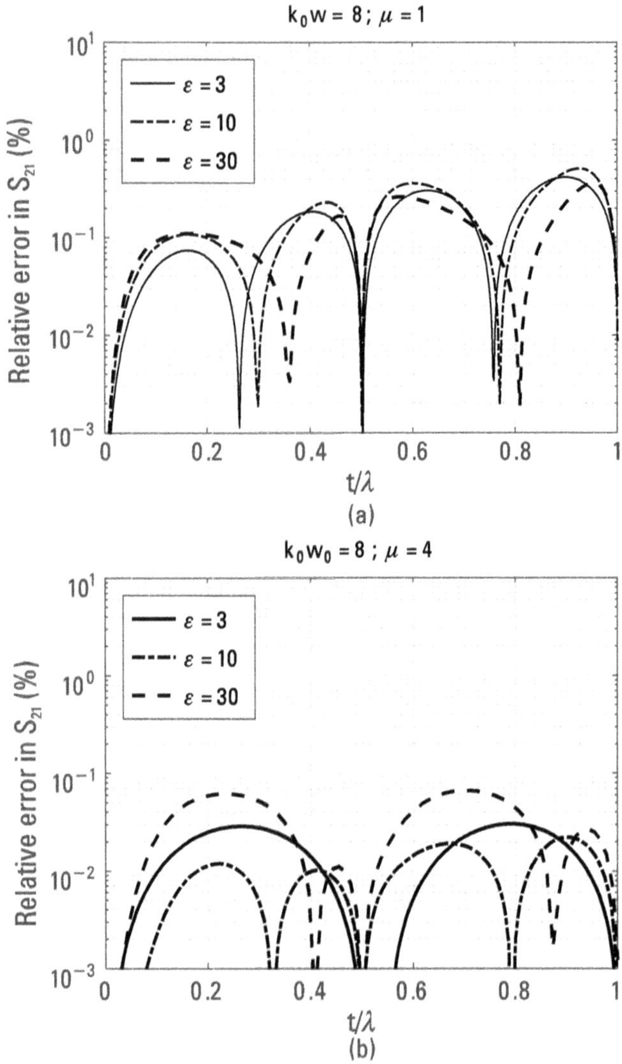

Figure 4.17 Relative error in plane-wave approximation for transmission when $k_0w_0 = 8$: (a) dielectric-only slabs, and (b) magnetic slabs.

focused-beam system. This is not necessary for lossy materials since the differences are very small. However, for low-loss dielectric materials, it is possible for the focusing error to have a more significant impact on the measured insertion loss. In particular, (4.39) can be solved numerically for a given thickness and

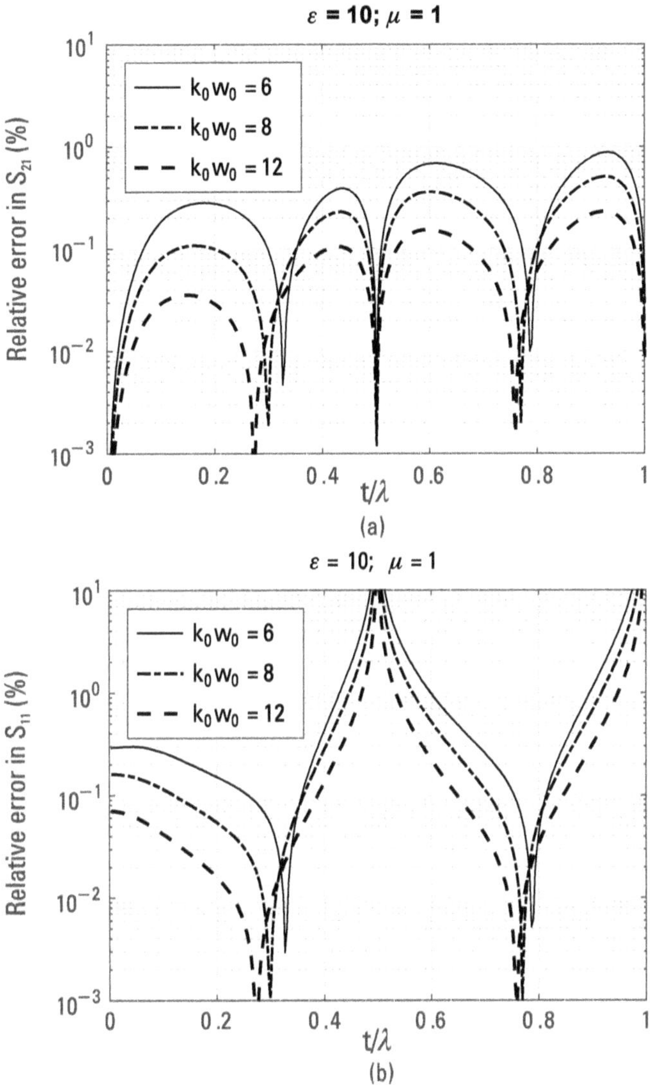

Figure 4.18 Relative error in plane-wave approximation with a dielectric slab for different beam diameters (k_0w_0).

permittivity to determine the systemic focusing error, and this can then be subtracted from the measured S_{21} as a correction. For example, Figure 4.19 shows the amplitude correction for a 6.3-mm-thick HDPE specimen measured with a focused-beam system that has a $k_0w_0 \sim 20$.

4.3.3 Beam-Shift Error

Another potential source of bias in low loss measurements is a beam shift from refraction through the sample. This effect is shown notionally in Figure 4.20 for the case of a diverging beam, such as the spot probes described in Chapter 3. Refraction by the specimen causes the beam to shift forward toward the receive side. This is similarly an effect that can occur in focused beams since insertion of a specimen interacts with the plane-wave distribution of the beam, also causing a net shift of the beam going to the receive horn.

The amount of shift, Δd, can be quantified by applying Snell's law to rays passing through a slab of thickness t and permittivity ε. For a directive beam, we assume the paraxial- or small-angle approximation, and the resulting beam shift is

$$\Delta d = t\left(1 - \frac{1}{\sqrt{\varepsilon}}\right) \tag{4.48}$$

The only other information needed to implement this correction is the dependence of transmission amplitude on distance between the transmit and receive antennas. Either numerical simulations or direct experimental measurements can be applied to create a table of S_{21} amplitude versus transmit/receive separation. With a shift calculated by (4.48), the S_{21} versus shift dependence

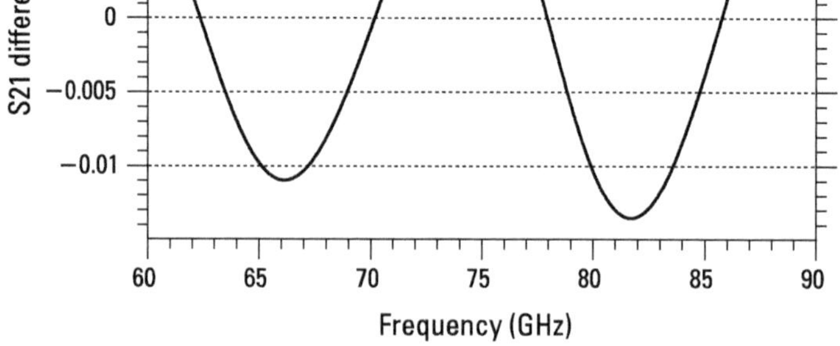

Figure 4.19 S_{21} amplitude-focusing correction for 6.3-mm-thick HDPE in an E-band focused-beam system.

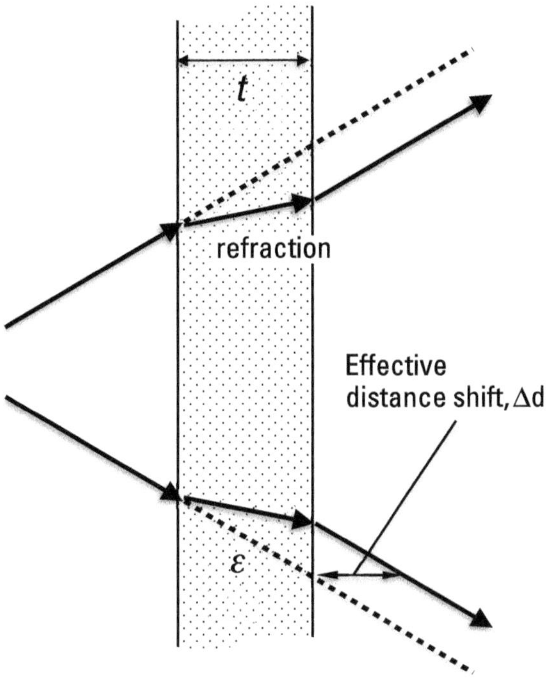

Figure 4.20 Sketch showing the beam-shift effect from a specimen with permittivity *e* and thickness *t*.

is then interpolated to correct the measured S_{21} amplitude. An example of this correction is shown in Figure 4.21 for a spot probe–measurement system. This data consists of measurements of a low-loss, high-density polyethylene (HDPE) specimen that is 6.3-mm-thick. The oscillation of the amplitude versus frequency is from constructive and destructive interference between the front and back surfaces. Without correction, the peak transmission goes above 0 dB, which is nonphysical. After beam-shift correction, the measured amplitude is no greater than 0 dB, as expected.

While Figure 4.21 shows the beam-shift correction applied to spot probes, it can also be applied to a focused beam. Historically, free-space methods have had a reputation for poor sensitivity when it comes to measuring the imaginary permittivity or loss tangent of low-loss dielectrics. Much of this reputation is due to errors such as this beam-shift effect. However, application of the

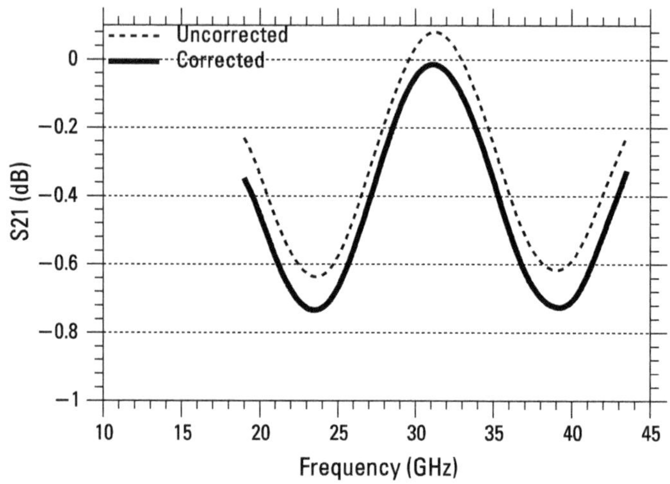

Figure 4.21 Measured S_{21} amplitude of 6.3-mm-thick HDPE, without and with beam-shift correction applied.

previous correction goes a long way toward improving accuracy and sensitivity of broadband free-space methods for these cases. For example, Figure 4.22 shows the dielectric loss tangent of polymethylmethacrylate (acrylic) measured in two different focused-beam systems and in a spot-probe system. This data had both the beam-shift and focusing corrections applied. Figure 4.22 also shows some literature-reported resonant cavity measurements of acrylic with comparable values [15, 23, 24]. Resonant measurement methods have good sensitivity for measuring low-loss tangents but can be inconvenient, since they limit the specimen thickness and only measure specific frequencies. The broadband free-space methods described here have no restrictions on specimen thickness or frequency and can have loss tangent measurement sensitivities as low as 0.002–0.0002 [25].

4.3.4 Specimen Position

Specimen-positioning error primarily affects S_{11} measurements. Ideally, the face of the specimen is either at the same position or at a known displacement from the reference plane set by the response calibration. However, if the calibration is not completely flat or the specimen is not flat, there can be uncertainty in this position. In this case, the transmission path length varies by that uncertainty so that there will be a phase offset. Thus, the uncertainty

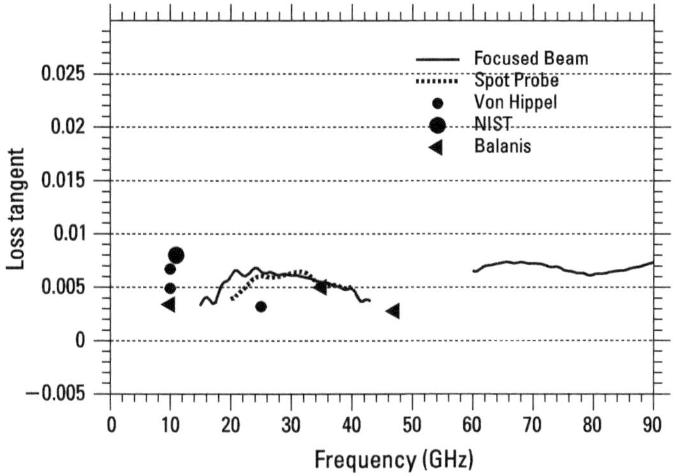

Figure 4.22 Dielectric loss tangent extracted for a 3.1-mm-thick acrylic specimen. The spot-probe data included a beam-shift correction while the focused-beam data had both beam-shift and focusing corrections applied [15, 23, 24].

in phase angle is expressed by the uncertainty in the specimen position multiplied by the propagation constant,

$$\delta\theta = \gamma_0 \delta L, \text{ where } \gamma_0 = \sqrt{k_c^2 - k_0^2} \tag{4.49}$$

When using appropriate inversion methods, this source of uncertainty is not normally a concern. For measuring nonmagnetic materials, it is typical to use the iterative S_{21} inversion algorithm, which is insensitive to the specimen position. For measuring materials where both permittivity and permeability are unknown, then it is possible to use the position-independent, four-parameter algorithm, which is also insensitive to specimen position.

4.3.5 Other Errors: Network Analyzer and Specimen

An overview of error sources in vector network analyzers is given by Ryting [26], and Wong provides a history of network analyzer calibration [27]. The error caused by limitations of the network analyzer is dependent on the equipment used, and manufacturers provide error estimates specific to their equipment. These errors will also depend on the calibration method and post-processing as well as network analyzer settings such as IF bandwidth and power levels. Network-analyzer noise levels can be measured directly by conducting repeatability measurements and calculating relevant standard deviations of the desired S-parameters.

Another, often dominant source of uncertainty lies in the material specimen itself. A materials-measurement laboratory may handle a wide variety of materials from various sources, including complex composite materials. The exact microstructure of these material samples may be unknown, and there may be inhomogeneities that violate the assumptions of the inversion algorithms outlined in Chapter 2. In engineered composites, the material specimens are constructed from multiple constituents, and there can be spatial variations within the specimen due to inconsistent distribution of these constituents. For example, fiber-reinforced composites may have variations in the fiber weave or trapped voids that lead to local variations in the dielectric properties. In mixtures like magnetic-absorber materials or artificial dielectrics, which consist of magnetic or dielectric pigments within a polymer binder, there may be particle settling that leads to varying concentrations of pigment particles within the material.

An illustration of specimen inhomogeneity is provided in Figure 4.23, which shows a simple two-layer model of a material specimen, in which one side has a greater concentration of magnetic pigment than the other. As a

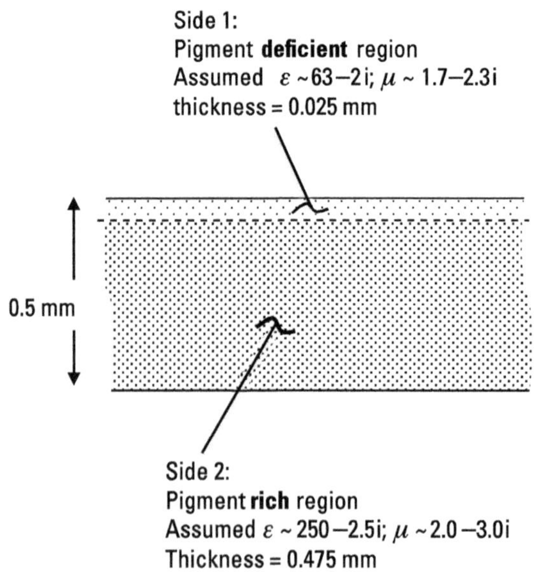

Figure 4.23 Diagram of representative inhomogeneity due to pigment particle settling in a composite mixture.

result, one side has a significantly higher permittivity and permeability than the other side. When this two-layer material specimen is inverted as if it is homogeneous, then the resulting apparent permittivity and permeability of Figure 4.24 results. In this case, the material, as specified in Figure 4.23, was inverted using the four-parameter method. Of particular note is the

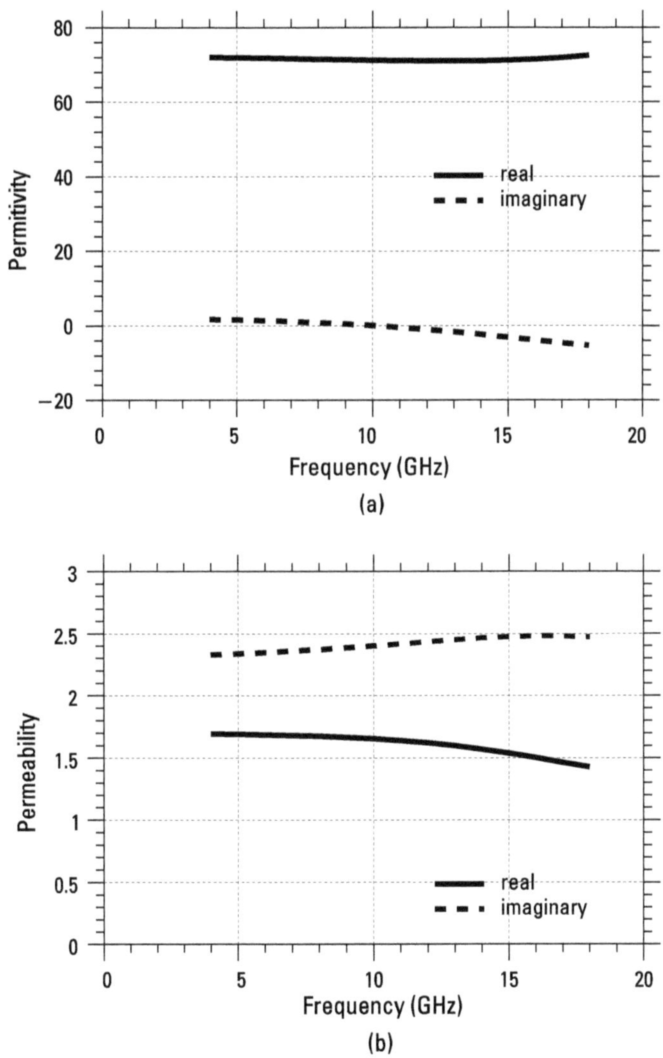

Figure 4.24 (a) Apparent permittivity and (b) permeability of an inhomogeneous specimen assumed to be homogeneous.

imaginary part of the total permittivity, which becomes negative at the higher frequencies of Figure 4.24. With the sign convention adopted in this book, imaginary permittivity and permeability should never be less than zero. So, the apparent intrinsic properties of Figure 4.24 are not physically realizable and are erroneous. Therefore, not only can the measurement fixture impose uncertainties on measured data, but so can the specimen itself, particularly under the assumption that it is homogeneous. Ideally, a materials measurement laboratory should have access to optical or electron microscopes so that specimen microstructures can be evaluated and correlated to the intrinsic microwave properties.

Another source of uncertainty associated with the specimen is its thickness. Thickness errors may arise from the uncertainty in thickness estimation (i.e., micrometer measurements), or from surface roughness or location-dependent thickness variations in the specimen itself. Specimen dimensions are typically measured with a caliper or micrometer, and uncertainty in these instruments is typically less than 0.01 mm (0.0005 inches). Since specimens may not be uniformly thick, it is important to measure the center of the specimen, and not just the edge; and this can be accomplished with a micrometer that has an extra deep yoke or throat. When a specimen is centered in the focused-beam system, most of the illuminating power is in the center of the specimen, so emphasis should be placed on the center thickness. Even when the center thickness is measured, there may be substantial variation in that thickness over the area of the beam diameter, which represents a source of uncertainty in thickness, t. Sometimes material specimens are elastomeric or soft. In this case the measured thickness may depend on how much pressure is exerted by the micrometer or caliper during the thickness measurement.

To illustrate the effect of thickness and network analyzer errors on the uncertainty in inverted properties, some representative cases are shown in Figures 4.25 and 4.26. This uncertainty data is shown as a function of the electrical-specimen thickness. The thickness and network-analyzer uncertainties were propagated using the RMS method of (4.33) and (4.34). In these plots, the thickness error was assumed to be ±0.0127 mm. The network analyzer error was assumed to be ±0.2 degrees in phase, and ±0.5% in amplitude plus an absolute amplitude error of 0.001. The absolute amplitude error is equivalent to a measurement dynamic range of 60 dB (e.g., a relative noise floor of −60 dB with respect to the maximum). For this illustration, no other uncertainty sources were included.

Figure 4.25 shows the estimated measurement uncertainty for a low-loss, low-dielectric material such as an unfilled polymer. The top plot is the uncertainty in the real part of the permittivity, and the bottom plot is the

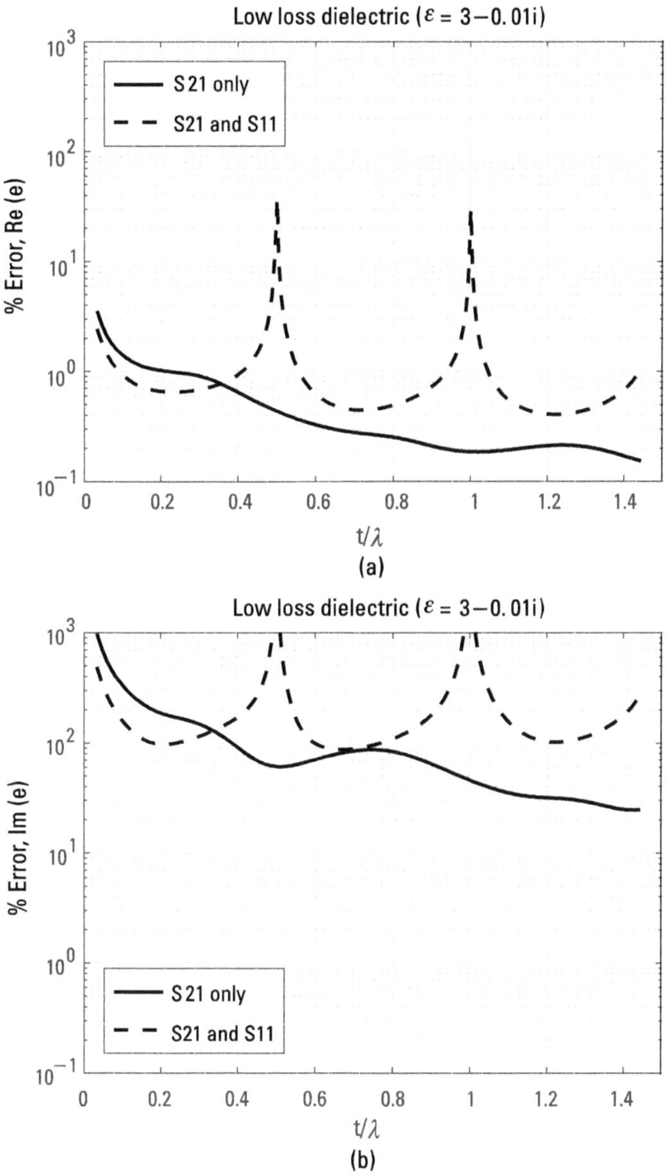

Figure 4.25 Representative measurement error for a low-loss dielectric specimen due to thickness and network analyzer uncertainties.

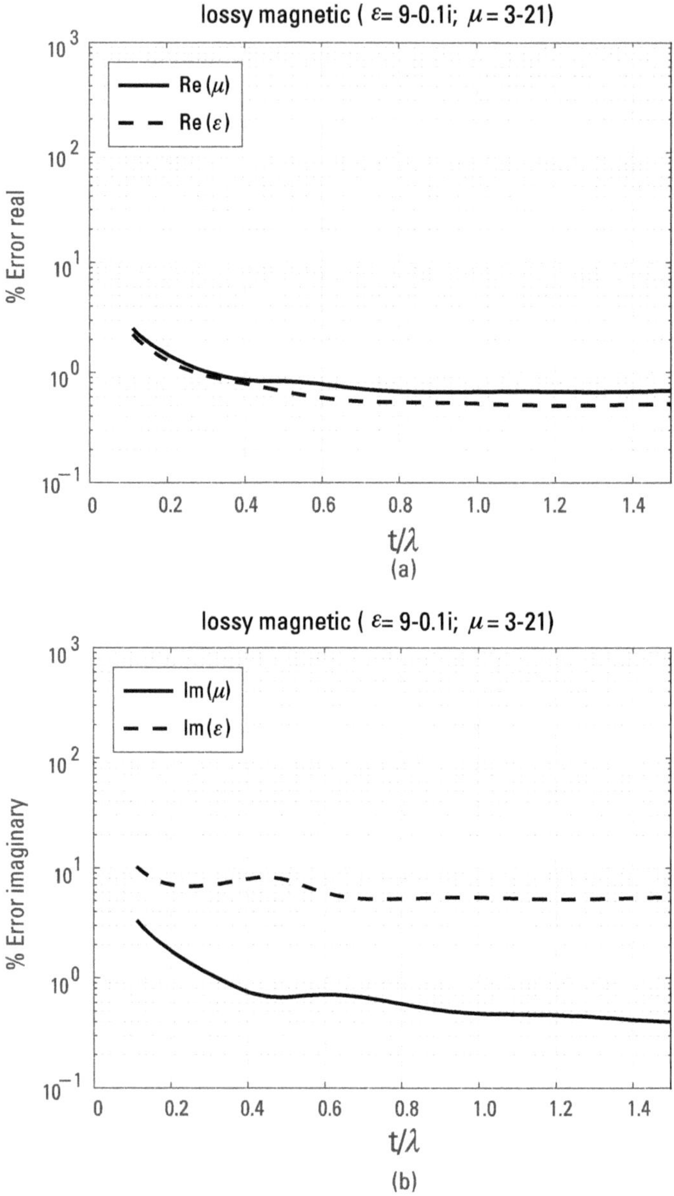

Figure 4.26 Representative measurement error for a lossy magnetic material due to thickness and network analyzer uncertainties.

uncertainty in the imaginary permittivity for this material. Curves for two different inversion algorithms are shown: (1) S_{21} iteration and (2) S_{21} and S_{11} iteration. The uncertainty calculated for the S_{21} and S_{11} iteration is the same as what would be calculated for the NRW inversion method. When the S_{21} iteration is used, the uncertainty generally decreases as specimen thickness increases. This follows since a thicker specimen has a greater effect on the transmitted phase, providing an improved signal to noise.

On the other hand, when the S_{21} and S_{11} iteration method is used, there is a very large uncertainty for specimen thicknesses corresponding to integral multiples of $\lambda/2$. While the uncertainty in the imaginary permittivity (bottom plot of Figure 4.25) is very large for this case, the imaginary permittivity itself is a small number to begin with ($\varepsilon'' = 0.01$). So, the absolute error for this example is small relative to the real part of the permittivity. This kind of uncertainty in low-loss dielectric-material measurements is similar to what is typically observed in other transmission-line methods, such as rectangular waveguide or coaxial airline fixtures.

A different example of uncertainties is shown for a lossy magnetic material in Figure 4.26. In this data, inversion via the four-parameter iteration is assumed since both permittivity and permeability must be evaluated. The error curves show the calculated uncertainties for both dielectric and magnetic properties as a function of electrical thickness. Because the values of the imaginary permeability and permittivity are larger than the low-loss dielectric, their relative uncertainties are correspondingly lower, even though their absolute uncertainties are similar.

4.4 Apertures

The uncertainty sources described in Section 4.3 assume that the specimen is large enough that the illumination on the edges is negligible. However, there are often constraints on the maximum size of a material specimen so that overillumination of the specimen cannot be ignored. This is particularly true at lower frequencies, where the large wavelength drives ever-larger illumination areas. In some cases, it might be possible to overcome this limitation by encompassing a too-small specimen, within a larger conductive ground plane. This concept of positioning an aperture around a specimen is illustrated notionally in Figure 4.27. The main idea is that the aperture is built into a conductive plane that is significantly bigger than the illuminating-beam diameter. This larger plane acts as a barrier so that energy does not travel around the specimen to the other side. Measurement with an aperture such as that shown in

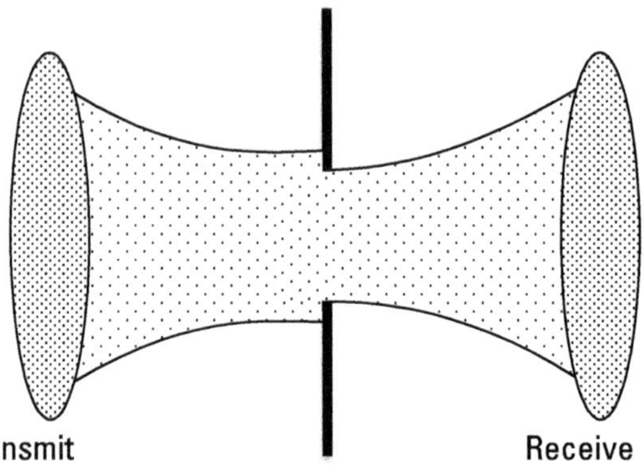

Figure 4.27 Notional sketch of aperture in a focused-beam system.

Figure 4.27 represents a compromise between accuracy and small specimen size since it does not exactly match the behavior of the material under test.

To illustrate the effects of apertures, Figure 4.28 shows the transmission coefficient of a thin resistive specimen measured by a simulated focused-beam system. This data was calculated by a full-wave FDTD solver. The focused beam was simulated by adding a series of weighted plane waves with a Gaussian distribution to create a beam waist of $0.8\lambda_0$ ($k_0 w_0 = 5$). This beam approximates the beam from a large experimental 1.8-m diameter lens system designed for use at UHF frequencies The transmission coefficient was determined by propagating the focused beam through the specimen sheet. The transmission coefficient data was computed as a ratio to the transmission of a clear site (no sample) to minimize computational errors and simulate what is done in actual focused-beam measurements.

The calculated amplitude transmission coefficients for a 150-Ω/square sheet are shown in Figure 4.28. Data is shown for an infinite sheet with no aperture (solid line), an apertured sheet calibrated to an infinite clear site, and an apertured sheet calibrated to an apertured clear site. The apertured sheet is 1.1×1.2m in size. At the lower end of the frequency band, the sample height and width are small enough relative to the incident beam to allow significant power to go around the specimen. Thus, the metal ground plane around the specimen limits the leakage around the specimen. This data shows that the beam begins overilluminating the sample below 1 GHz, but that calibrating to an apertured clear site is an effective way to correct for this overillumination.

When properly calibrated, the apertured transmission amplitude shows negligible errors from overillumination; however, significant phase variations

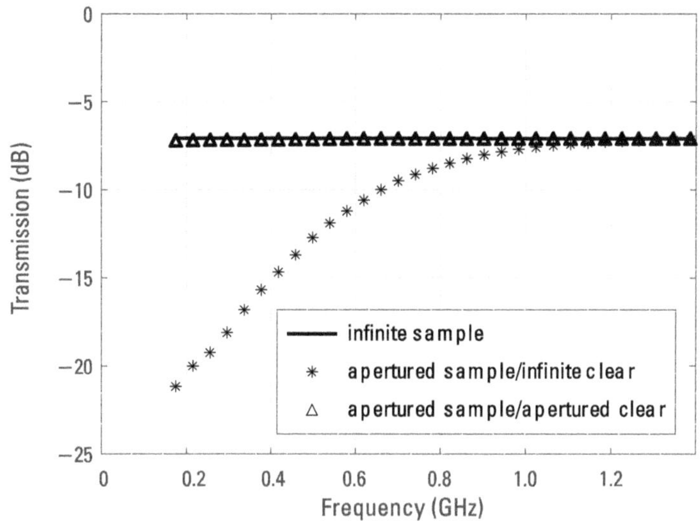

Figure 4.28 FDTD-calculated transmission coefficient of a 150-Ω/square sheet illuminated by a Gaussian beam.

occur from the aperture-edge scatter. These errors are more intuitively viewed in terms of effective measured impedance. The effective sheet impedance, Z, is determined from the transmission coefficient, S_{21}, by (2.51). The calculated impedances from an apertured 150-Ω/square sheet calibrated to an apertured clear site are shown in Figure 4.29. Two cases of apertured specimen are shown: (1) no gap, where the impedance sheet is electrically connected to the ground plane; and (2) a 2-cm gap, between the periphery of the sheet and the edge of the aperture. When there is good electrical contact between the sample and the ground plane, the real impedance is close to the actual impedance of the sheet, but a small inductance (positive imaginary impedance) occurs at the lowest frequencies. With an airgap between the sample edges and the ground plane, the real impedance increases at lower frequencies, and the imaginary impedance shows a significant capacitance (negative imaginary impedance) from interaction between the sample and ground plane edges across the gap.

The inductive and capacitive effects caused by the aperture edges are confirmed by the model-measurement comparison in Figure 4.30. The FDTD-simulated data of these plots was calculated for a 440 Ω/square sheet. The measured sample was a carbon filled polyimide sheet that used a 1.8-m-diameter lens system previously constructed by myself and others [28]. Both measured and simulated data were processed with a 4-ns time-domain gate, and this data shows qualitative agreement. However, there are some quantitative differences

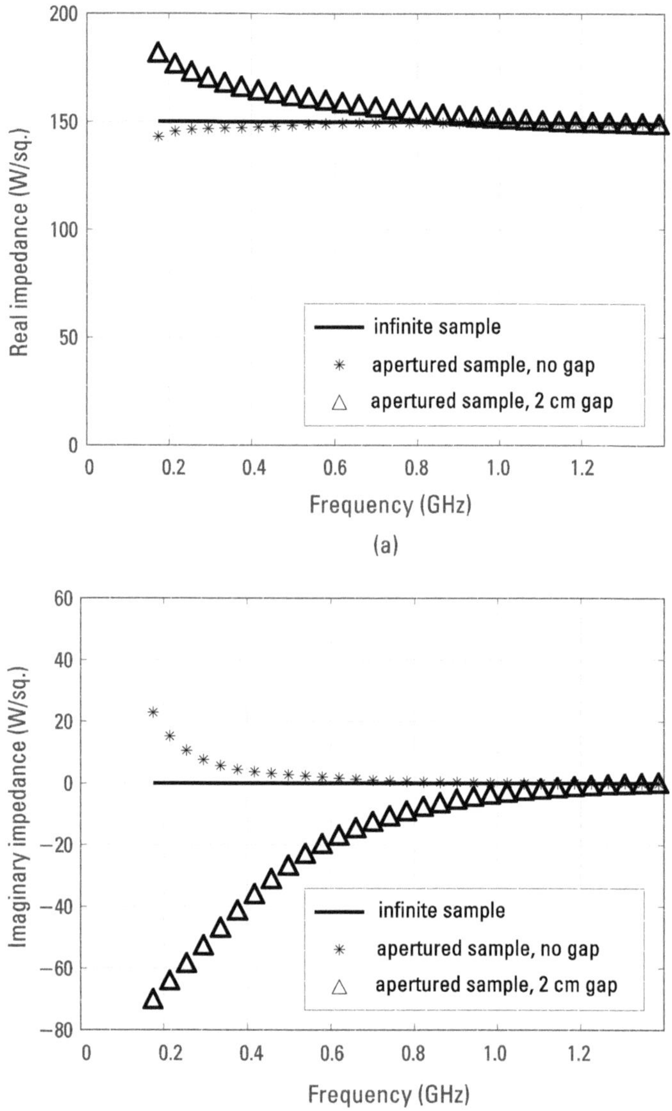

Figure 4.29 Effect of gap on real and imaginary sheet impedance calculated for 150-Ω/square sheets in an aperture.

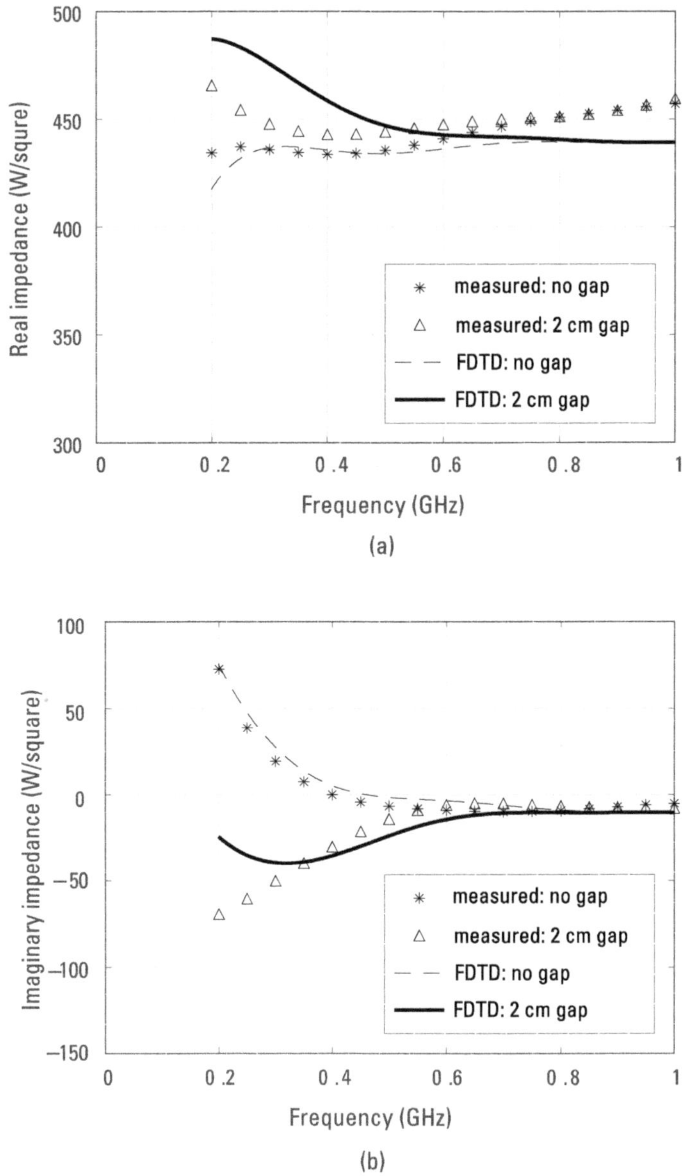

Figure 4.30 Experimentally measured and FDTD-calculated impedance with and without 2-cm gaps.

due to frequency dispersion in the actual sample—the FDTD model assumes a constant impedance for all frequencies. In addition, the beam was assumed to be a symmetrical Gaussian beam with a constant k_0w_0 in the FDTD simulations; however, the actual measurement system has a slightly elliptical beam with some frequency dispersion in k_0w_0.

These results show that in some circumstances and with appropriate calibration, a conductive aperture can extend the focused-beam methodology to samples that are only a wavelength across. The conductive aperture is effective in this case because the materials are semiconductive themselves. In other words, the difference between these resistive specimens and metal is smaller than the difference between resistive specimens and air or no aperture. On the other hand, if the materials under test are low-dielectric materials, then a metallic aperture can create a larger error than if no aperture was used. So, the choice of an aperture depends on the types of materials under test. The best material for an aperture is a material that is not too different from the material under test. For example, if a moderate- or high-dielectric material is being measured, then ideally the aperture should be constructed from materials with similar properties.

In some cases, material-manufacturing limitations may further limit specimen size, driving a need for techniques that allow even smaller specimens. This can be accomplished by leveraging the idea of using similar materials for the aperture. Continuing our example of resistive materials, specimen-size reduction could be realized by filling the larger metallic aperture describe above with subsized strips of the material under test, combined with previously measured known materials. An example geometry for this idea is shown in Figure 4.31, which includes a metallic aperture with a small strip of the unknown material Z_2 surrounded by strips of a known impedance material, Z_1. This geometry is useful when specimens for testing can only be obtained in narrow strips.

The transmission coefficient from this three-strip fixture will result in an effective impedance that is an average of the two known and one unknown strips. It can be modeled as a weighted-area average of the voltage transmission coefficients of each impedance region,

$$T_{\text{effective}} = \mathrm{W}_{Z_2} T_2 + \left(1 - \mathrm{W}_{Z_2}\right) T_1 \tag{4.50}$$

where T_1 is the voltage transmission coefficient of the known impedance strips, and T_2 is for the unknown. The weight for the middle (unknown) impedance strip is proportional to the voltage across the strip, which is calculated by integrating the Gaussian beam taper over the middle strip width ($2r$),

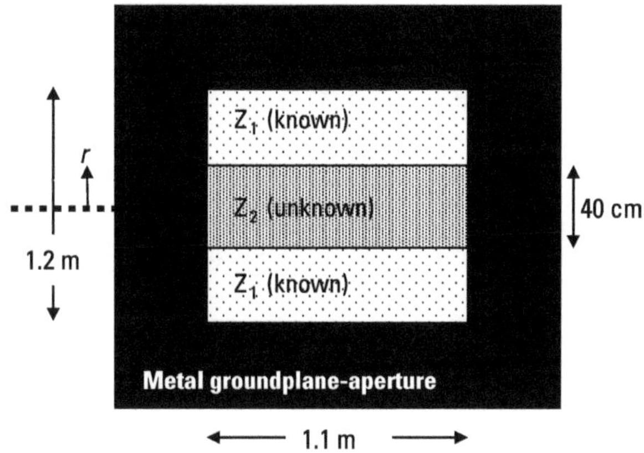

Figure 4.31 Geometry for measuring impedance of a narrow strip sample, Z_2. The rest of the aperture is filled with a known impedance sheet, Z_1.

$$W_{Z_2} = \frac{\int_0^r e^{-x^2/w_0^2} dx}{\int_0^\infty e^{-x^2/w_0^2} dx} = \frac{w_0 \sqrt{\pi/4}\, \mathrm{erf}\left(r/w_0\right)}{w_0\sqrt{\pi/4}} = \mathrm{erf}\left(\frac{r}{w_0}\right) \tag{4.51}$$

Because the aperture is smaller than the total beam, this weight is also normalized to the total field going through the aperture. After including an empirical multiplicative factor, f_1, to account for diffraction and aperture shape effects, the resulting weight is

$$W_{Z_2} = \frac{\mathrm{erf}\left(f_1 r/w_0\right)}{\mathrm{erf}\left(f_1 R/w_0\right)} \tag{4.52}$$

Figure 4.32 shows an example of this weighted-average model compared to FDTD calculations of effective impedance. These calculations are for an assumed aperture size of 1.1 × 1.2m, where the width of each strip is 0.4m. The two known strips have sheet impedances of 150 Ω/square. The FDTD calculations and the semi-empirical model in Figure 4.32 show agreement even when Z_1 and Z_2 differ by a factor of two. Therefore, the semi-empirical model can be used to supplement the simulations, thereby reducing the needed computational effort. That said, these various aperture approaches add complication to the measurement and data analyses, and care must be taken to evaluate them before use to ensure that errors from undesired diffraction are minimized.

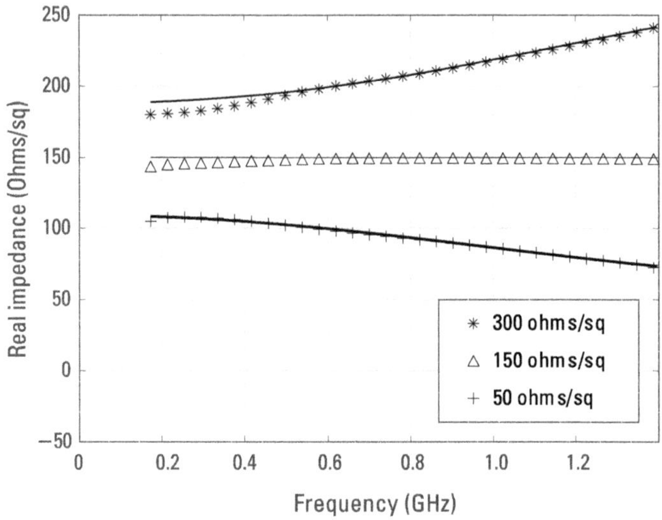

Figure 4.32 Effective impedance measured with aperture fixture of Figure 4.31 for different Z_2 values. Symbols are FDTD calculations, and lines are semi-empirical model.

References

[1] Hill, L. D., "A Quasi-Optical Microwave Focused-beam System," *Proceedings of Antenna Measurement Techniques Association Symposium*, Denver, CO, Oct. 21–26, 2001.

[2] Iyer, S., et al., "Compact Gaussian Beam System for S-Parameter Characterization of Planar Structures at Millimeter-Wave Frequencies," *IEEE Trans. Instrumentation and Measurement*, Vol. 59, No. 9, Sept. 2010, pp. 2437–2444.

[3] Goldsmith, P. F., *Quasioptical Systems, Gaussian Beam Quasioptical Propagation and Applications*, Piscataway, NJ: IEEE Press, 1998.

[4] Boyd, R. W., *Nonlinear Optics* (Fourth Edition), San Diego, CA: Academic Press, 2020.

[5] Goldsmith, P. F., T. Itoh, and K. D. Stephan, "Quasi-Optical Techniques," in *Handbook of Microwave and Optical Components*, Volume 1, K. Chang (ed.), Hoboken, NJ: Wiley, 1989, pp. 344–363.

[6] Musil, J., and F. Zacek, *Microwave Measurement of Complex Permittivity by Free Space Methods and Their Applications*, Amsterdam, Netherlands: Elsevier, 1986.

[7] Marcuse, D., *Light Transmission Optics* (Second Edition), New York, NY: Van Nostrand Reinhold, 1982.

[8] Peatross, J., and M. Ware, *Physics of Light and Optics*, 2015 Edition, August 9, 2022, Revision Provo, UT: Brigham Young University.

[9] Bruce, I., "ABCD Transfer Matrixes and Paraxial Ray Tracing for Elliptic and Hyperbolic Lenses And Mirrors," *European J. Phys.*, Vol. 27, 2006, pp. 393–406.

[10] Gangopadhyay, S., and S. Sarkar, "ABCD Matrix for Reflection and Refraction of Gaussian Light Beams at Surfaces of Hyperboloid of Revolution and Efficiency Computation for Laser Diode To Single-Mode Fiber Coupling by Way of a Hyperbolic Lens on the Fiber Tip," *Applied Optics*, Vol. 36, No. 33, 1997, pp. 8582–8586.

[11] Trabelsi, S. S., and O. Nelson, "Nondestructive Sensing of Physical Properties of Granual Materials by Microwave Permittivity Measurement," *IEEE Trans. Instrumentation & Measurement*, Vol. 55, No. 3, June 2006, pp. 953–963.

[12] Kocharyan, K. N., M. N. Afsar, and I. I. Tkachov, "Millimeter-Wave Magnetooptics: New Method for Characterization of Ferrites in the Millimeter-Wave Range," *IEEE Trans. Microwave Theory and Techniques*, Vol. 47, No. 12, Dec. 1999, pp. 2636–2643.

[13] Zhu, D. Z., et al., "Fabrication and Characterization of Multiband Polarization Independent 3-D-Printed Frequency Selective Structures with UltraWide Fields of View," *IEEE Trans. Antennas and Propagation*, Vol. 66, No. 11, November 2018, pp. 6069–6104.

[14] Hilario, M. S., et al., "W-Band Complex Permittivity Measurements at High Temperature Using Free-Space Methods," *IEEE Trans Components, Packaging, and Manufacturing Tech.*, Vol. 9, No. 6, June 2019, pp. 1011–1019.

[15] Riddle, B., J. Baker-Jarvis, and J. Krupka, "Complex Permittivity Measurements of Common Plastics over Variable Temperatures," *IEEE Trans. On Microwave Theory and Techniques*, Vol. 51, No. 3, March 2003, pp. 727–733.

[16] Kirby, R. K., "Thermal Expansion of Polytetrafluoroethylene (Teflon) from −190 to +300C," *Journal of Research of the National Bureau of Standards*, Vol. 57, No. 2, August 1956, pp. 91–94.

[17] Xiu, D., "Fast Numerical Methods for Stochastic Computations: A Review," *Comm. In Computational Phys.*, Vol. 5, No. 2–4, 2009, pp. 242–272.

[18] Taylor, J. R., *An Introduction to Error Analysis*, Oxford, United Kingdom: Oxford University Press, 1982.

[19] Baker-Jarvis, J., "Transmission/Reflection and Short-Circuit Line Permittivity Measurements," NIST Technical Note 1341, 1990.

[20] Baker-Jarvis, J., et al., "Transmission/Reflection and Short-Circuit Line Methods for Measuring Permittivity and Permeability," NIST Technical Note 1355, 1992.

[21] Petersson, L. E. R., and G. S. Smith, "An Estimate of the Error Caused by the Plane-Wave Approximation in Free-Space Dielectric Measurement Systems," *IEEE Trans. Antennas and Propagation*, Vol. 50, No. 6, 2002, pp. 878–887.

[22] Petersson, L. E. R., "Analysis of Two Problems Related to a Focused Beam Measurement System," Ph.D. dissertation, Georgia Institute of Technology, Nov. 2002.

[23] Von Hipple, A. R., *Dielectric Materials and Applications*, New York: Wiley, 1954.

[24] Balanis, C. A., "Measurements of Dielectric Constants and Loss Tangents at E-Band Using a Fabry-Perot Interferometer," NASA Technical Note D-5583, 1969.

[25] Schultz, J. W., R. Geryak, and J. G. Maloney, "New Methods for Improved Accuracy of Broad Band Free Space Dielectric Measurements," *2020 50th European Microwave Conference (EuMC)*, Utrecht, Netherlands, Jan. 12–14, 2021.

[26] Rytting, D. K., "Network Analyzer Accuracy Overview," *ARFTG Conference Digest—Fall*, 58th, Vol. 40, 2001, pp. 1–13.

[27] Wong, K., "Network Analyzer Calibrations—Yesterday, Today and Tomorrow," Symposium Digest, *IEEE MTT-S International*, 2008, pp. 19–25.

[28] Schultz, J. W., "Numerical Analysis of Transmission Line Techniques for RF Material Measurements," *Proceedings Antenna Measurement Techniques Association*, Irvine CA Oct. 19–24, 2003.

5

Transmission Line Methods

5.1 Waveguides

Chapters 2–4 focused on free-space methods for characterizing electromagnetic materials. While free-space methods have many advantages, they require specimens that are electrically large—on the order of at least a couple of wavelengths across. However, when materials are difficult or expensive to manufacture it can be advantageous to use a measurement method that works with smaller specimens. This is possible with waveguide, which is a structure that guides electromagnetic energy from one location to another. The idea of electromagnetic propagation within a hollow conductive pipe was conceived by Rayleigh around 1897 but then received little attention again until the 1930s [1]. The use of waveguides to measure material properties dates to at least the 1940s [2, 3].

As a general concept, waveguides can be in the form of hollow pipes or have multiple conductors. Multiconductor transmission lines are discussed later in Sections 5.2 and 5.3. This section focuses on hollow pipes, which are usually just referred to collectively as waveguide. Specifically, this section reviews the rectangular waveguide, which is the most common waveguide method used in RF material measurements. Circular, corrugated, and ridged waveguides, which can also be used for material measurements, are briefly described at the end of this chapter in Section 5.4. For material measurements, rectangular waveguide is often preferred, because the orientation of the electric field is well established by the dimensions of the waveguide. This well-defined E-field is important when materials are anisotropic. Rectangular specimens are also more convenient to cut or machine than samples for fixtures with curved sides.

A coordinate system for a rectangular waveguide is shown in Figure 5.1, with propagation of the guided wave in the z-direction. Tangentially oriented E-fields cannot exist adjacent to the conductive boundaries of a waveguide. Thus, there are a limited set of electric and magnetic field configurations that can occur within the rectangular pipe. In free space, a propagating plane wave has both the electric and magnetic fields tangential to the propagation direction, but this is not the case within a rectangular waveguide, which may have transverse-electric or transverse-magnetic fields, but not both. For the dominant mode, the allowed fields within the waveguide can be derived by superimposing a pair of plane waves at complementary angles with respect to the xz plane such that they satisfy the vanishing tangential E-field at the conductive boundaries [4]. The E-field within the waveguide is given by

$$E_x = E_0 \sin\left(k_c y\right) e^{-\sqrt{k_c^2 - k^2}\, z} \tag{5.1}$$

where k is the wavenumber, $k = 2\pi f/c$, f is the frequency, and c is the speed of light. E_x vanishes at both tangential boundaries, $y = 0$ and $y = a$, as long as

$$k_c = \frac{n\pi}{a}, \quad n = 1,2,3,\ldots \tag{5.2}$$

The variable k_c is also known as the cutoff wavenumber, and by association a cutoff frequency can be defined by $f_c = nc/(2a)$. Evaluation of (5.1) indicates that a wave can propagate only when $k_c^2 - k^2$ is negative so that E_x

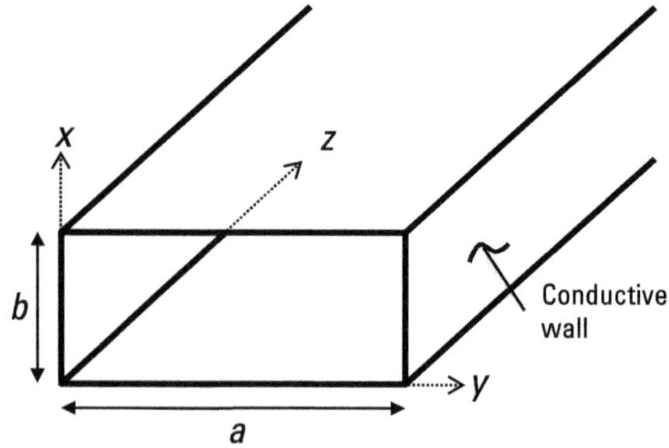

Figure 5.1 Rectangular waveguide coordinates and dimensions.

does not decay as a function of z. Different cutoffs exist for different values of n, which are called modes. Modes that are allowed in a rectangular waveguide can have either a TE or a TM field configuration, and the modes described by (5.1) are TE. The mode with the lowest cutoff, the TE, $n = 1$ mode, is sometimes called the dominant or fundamental mode. When using a waveguide to measure material properties, the existence of higher-order modes significantly complicates the analysis, so only this lowest mode is considered, and $n > 1$ or TM modes are avoided. For this reason, the frequency range over which a rectangular waveguide can be used for materials characterization is limited to between the $n = 1$ cutoff frequency and the cutoff for the next higher-order mode.

Equation (5.2) is only true for certain TE modes, and a more general expression for calculating mode cutoffs is given by [5]

$$\left(k_c\right)_{n,m} = \pi\sqrt{\left(\frac{m}{b}\right)^2 + \left(\frac{n}{a}\right)^2}, \quad m,n = 0,1,2,\ldots \tag{5.3}$$

The lowest-order mode is TE and when $m = 0$ and $n = 1$ is designated as TE_{10}. This notation is not fully universal, however, and some references have m and n (or a and b) switched [4]. Notice that if a rectangular guide is square, $a = b$, and multiple or degenerate modes exist even at the lowest frequencies. Therefore, rectangular guides are usually constructed with $b/a > 1$. Standard waveguide sizes exist, and one that is common is called the X-band waveguide, which operates from 8.2 to 11.4 GHz. Other letter designations are commonly used to denote other frequency bands.

Another standard way to designate waveguide is based on its width (a in Figure 5.1) and an X-band waveguide is also called WR-90, since it is 0.9-in-wide (22.86 mm). While based on the older unit of inches, this designation is useful since it relates directly to the waveguide width, and the dominant mode cutoff can be easily derived. Commonly used waveguide sizes also have standard heights. For the WR-90 waveguide, the standard height is 0.4 in (10.16 mm). Other standard waveguide sizes have similar, but not exactly the same, width-to-height ratios. For material measurements at frequencies below 1 GHz, it is not uncommon to have larger width-to-height ratios to reduce the overall size of the material measurement specimen. For example, VHF waveguide measurement systems have been constructed with width/height ratio of 4 or 8 or even higher, versus the width/height of 2.25 for the standard WR90 waveguide.

While the E-field in the dominant mode of a rectangular waveguide is x-oriented in Figure 5.1, the magnetic field is oriented in both the y and z

directions. This is an important consideration when measuring anisotropic materials. For example, a waveguide can be used to separately measure anisotropic dielectric permittivity components by constructing multiple samples and orienting them appropriately in the waveguide. The inverted permittivity corresponds to the *x*-direction in Figure 5.1. However, when magnetic permeability is also anisotropic, the inverted permeability will be an average of the properties in the *y* and *z* directions based on the orientations of the magnetic field lines.

5.1.1 Waveguide Calibration

To obtain quantitative S-parameters from a waveguide fixture, it must first be calibrated. The relevant components of a waveguide measurement fixture include (1) the waveguide itself, (2) transitions that convert from RF cables to the rectangular waveguide, (3) RF cables that connect the waveguide to the microwave analyzer, and (4) the microwave VNA, which has oscillators and mixers for generating and quantifying the ingoing and outgoing electromagnetic waves. The VNA includes internal components such as switches, directional couplers, sources, and receivers. Measurement errors can come from imperfections in all these components and detailed error models can be used to describe each of the dominant error sources [6]. Accounting for these errors involves measurement of known standards. Ideally the number of standards measurements should be the same as the number of error terms to be dealt with.

Waveguide measurements can use a variety of different calibration standards [7]. Probably the most convenient and commonly employed for a two-port waveguide fixture is the TRL combination [8]. The thru is a measurement of the waveguide fixture without any specimen, and all four scattering parameters are captured. The short is a measurement of the waveguide separated in the middle and capped with conductive ends, which act as electrical short circuits. In this case, two more measurements are captured: S_{11} and S_{22}. The line standard is a length of extra empty waveguide that is inserted into the fixture. It provides an additional two independent measurements of transmission in each direction with a well-controlled phase shift. In particular, the phase delay in radians (ϕ) provided by this line can be calculated from,

$$\phi = lk_0\sqrt{1-\left(f_c/f\right)^2} \tag{5.4}$$

where l is the length of the inserted guide. This contrasts with a phase shift from a wave traveling a distance l in free space, which would only be $\phi = lk_0 =$

$2\pi l/\lambda_0$. The effect of a waveguide with a given cutoff frequency is a frequency-dependent phase velocity [4].

For the line standard to be effective it needs to provide a substantially different phase to the measured transmission signals than the thru standard. The usual rule of thumb is that the phase delay from the line standard is greater than 20 degrees at the lowest frequency of interest and no more than 160 degrees at the highest frequency. Getting too close to 180 degrees would create a phase ambiguity in the calibration standards. Another way to determine the correct line length is to set the phase delay to be equivalent to a quarter-wavelength or 90 degrees in the middle of the band of interest [9].

A waveguide system includes a pair of transition sections that convert from the RF cable to the waveguide and then one or more sections of rectangular waveguide between the two transitions. Under ideal conditions, there should be about two wavelengths of waveguide between the transition and the specimen location. The reason for this is to ensure that evanescent energy from the transition is sufficiently decayed so that it doesn't interact with the material specimen. This is, however, not always practical, and a waveguide that operates at VHF frequencies may end up being extremely long. For example, a WR-4200 waveguide has a cutoff frequency of about 140 MHz, and the wavelength at 140 MHz is a little over 2m. By the above rule of thumb, the sample position should be about 4m from the transition, making the whole waveguide system probably about 9m-long (assuming two 4-m sections and two 0.5-m transitions). In practice, that may be too long to fit in a laboratory space. Fortunately, it is possible to operate with a shorter system. For example, an operational WR-4200 waveguide currently in use is closer to 3.7m in total length with no significant systematic error from the reduced length. In cases like this, when the overall waveguide is shorter than desired, the line length used for TRL calibration can also be negative, where an appropriate section of waveguide is removed rather than added relative to the thru. This is helpful at low frequencies where waveguide systems are electrically short for practical reasons and the specimen under test needs to be as far away from the transitions as possible.

In some waveguide configurations, a short section of waveguide is used to hold the specimen and is clamped between longer waveguide sections connected to the transitions. Calibration can be done without this extra sample-holder waveguide; however, the phase shift added by this section will need to be extracted from the measured S-parameter data before inverting the data. A better method is to include all the waveguide employed when the specimen is loaded as part of the thru standard and to ensure that the reflection calibration standard (short) also includes the specimen waveguide section for one of the directions. This avoids having to precisely know its length, since it is inherently included in the calibration. Fortunately, most vendors of VNAs provide easy-to-use software for setting up and conducting waveguide calibrations for a variety of setups. Ready-made waveguide calibration kits are also commercially available for the most common standard waveguide sizes.

5.1.2 Waveguide Property Inversion

Dielectric or magnetic property measurements can be done with waveguide data in a manner very similar to that used in free space. In free space, the inversions are derived by assuming a wave that interacts with a slab specimen and calculating the transmission and reflection behavior at interfaces and in the middle of the material. As described in Chapter 2, microwave network analysis is used to cascade these different effects together to determine the total reflection or transmission behavior of a slab. A key difference, however, is that the intrinsic impedance of free space is a constant ($Z_0 = \sqrt{\mu_0/\varepsilon_0} \approx 377\Omega$), while a wave propagating within a waveguide experiences a frequency-dependent impedance. Specifically, the wave impedance depends on the cutoff frequency for the propagating mode within the air-filled waveguide and is calculated by [4]

$$Z_0^{wg} = \frac{Z_0}{\sqrt{1-\left(f_c/f\right)^2}} \tag{5.5}$$

Similarly, the propagation constant in an air-filled waveguide is also a function of the cutoff frequency,

$$\gamma_0^{wg} = \sqrt{k_c^2 - k_0^2} = ik_0\sqrt{1-\left(f_c/f\right)^2} \tag{5.6}$$

Within a material specimen that fills the inside cross-section of a waveguide we have

$$\gamma^{wg} = \sqrt{k_c^2 - k^2} = \sqrt{k_c^2 - \varepsilon\mu k_0^2} = ik_0\sqrt{\varepsilon\mu - \left(f_c/f\right)^2} \tag{5.7}$$

Armed with (5.5) to (5.7), we can then calculate reflection and transmission coefficients necessary for material inversions. These expressions are derived elsewhere [9] and are simply summarized in the following. The reflection coefficient at an interface between air and a material under test is

$$\Gamma = \frac{\mu\gamma_0^{wg} - \gamma^{wg}}{\mu\gamma_0^{wg} + \gamma^{wg}} = \frac{\mu\sqrt{1-(f_c/f)^2} - \sqrt{\varepsilon\mu - (f_c/f)^2}}{\mu\sqrt{1-(f_c/f)^2} + \sqrt{\varepsilon\mu - (f_c/f)^2}} \tag{5.8}$$

The transmission coefficient through a thickness, t, of material is

$$T = e^{-t\gamma^{wg}} = e^{-ik_0 t\sqrt{\varepsilon\mu - (f_c/f)^2}} \tag{5.9}$$

Notice that when the cutoff frequency is zero, (5.8) and (5.9) reduce to the free-space case (2.24).

Substituting these into the various expressions in Chapter 2 for material inversion, we can then calculate the equivalent inversion equations for the waveguide. In other words, S_{11}-only, S_{21}-only, and combined S_{11} plus S_{21} property inversions in a waveguide use the same S-parameter relationships as free space (2.20),

$$S_{11} = \frac{\Gamma(1-T^2)}{1-\Gamma^2T^2} \quad \text{and} \quad S_{21} = \frac{T(1-\Gamma^2)}{1-\Gamma^2T^2} \tag{5.10}$$

but with the waveguide equivalent Γ and T provided above. As in the free-space case, the Newton's iterative root-finding method can be used as well, and the derivative expressions in (2.25) to (2.28) are applicable, as long as they also incorporate the derivatives of (5.8) and (5.9). Similarly, the four-parameter method can be done with the waveguide equivalent Γ and T. In this case the equations that combine the S-parameters (2.46) and (2.47) are replaced by,

$$S_{11}^{\text{cal}} S_{22}^{\text{cal}} e^{-2\gamma_0^{wg}(t_s - t_m)} - S_{21}^{\text{cal}} S_{12}^{\text{cal}} e^{-2\gamma_0^{wg}(t_s)} = \frac{\Gamma^2 - T^2}{1-\Gamma^2T^2} \tag{5.11}$$

$$e^{-\gamma_0^{wg} t_s} \frac{S_{21}^{\text{cal}} + S_{12}^{\text{cal}}}{2} = \frac{T(1-\Gamma^2)}{1-\Gamma^2T^2} \tag{5.12}$$

The NRW inversion can also be adapted to the waveguide; in fact, it was originally derived for waveguide measurements. Equations (2.15) to (2.17) are applicable, and the expressions for calculating permeability and permittivity (2.18) and (2.19) are modified to be [10, 12]

$$\mu = \frac{2\pi}{\Lambda\sqrt{k_0^2 - k_c^2}}\left(\frac{1+\Gamma}{1-\Gamma}\right) \tag{5.13}$$

$$\varepsilon = \frac{1}{\mu k_0^2}\left(\frac{4\pi^2}{\Lambda^2} + k_c^2\right) \tag{5.14}$$

For a shorted S_{11} measurement, where the specimen is placed against a conductive ground plane, the free-space expression, (2.37), is replaced by a more general form,

$$S_{11} = \frac{\gamma_0^{wg}\tanh\left(\gamma^{wg}t\right) - \gamma^{wg}}{\gamma_0^{wg}\tanh\left(\gamma^{wg}t\right) + \gamma^{wg}} \tag{5.15}$$

Finally, when resistive sheets are measured, the free-space expressions (2.51) and (2.52) are modified by replacing the free-space impedance by the frequency-dispersive impedance of (5.5),

$$Z_s = \frac{Z_0 S_{21}}{2\left(1 - S_{21}\right)}\frac{1}{\sqrt{1-\left(f_c/f\right)^2}} \tag{5.16}$$

$$Z_s = \frac{-Z_0\left(1 + S_{11}\right)}{2S_{11}}\frac{1}{\sqrt{1-\left(f_c/f\right)^2}} \tag{5.17}$$

5.1.3 Waveguide Air-Gap Correction

Waveguide measurements can be complicated by differences between the specimen dimensions and the internal size of the waveguide. Air gaps between the sample material and the waveguide walls contribute to errors in the measured properties. For dielectric slab samples with gaps that are small relative to the waveguide dimensions, a first-order correction is often applied [14],

$$\varepsilon^{\text{corrected}} = \frac{b - g}{\dfrac{b}{\varepsilon^{\text{uncorrected}}} - g} \tag{5.18}$$

where b and g are defined in Figure 5.2. This correction is equivalent to a lumped-circuit model of a dielectric capacitor in series with an air capacitor.

Other expressions for air gap correction are described in [15–17]. For small gaps, these alternate formulations provide similar corrections. However, when gaps are large or the dielectric permittivity is very large, air-gap

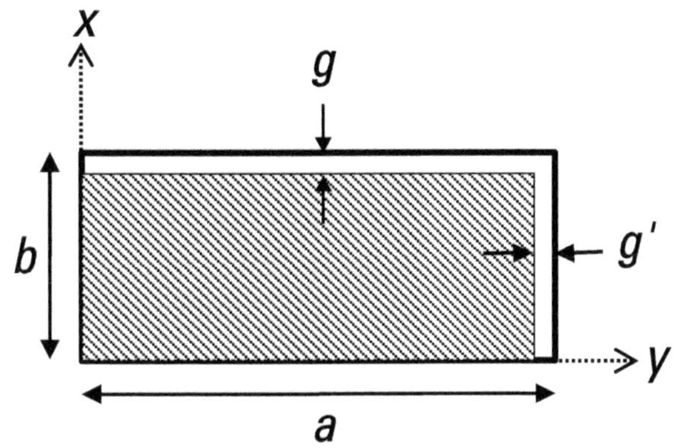

Figure 5.2 Geometry of air gaps between a material specimen and the waveguide walls.

corrections are too approximate and may become invalid. They also rely on the ability to accurately know the dimensions of the gap, which can be a significant source of uncertainty. Sample manufacturing may also make it difficult to have a uniform gap, further adding to the measurement uncertainty. Figure 5.2 shows gaps in the short and long directions of the waveguide. However, for the dominant mode, the E-field is oriented only in the short direction and goes to zero at the $y = 0$ and $y = a$ boundaries. So, small gaps at these boundaries do not significantly impact the measured permittivity.

When measuring magneto-dielectric materials, gaps between the specimen and waveguide walls can also affect inverted magnetic permeability. Gap corrections for permeability can be similarly derived using circuit models [17]. For gaps in the x-dimension, as shown in Figure 5.2, they can be modeled as two inductances in series, one for the material under test and the other for the air gap. This series model then results in a corrected permeability of

$$\mu^{\text{corrected}} = \frac{b}{b-g}\mu^{\text{uncorrected}} - \frac{g}{b-g} \tag{5.19}$$

where the relative permeability of the air gap is assumed to be 1. Unlike the case for permittivity, permeability can be affected by gaps in both directions of a waveguide. In the y-dimension case the magnetic field lines cross the boundary between the specimen and the gap, so a parallel-inductor model is more appropriate. Inductors in parallel follow the same form as capacitors in series, so this correction is similar to the permittivity correction described above, and for μ is given by

$$\mu^{\text{corrected}} = \frac{a - g'}{\dfrac{a}{\mu^{\text{uncorrected}}} - g'} \tag{5.20}$$

where g' is the gap width in the y-dimension and a is the waveguide inside width in that direction.

To demonstrate the effect of gaps in typical materials, Figure 5.3 shows an example calculation of the apparent permeability and permittivity that would be measured as a function of the size of the gap in a standard WR-90 waveguide fixture, 0.4-in-tall by 0.9-in-wide (10.16 by 22.86 mm). This data is based on the gap models in (5.18) to (5.20), and the WR-90 waveguide operates from 8.2 to 12.4 GHz. This data also assumes a relative permittivity of 25-1i and permeability of 3-2i, which is typical for a commercial high-performance microwave-absorber material in that frequency range. The dielectric permittivity varies rapidly with gap dimension and shows that the presence of a gap generally causes the apparent permittivity to be lower than the actual. The rapid decrease in permittivity or permeability is also influenced by the contrast between the material and air. So, a higher permittivity or permeability will be more strongly affected by air gaps than a lower permittivity or permeability material.

An analogous edge-capacitance model can also be developed for air gaps in waveguide measurements of resistive sheets [18]. The two-dimensional nature of the resistive sheets makes this a bit more complicated than the simple circuit models described above. Instead, a model based on Laplace's equation can be developed. Laplace's equation is a way to express the electrical potential in a space with no charges, such as the gap region between the resistive sheet sample and the waveguide walls. The edge capacitance is obtained from the electric flux ending on the sample, which can be calculated by a conformal transformation for Laplace's equation. Conformal transformations are mathematical operations that translate a complex geometry into one that can be more easily solved. For the geometry of Figure 5.4, a conformal mapping using an analytic inverse-cosine transformation can be applied; this maps from the x,z coordinate system to transformed u,v coordinates [5],

$$u + \mathrm{i}v = V_0\left(1 - \frac{2}{\pi}\cos^{-1}\left(\frac{x + iz}{g}\right)\right) \tag{5.21}$$

where u is the potential function, $-\varepsilon v$ is the flux function, and g is the width of the gap. Recognizing that $\cos x = i \cosh x$, (5.21) can be used to write the flux function,

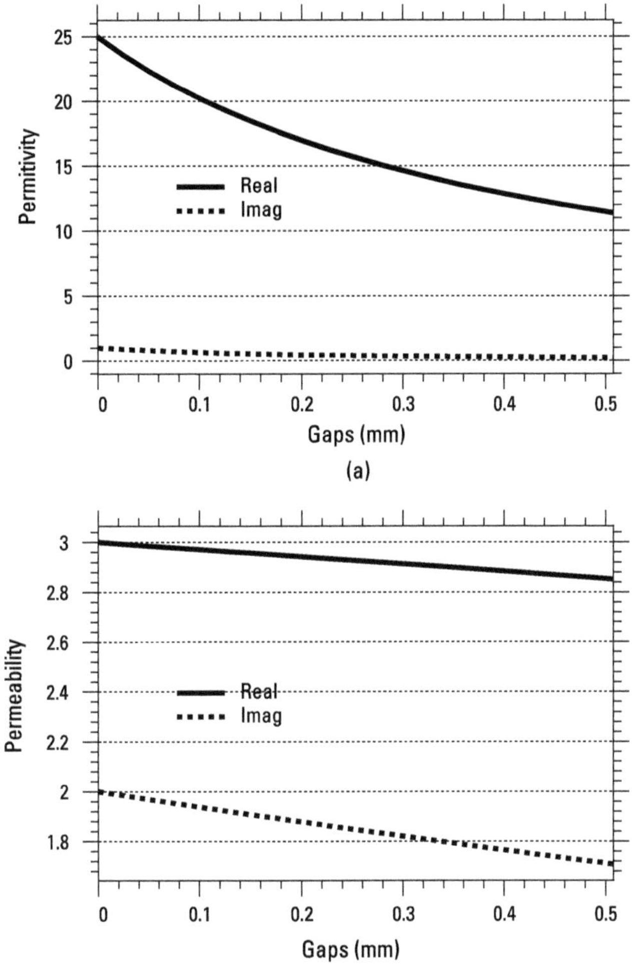

Figure 5.3 Calculated apparent (a) permittivity and (b) permeability that would be inverted from a typical magneto-dielectric specimen within a WR-90 waveguide fixture as a function of air-gap size.

$$\Psi = \text{Im}(-\varepsilon v) = \frac{2\varepsilon_0 V_0}{\pi}\cosh^{-1}\left(\frac{x}{g}\right) \tag{5.22}$$

While this map assumes a semi-infinite half-plane at a potential of V_0, the resistive sheet within the waveguide has a finite width, w. For $g << w$, the capacitance can be estimated by calculating the flux function at the edge of

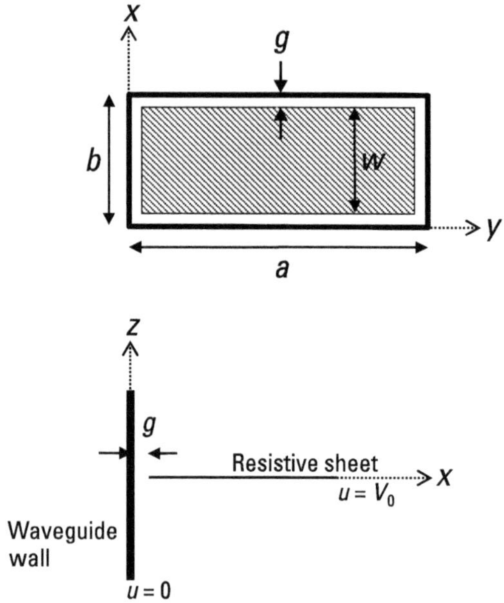

Figure 5.4 Geometry of a resistive sheet in a waveguide fixture with known air gaps between the specimen and the walls.

the sample at $x = w + g$. From Gauss's law, the charge on the sheet is equal to the total flux ending on it, and capacitance per unit length is then

$$C = \frac{q}{V_0} = \frac{2\varepsilon_0}{\pi}\cosh^{-1}\left(\frac{w+g}{g}\right) \tag{5.23}$$

The total edge capacitance is estimated by multiplying (5.23) by the length of the resistive sheet within the waveguide. There is significant E-field only on the top and bottom edges of Figure 5.4; thus, the effect of the gap on the left and right sides can be neglected. In addition, the E-field along the top and bottom varies as a function of position. For the fundamental TE mode, the E-field amplitude goes as $E_x = E_0\sin(\pi y/a)$, thus the effective length of the edge capacitor, l, is estimated as the half-width of this field distribution, which is $2a/3$ for each side (top and bottom). The total sheet impedance, Z, measured by the waveguide includes the resistive sheet and gap capacitance in series,

$$Z_{\text{total}} = Z_{\text{sheet}} + 2Z_{\text{edge}} = Z_{\text{sheet}} + \frac{2}{i\omega C_{\text{edge}}} \tag{5.24}$$

where C_{edge} is computed from (5.23) multiplied by the effective length, $2a/3$.

The edge-capacitance model of (5.23) and (5.24) affects only the imaginary component of the measured sheet impedance. However, the real sheet impedance will also increase due to gaps and must be accounted for. Specifically, there will be a difference in the apparent impedance measured due to increased transmission from the area of air (infinite sheet impedance) material within the sample plane. The total transmission coefficient is better modeled as a weighted sum of the transmission through the sample sheet and air, where the weights are equal to the relative area of each component. Because the dominant TE mode has a negligible effect from the gaps on the waveguide sides they can be neglected. Thus, a simplified area average model of the transmission amplitude (in linear units) is

$$T_{\text{total}} = \frac{b-2g}{b} T_{\text{sheet}} + \frac{2g}{b} T_{\text{air}} \tag{5.25}$$

where $T_{\text{air}} = 1$. Solving this equation for T_{sheet} in terms of T_{total} and T_{air} then provides a way to correct for the air gap area.

To demonstrate the effectiveness of these air-gap corrections for resistive sheets, Figure 5.5 shows a series of computational electromagnetic simulations using a full-wave FDTD solver. These simulations were of a WR-2300 waveguide excited from 320–540 MHz, and the symbols data are the inverted sheet impedance of the 30, 75, 150, and 300Ω/square resistive sheets. The top plot shows that the inverted real impedance is 10 to 20% higher than the actual resistive sheet simulated, and this is because a 10-mm air gap was also included. The 10-mm air gap was about 1.7 percent of the long dimension of the waveguide and was included on all four sides of the resistive sheets in these simulations. The solid lines in the top plot of Figure 5.5 show the apparent sheet impedance that is calculated with (5.25), which agrees well with the FDTD simulated results. Thus, this simple transmission model provides a reasonable approximation for correcting the real impedance.

The bottom plot shows the imaginary impedance calculated by FDTD as symbols, and the analytical model of (5.23) and (5.24) as a solid line. The original simulation was of purely real resistive sheets with no reactive component. So, the nonzero imaginary impedance predicted by the FDTD simulations is due to the air gap. The solid line is the capacitance model described above, which accounts for most of the observed capacitance in the FDTD calculations. In this approximate model, there is no dependence of the gap capacitance on the sample impedance. This is mostly consistent with the FDTD simulations, which show only a weak dependence.

Another way to minimize the effects from air gaps in resistive sheet measurements is to mount the specimens so they are pinched between two

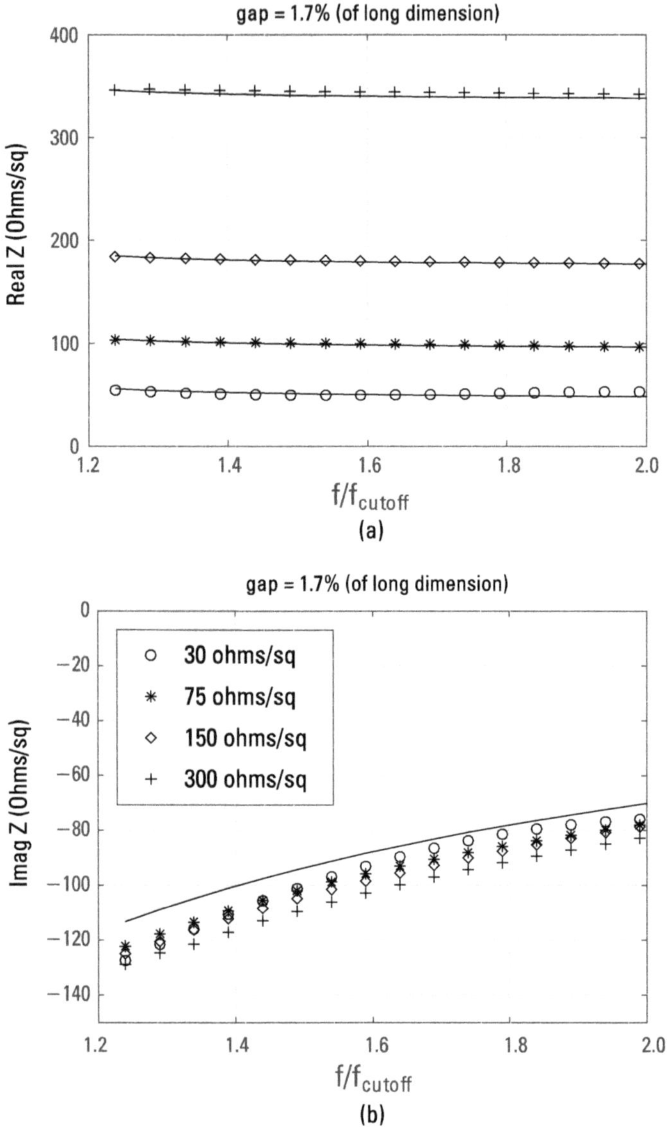

Figure 5.5 FDTD calculated (a) real and (b) imaginary sheet impedance from waveguide transmission of 30, 75, 150, and 300Ω/square sheets with gap widths of 1.7% (relative to *a*) around their perimeter.

waveguide flanges. This is shown schematically in Figure 5.6 and effectively eliminates the air gap shown in Figure 5.3. However, resistive card materials are usually manufactured from a conductive layer supported on a dielectric substrate, making one side not conductive. Sometimes there are protective layers on the other side as well so that when clamped, the resistive sheet does not actually make electrical contact with the metal flanges of the waveguide. Since there is a thin layer of nonconductive material that separates the conductive sheet from the flange, electromagnetic energy can propagate into the flange region in that thin layer. For this reason, it is important to ensure there is a way to choke off energy trapped between the flanges so that it doesn't radiate outside of the waveguide. This electrical choking can be done with thru bolts as are shown in Figure 5.6. Furthermore, enough bolts should be used around the perimeter of the flange so that microwave energy doesn't leak between them. Ideally the bolts should be spaced well under a fourth of a wavelength apart at the highest use frequency. Bolt turning in large waveguides can be tedious, and some laboratories use pneumatic systems to clamp the waveguides together. When this is done, it is still important to ensure that a direct electrical connection is maintained between the flanges to choke off RF energy that may leak through gaps between the flanges.

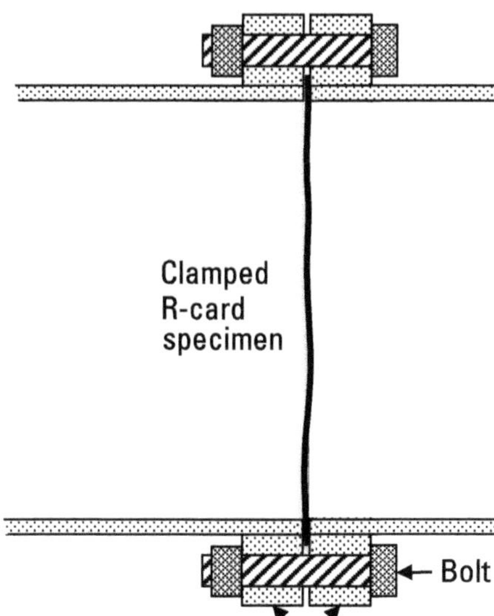

Figure 5.6 Resistive sheet specimen clamped between flanges with thru bolts for waveguide measurement.

5.2 Coaxial Air Lines

The coaxial line is a type of transmission line that includes a conductor within another conductor, separated by air or a dielectric. The coaxial airline can be used to characterize materials by inserting a toroidal shaped specimen to fill the space between the inner and outer conductor within a length of coaxial line. This transmission line is then connected to a VNA to determine the transmission and/or reflection from the known length of material specimen. The most common fixture for this is the 7-mm coaxial airline, which has been a standard size since the 1960s [11]. Like the hollow waveguides, the coaxial line can propagate energy in different modes depending on frequency. Unlike the waveguides, the dominant mode in a coaxial line has a cutoff frequency of $f_c = 0$. The coaxial line in this mode is not dispersive and instead has a constant wave velocity versus frequency just like free space. Additionally, the characteristic impedance of a coaxial line is constant and can be calculated from the dimensions of the conductors [5],

$$Z_{\text{coax}} = \frac{\ln(b/a)}{2\pi}\sqrt{\frac{\mu\mu_0}{\varepsilon\varepsilon_0}} \tag{5.26}$$

where a and b are the radii of the inner and outer conductors as shown in Figure 5.7. For a 7-mm coaxial line, the 7-mm refers to the inside diameter of the outer conductor ($2b$). Coaxial lines are also usually standardized to have a 50-Ω characteristic impedance. Using (5.26), this leads to an inner conductor diameter of 3.04 mms.

Like a rectangular waveguide, a coaxial line does have an upper frequency where the next higher-order mode can also propagate. These higher-order modes are dispersive and will complicate data interpretation in material measurements, and so they are to be avoided. Determining higher modes in coaxial lines requires solution of a transcendental equation; however, it can be estimated by reasoning that the higher-order cutoffs will occur when the circumference of transmission line, midway between the inner and outer conductors, is an integral multiple of a wavelength [5]

$$n\lambda_c \approx 2\pi\left(\frac{a+b}{2}\right)\sqrt{\mu\varepsilon}, \quad n = 1,2,3\ldots \tag{5.27}$$

For a 7-mm-diameter coaxial line, this translates to about 19 GHz. Therefore, the upper limit for 7-mm coaxial measurements is usually said to be 18 GHz. It may in fact, be lower than that depending on how high the

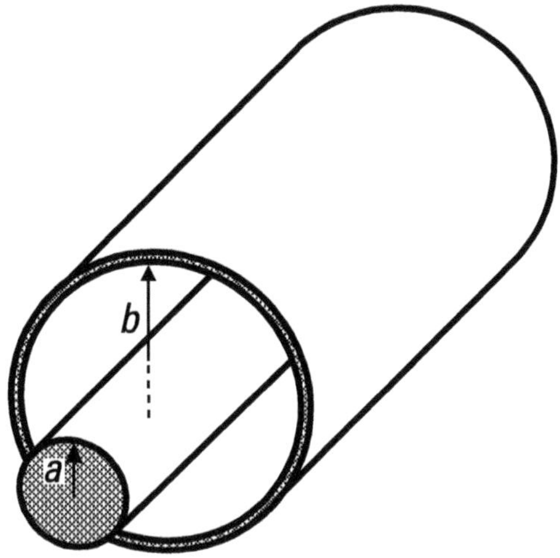

Figure 5.7 Drawing of coaxial transmission line showing inner and outer conductors.

ε and μ of the material under test is. In other words, the cutoff for the next higher mode is lowered within the material because of the intrinsic properties of that material. Any small inhomogeneity in the material or imperfection in its shape or flatness can then excite that next mode resulting in increased measurement error at those higher frequencies. In some cases, larger diameter coaxial lines are used for material measurements. The upper use frequency of these larger fixtures can also be estimated by (5.27).

Another important difference between coaxial air lines and waveguides is that the dominant coaxial mode has both the electric and magnetic fields transverse to the propagation direction. The orientation of these fields is shown schematically in Figure 5.8. The E-field lines are oriented radially (solid lines), while the magnetic field lines are circumferential (dashed lines). If a material under test is anisotropic, the measured permittivity and permeability will both be averaged over the in-plane directions.

5.2.1 Coaxial Calibration and Material Inversion

The calibration used in coaxial airlines is similar to that of waveguides. There are a variety of measurement standards that can be used, and ready-made calibration kits are easily obtained for standard sizes such as the 7-mm coaxial airline. The TRL calibration method described in Section 5.1.1 is also sometimes recommended for coaxial lines. The biggest difference, however, is that while rectangular or circular waveguides have bandwidths limited to less than

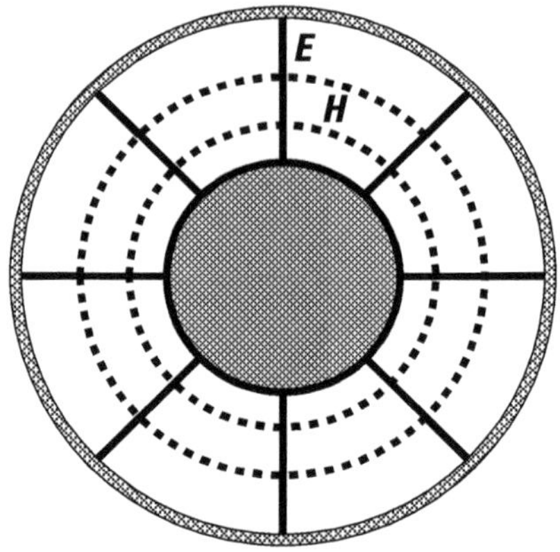

Figure 5.8 Coaxial transmission line cross-section showing orientation of E-field lines (solid) and magnetic field lines (dashed).

2:1, a coaxial airline can span multiple decades of frequency. For example, the 7-mm coaxial line described in Section 5.2 has an upper frequency limit of about 18 GHz. However, at the lower frequency, there is technically no restriction due to the coaxial line. With the cutoff frequency being 0, the only restriction on how low in frequency a material can be measured is due to the microwave analyzer and specimen thickness.

The TRL calibration uses thru, reflection, and line standards that are measured to calculate the calibration coefficients. As described in Section 5.1.1, the line standard establishes a known phase shift, which is set by the physical length of the standard, l. For the coaxial airline, that phase shift is $\phi = lk_0 = 2\pi l/\lambda_0$. However, if that phase shift is too close to integral multiples of a half-wavelength (180 degrees), then the phase difference from the thru will be insufficient to act as a useful standard. Additionally, if measurements need to be done over extremely wide bandwidths, a single line standard is insufficient, and multiple line standards need to be used to fully span the frequency range. The recommended phase range over which a line standard is valid is from 20 to 160 degrees of phase delay. For example, a 7.4-mm-long line standard will achieve the 160-degree phase delay criterion at 18 GHz but will hit the other limit of 20 degrees at about 2.25 GHz. So, additional line standards of increasing length are required to extend the measurement below that frequency. Depending on how low in frequency measurement data is desired, line standards may become impractically long. When this happens, it is possible to use a thru, reflect, and match (TRM) method to further decrease the lowest measurement frequency [13].

Once a calibrated set of network scattering parameters is obtained, the intrinsic properties of a material specimen can be inverted. Because the dominant-mode cutoff frequency is zero, the propagation of the wave through the waveguide follows the same behavior as free space. For that reason, the inversion equations described in Chapter 2 for free-space material property inversion also apply to coaxial line data without any modifications.

An example of an iterative S_{21} inversion of a 12-mm-thick acrylic specimen is shown in Figure 5.9. Note that this measurement used a 76.2-mm-diameter coaxial airline, which has an upper frequency limit of about 2 GHz. This data was calibrated with the TRL method, and a significant blip in the inverted data is evident at 1.226 GHz. The line standard was about 122.5-mm-long, which is half a wavelength at that same frequency. Thus, this blip is where there is a phase ambiguity in the line standard of the TRL calibration. This artifact would be eliminated if the calibration had included another line standard of a different length. Excluding this artifact, the mean permittivity of the specimen across the measurement band is 2.64-.025i, which is about the expected permittivity of acrylic. The acrylic specimen was machined with very good precision so that the air gaps between the specimen and the inner and outer radii of the coaxial airline were less than 0.05 mm. As discussed in Section 5.2.2, correcting for this small of an air gap in a low-dielectric polymer such as acrylic results in only a few percent shift of the inverted permittivity.

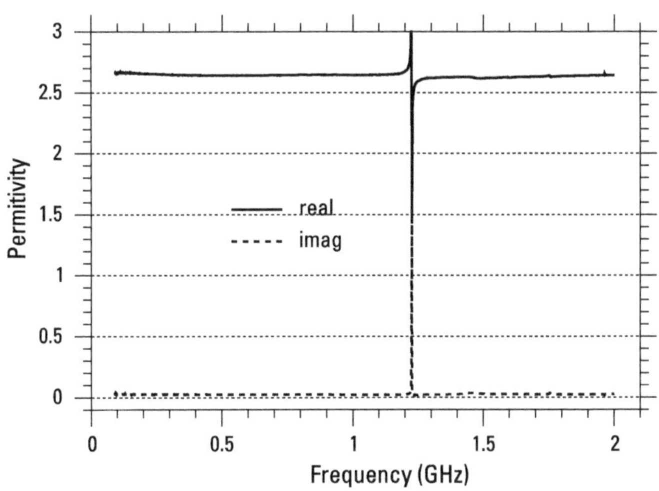

Figure 5.9 Coaxial airline measurement of 12-mm-thick acrylic specimen, calibrated with a TRL calibration method.

Another measurement example is in Figure 5.10, which shows the inverted properties from a commercial microwave absorber consisting of aligned sendust platelet particles in a polymer matrix. Absorbers such as this are often used for electromagnetic interference mitigation in consumer electronics and have extremely high dielectric permittivity and excellent magnetic loss at VHF and UHF frequencies. This absorber normally comes in 0.5-mm-thick sheets, and

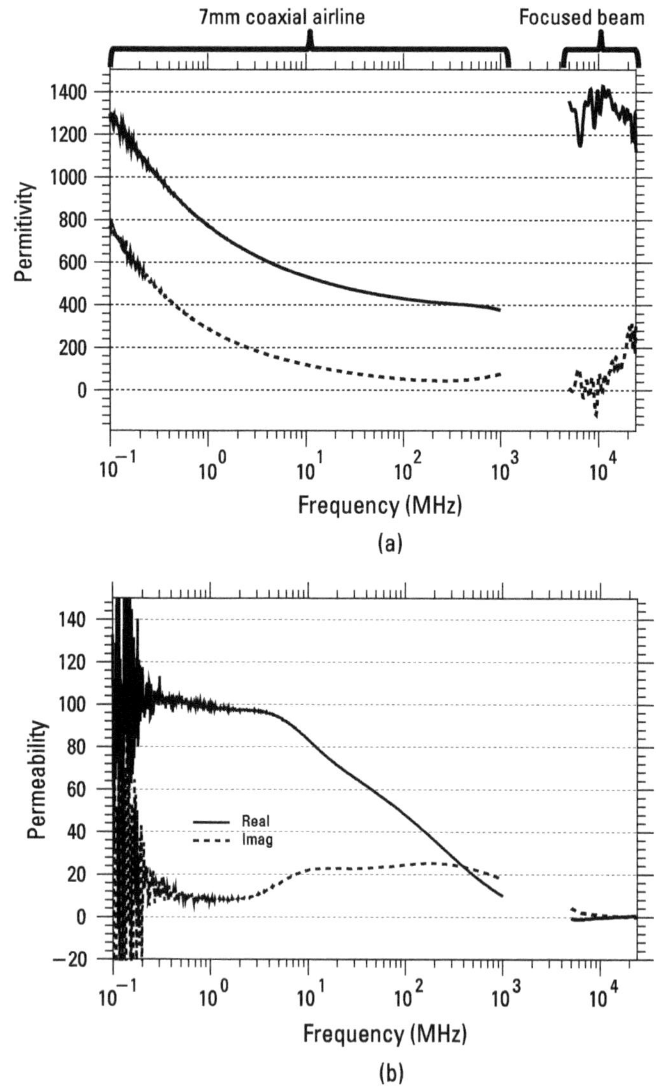

Figure 5.10 (a) Permittivity and (b) permeability of a commercial absorber material made from aligned sendust particles in a polymer matrix.

three of these sheets were stacked together to make the specimen thicker to improve the lower frequency sensitivity. In contrast to the acrylic specimen, this absorber measurement was done in a 7-mm coaxial airline, and the calibration method was a TRM load so that it would be accurate to much lower frequencies. To provide some reference information, high-frequency free-space focused-beam measurement data of this material is also included on the plots.

Looking first at the dielectric permittivity data on the top of Figure 5.10, there is large discrepancy between the coaxial airline measurements and the free-space data. This difference is due to limitations from air gaps in the coaxial specimens. First, absorber materials such as this can be difficult to cut cleanly into the 7-mm toroid shapes. Second, the cutting action can also result in local damage of the magnetic platelets in the material edges due to shear, changing the local particle alignment at the toroid edges and effectively creating a lower-dielectric barrier around the perimeter. So, the platelet deformation combined with the extremely high dielectric permittivity of these materials creates a large measurement error even when the gaps between the specimen and the coaxial line are minimal. In this case, the materials fit tightly into the coaxial airline, but the apparent permittivity is still much lower than that measured by the high-frequency focused-beam method. The high-frequency permittivity data here is more reliable, because free-space methods do not have air gaps in the first place. This large gap error is also beyond the ability of the below-described gap corrections (Section 5.2.2) to fix. Materials such as this, with permittivities in the range of hundreds to thousands, are often impractical to measure in a 7-mm coaxial airline.

While the permittivity cannot be effectively determined with this coaxial line method, the magnetic permeability on the lower plot of Figure 5.10 is still reliable. In particular, the circumferential magnetic field lines of the fundamental mode do not cross a material boundary into air, so there is no strong demagnetization error. In other words, even when permeability is high, the permeability errors from small gaps in the coaxial airline are small, and the data in the lower plot of Figure 5.10 is consistent between the coaxial airline measurements and the higher-frequency focused-beam results. Moreover, the magnetic permeability results in Figure 5.10 are consistent with other measurement methods, including the CEM methods and impedance analysis techniques described later in this book (Chapters 7 and 8).

The data in Figure 5.10 also shows a lower limit to measurement accuracy for magnetic permeability of this material, which can be related to the sensitivity of the VNA and the thickness of the specimen. At or below 1 MHz, the inverted permeability becomes noisier, and below about 200 KHz, the noise level in the inverted permeability is too large to provide accurate results.

The specimen is 1.5-mm-thick, and the lower-frequency performance could be improved by increasing the thickness of the specimen, such as by stacking up more layers of the material. That said, this data does not show that the intrinsic properties can be measured by a coaxial airline over more than three decades of bandwidth.

5.2.2 Air Gap Corrections in Coaxial Airlines

Usually, the largest source of measurement uncertainty in the coaxial airline is from air gaps between the sample and fixture. As Figure 5.10 shows, these gap errors limit the usefulness of coaxial measurements, particularly for materials and absorbers that have medium or high permittivities and conductivities. Similar to the rectangular waveguide fixtures, analytic expressions can be derived to estimate the effect of air gaps in coaxial airlines. The usual approximation is based on a capacitor model where the air gap and material form layers of a capacitor [19].

$$\varepsilon_{\text{corrected}} = \frac{L_2 \varepsilon_{\text{uncorrected}}}{L_3 - L_1 \varepsilon_{\text{uncorrected}}} \tag{5.28}$$

where $L_3 = \ln(r_o/r_i)$, $L_2 = \ln(r_{os}/r_{is})$, and $L_1 = L_3 - L_2$, and the various radii are defined in Figure 5.11. As a simple circuit model, (5.28) is frequency-independent and only valid at lower frequencies where the gaps are electrically small.

For magnetic permeability, a similar circuit model analyzes the sample and gaps in terms of inductances [17, 19],

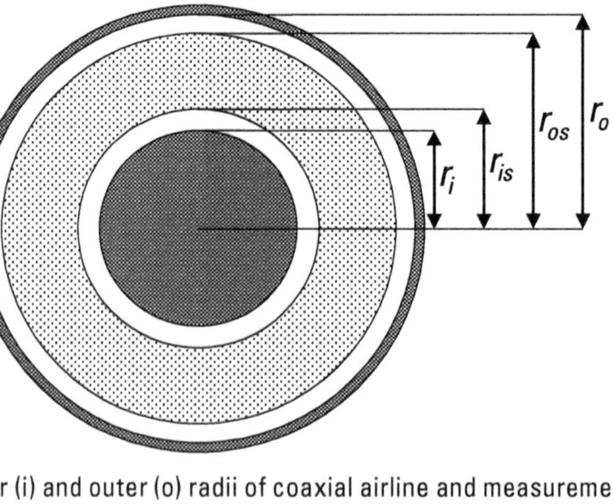

Figure 5.11 Inner (i) and outer (o) radii of coaxial airline and measurement specimen.

$$\mu_{\text{corrected}} = \frac{\ln\left(r_o/r_i\right)\mu_{\text{uncorrected}} - \left(\ln\left(r_{is}/r_i\right) + \ln\left(r_o/r_{os}\right)\right)}{\ln\left(r_{os}/r_{is}\right)} \tag{5.29}$$

Because the electric field lines cross the gap, while the magnetic field lines are parallel to it, the effect of a gap is significantly stronger for the dielectric permittivity. This can be seen in Figure 5.12, which shows the application of (5.28) and (5.29) to intrinsic properties representative of a commercial carbonyl iron–filled magnetic absorber in the 2–18 GHz range. These plots show the effect of varying inner and outer gaps when the material has a permittivity of 25-1i and permeability of 3-2i. In both cases the inner gap has a more significant effect.

Markuvitz derived a higher-order approximation for coaxial airlines that treats the coaxial cross-section as a radial transmission line with two regions: gap and specimen [16, 19]. The boundary between the gap and the specimen then represents an equivalent microwave network of two radial transmission lines with electrical shorts. From this, a relationship between the material and gap regions can be derived,

$$\text{ct}\left(k_1 r_b, k_1 r_i\right) = \frac{k_1 \varepsilon_2}{k_2 \varepsilon_1} \text{ct}\left(k_2 r_b, k_2 r_o\right) \tag{5.30}$$

where we are assuming for simplicity that just one gap, r_b, is the radius of the boundary between the gap and the material; k_1, k_2 are the wavenumbers in the inner and outer radial regions respectively; and ε_1, ε_2 are the complex permittivities of the two regions. The ct function is defined in terms of Bessel functions of the zeroth and first order of the first and second kind,

$$\text{ct}\left(x, y\right) = \frac{J_1(x)N_0\left(y\right) - N_1\left(y\right)J_0(x)}{J_0(x)N_0\left(y\right) - N_0(x)J_0\left(y\right)} \tag{5.31}$$

The wavenumbers of the two regions are given by

$$k_1 = \frac{2\pi f}{c}\sqrt{\varepsilon_1 - \varepsilon_{\text{uncorrected}}} \text{ and } k_2 = \frac{2\pi f}{c}\sqrt{\varepsilon_2 - \varepsilon_{\text{uncorrected}}} \tag{5.32}$$

Equations (5.30) to (5.32) are a transcendental set that must be solved numerically, but they can provide an improved correction, particularly for higher frequencies.

Examples of cases where the Marcuvitz correction deviates from the simpler capacitor model are shown in Figure 5.13. This data is for an air gap

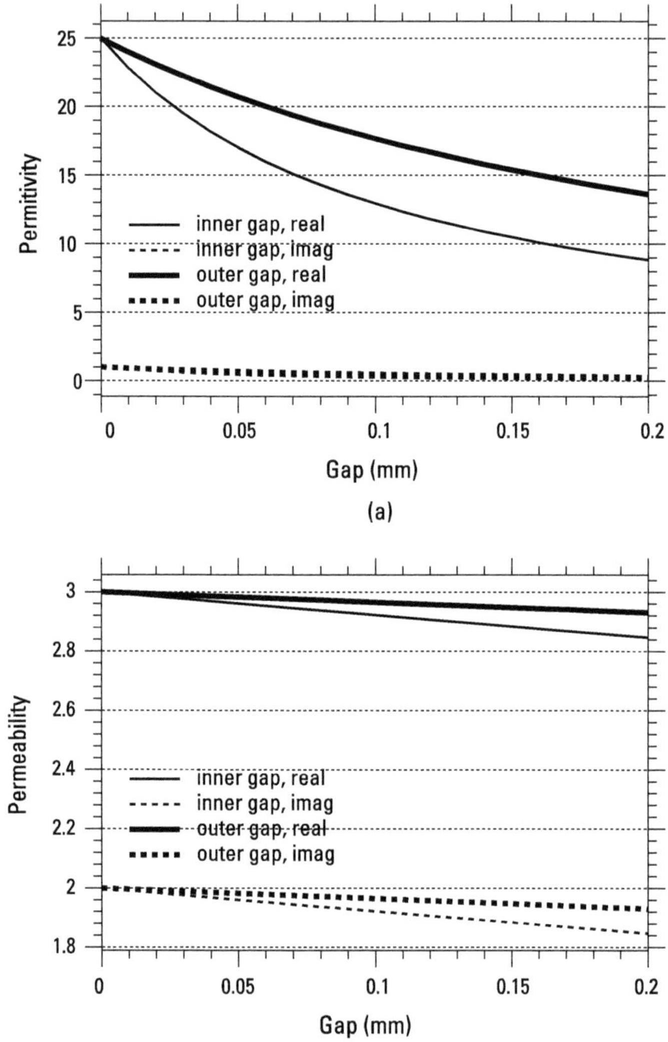

Figure 5.12 (a) Permittivity and (b) permeability (bottom) that would be measured in a 7-mm airline fixture for a magneto-dielectric material with intrinsic properties of $\mu = 3 - 2i$ and $\varepsilon = 25 - 1i$ as a function of inner or outer gap size.

of 0.1 mm between the specimen and the inner conductor, and three different sets of curves are shown for uncorrected real permittivities of 3, 10, and 15. The solid lines are the result after the capacitance correction is applied via (5.28), and the dashed lines are the corrected permittivity after the iterative

Marcuvitz correction is applied. For lower dielectric materials, there is little difference between the two gap correction models, and neither of them significantly differ from the uncorrected value of $\varepsilon' = 3$. For the cases when the uncorrected permittivity is higher, there are significant differences between the Marcuvitz correction and the capacitor model as frequency is increased. Thus, at higher frequencies and higher permittivities, the capacitor model overcorrects the measured data.

Both of the gap corrections we've discussed are based on assumptions that are approximate. For extremely high permittivity cases such as the measured data in Figure 5.10, even the Marcuvitz model becomes invalid. In fact, application of the Marcuvitz model to the Figure 5.10 data fails to find convergence. To more rigorously calculate the effects of air gaps on coaxial airline measurements, a full-wave computational simulation code is necessary for cases of higher permittivity or larger airgaps. The fundamental mode in a coaxial airline is circularly symmetric. Thus, body-of-revolution (BOR) symmetry can be assumed, which converts a three-dimensional geometry into a two-dimensional calculation. This dramatically speeds up calculations making BOR simulations a feasible way to better evaluate errors from gaps [20].

Ideally, the best way to correct for gaps is to not have them in the first place. This can be exceedingly difficult as many materials are difficult to machine or cut to sufficient precision. So, an alternative that some have used

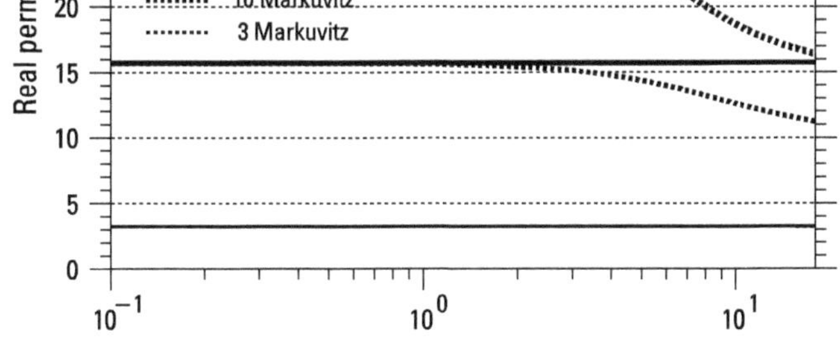

Figure 5.13 Corrected real permittivity of a series of specimens with uncorrected permittivities of 3, 10, and 15, using both capacitive and Marcuvitz gap correction methods.

is to fill the gap with a conductive or high-dielectric material. When the material under test is a higher dielectric, then a high-dielectric or conductive paste will have less contrast than an air gap, and the gap effect will be less [19]. Electroplating the curved parts of the toroidal specimens is another way to improve the contact with the coaxial transmission line boundaries [21]. Finally, a method that also reduces the errors from gaps is to use a larger-diameter coaxial airline. For a given achievable precision, the size of the gap relative to the radius of the specimen is smaller, making the error also smaller. The disadvantage is that the higher-order mode cutoff is at a lower frequency for a larger coaxial airline, so that it cannot measure as high in frequency.

5.2.3 Wrapped-Coaxial Airline Method

Thin magnetic sheets or films are sometimes used as constituents in other composite materials. Since these films are usually conductive, they cannot be measured using either the traditional waveguide or coaxial airline methods. When materials are too conductive, they don't have sufficient transmissivity to be measured accurately. However, a variation of the coaxial airline method can be used to characterize their magnetic permeability [22, 23]. This alternative method reorients the films relative to the traditional specimen orientation by wrapping them into a toroidal configuration within the coaxial airline. This is shown schematically in Figure 5.14, where the thin film is deposited on or coated with a thin dielectric substrate which is then wound about an axis to make a toroid.

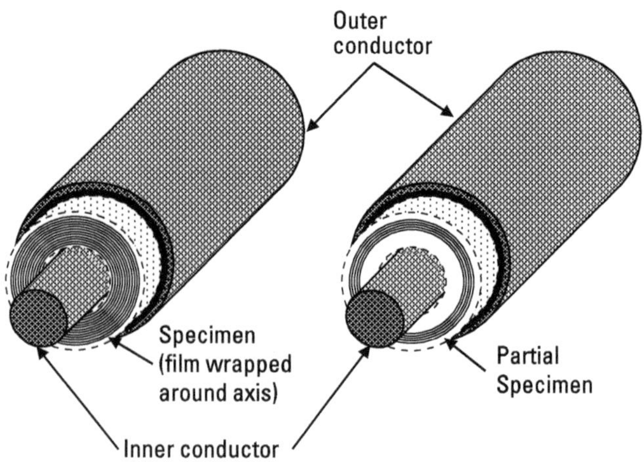

Figure 5.14 Wrapped magnetic thin film specimens to determine magnetic permeability of conductive films in a coaxial line.

This wrapping idea takes advantage of the field-vector orientations within the coaxial geometry. The magnetic field is oriented circumferentially while the electric field is radially directed. Thus, the measurement determines an apparent permeability in-plane and an apparent permittivity out-of-plane, relative to the wrapped thin film. For the configuration shown on the left side of Figure 5.14, a standard two-port measurement is done, and the usual four-parameter inversion method is used to determine in-plane permeability and out-of-plane permittivity of the wrapped specimen. Since the sample consists of alternating magnetic conductive and dielectric layers in the electric field line direction, the effect of air-gap errors is minimized.

If out-of-plane dielectric permittivity is of low importance, then the configuration on the right side of Figure 5.14 can be used to determine just the in-plane permeability of the layered magnetic/dielectric composite. In this configuration the film is wrapped onto a foam or other dielectric toroid, which supports and centers it within the coaxial fixture. The standard S-parameter inversion techniques are used to determine the apparent permeability of the coaxial cross-section. The actual permeability of the composite sample is then determined by modeling it as an inductor in parallel with concentric nonmagnetic inductors,

$$\mu_{\text{measured}} \ln\frac{R_4}{R_1} = \mu_1 \ln\frac{R_2}{R_1} + \mu_2 \ln\frac{R_3}{R_2} + \mu_3 \ln\frac{R_4}{R_3} \tag{5.33}$$

where the radii and relative permeabilities are defined in Figure 5.15. Furthermore, (5.33), can be expanded to model each individual film within the wrapped specimen so that the film permeability can be determined. Sometimes, stress induces changes in the film permeability, so a larger-diameter

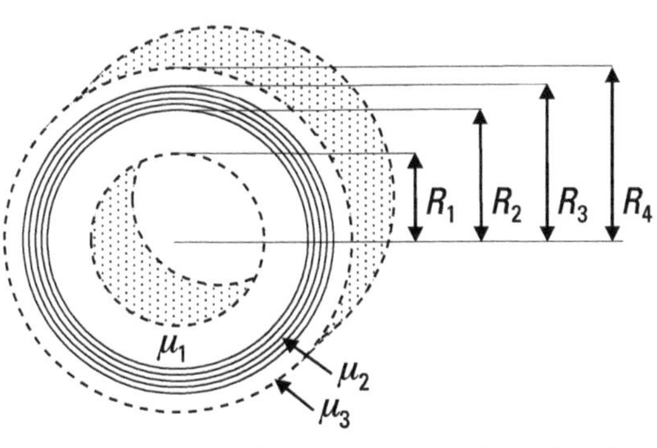

Figure 5.15 Radii of different magnetic and nonmagnetic regions in a wrapped coaxial line specimen.

coaxial air line should be used in this case to minimize the curvature of the films and decrease stress.

5.2.4 Square Coaxial Airline

Absorber materials used in anechoic chambers are typically manufactured in sizes that are 24 by 24-in (0.6 by 0.6-m) squares. These materials can be ferrite tiles or pyramidal-shaped carbon-loaded foam designed to line the walls of a test facility and minimize specular reflections. At high frequencies, free-space methods such as those described in Chapter 2 are used to characterize the performance of these types of absorbers. At low frequencies or long wavelengths, the required material specimen size and antenna aperture dimensions are too large to be practical for free space. Therefore, an alternative method is needed, and coaxial transmission lines have been adapted for this purpose [24]. In this coaxial fixture, the inner and outer conductor are both square and are sized to accommodate specimens that are 0.6 by 0.6-m. This is accomplished when the dimensions shown in Figure 5.16 are $d = 0.6$ and $D = 3d$. In this case eight different pieces of the absorber can be tiled within the square coaxial fixture to fill the space between the inner and outer conductors.

The cylindrical coaxial fixtures discussed above are dimensioned so that their intrinsic impedance is 50Ω, since that is the standard impedance of RF cables and VNA ports. While there is a simple analytical formula to calculate the impedance of a cylindrical coaxial line, that is not the case for a square coaxial line. To determine the intrinsic impedance of a square coaxial

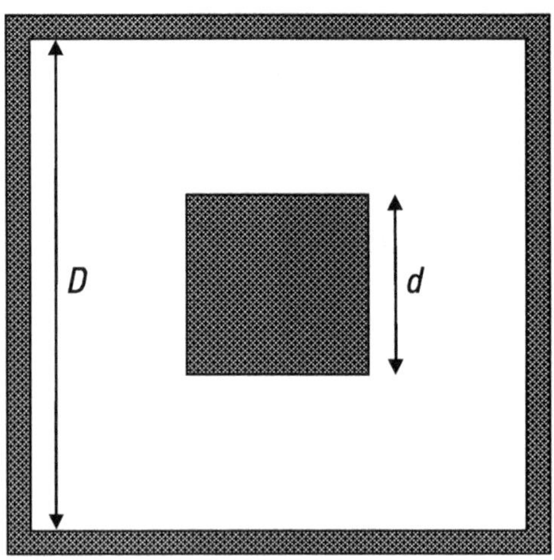

Figure 5.16 Dimensions of a square coaxial airline.

geometry, a mathematical trick called conformal mapping is required and results in a set of equations that have to be solved iteratively. There is, however, an approximate equation that provides an estimate of intrinsic impedance for a square coaxial line with better than 1% accuracy when $d/D \geq 0.25$ [25],

$$Z_{\text{square coax}} = Z_0 \left(\frac{1}{4\left(\frac{2d}{D-d} + 0.558 \right)} \right) \tag{5.34}$$

where Z_0 is the intrinsic impedance of free space (~ 377Ω).

Plugging in the dimensions for anechoic chamber absorber (0.6m) results in an intrinsic impedance of about 60Ω. Since this impedance is 10-Ω-higher than the impedance of the analyzer, a resistor can be added in series with the center conductor to improve the impedance match and reduce the reflections at the transition between the 50-Ω cable and the 60-Ω square coax. Since the fundamental mode is not frequency-dispersive, and if the coaxial system is made long enough, the mismatch reflection can also be subtracted by time-domain gating methods such as those used in free-space calibration [26].

The other applicable rule of thumb is the upper frequency where higher-order modes may exist and complicate the measured performance. In a coaxial line, this first higher-order mode occurs about where the mean circumference of the guide, midway between the inner and outer conductors, is about a wavelength [5]. For a guide where $d = 0.6$m and $D = 1.8$m, the mean circumference is 4.88m, and the cutoff frequency for that mode is therefore in the neighborhood of 60 MHz. However, in practice, that first higher-order mode may not be excited when specimens are symmetrical within the coaxial line, and the 1.8-m square coax can successfully operate up to 100 MHz [24]. Others have found that this 1.8-m coaxial line can take reasonable reflection data even as high as 500 MHz [26].

The propagation of the fundamental mode in the square coax is similar to a cylindrical coaxial airline with a cutoff frequency of 0. Therefore, the equations for material property inversion presented in Chapter 2 for free space also apply to the square coax fixture. Other smaller-sized square coaxial airlines can also be constructed to measure a variety of materials [27]. A square coax allows specimens to be machined into rectangular or trapezoidal shapes for insertion into the line, which in some cases is easier than machining curved toroids. Also, when not restricted to measuring certain sized square specimens, the inner and outer conductor dimensions can be set so that the intrinsic impedance of the square coax meets the 50-Ω input impedance of the microwave analyzer.

5.3 Stripline Methods

Another type of waveguide that can be used for material measurements with transmission and reflection is the stripline. The stripline consists of two outer conductors with an inner conductor midway between, as shown in Figure 5.17. Unfortunately, there is no simple exact expression for calculating the impedance, and a mathematical method such as conformal transformation must be used to develop expressions for calculating the intrinsic impedance. Expressions for the characteristic impedance of a number of different geometries are documented in [28], and the expression for a rectangular coaxial line is given by

$$Z_{r\text{coax}} = \frac{Z_0}{4} \frac{1}{\dfrac{w}{b-t} + \dfrac{1}{\pi}\left[\dfrac{b}{b-t}\ln\left(\dfrac{2b-t}{t}\right) + \ln\left(\dfrac{t(2b-t)}{(b-t)^2}\right)\right] \dfrac{\ln\left(1+\coth\left(\dfrac{\pi a}{2b}\right)\right)}{\ln 2}} \tag{5.35}$$

where a is the total width of the rectangular coaxial line and the other dimensions are as defined in Figure 5.17. Here, we are interested not in a rectangular coaxial line, but in a stripline, where the side walls effectively are moved to infinity, $a \to \infty$. In this case, (5.35) is simplified to [29]

$$Z_{r\text{coax}} = \frac{Z_0}{4} \frac{1}{\dfrac{w}{b-t} + \dfrac{1}{\pi}\left[\dfrac{b}{b-t}\ln\left(\dfrac{2b-t}{t}\right) + \ln\left(\dfrac{t(2b-t)}{(b-t)^2}\right)\right]} \tag{5.36}$$

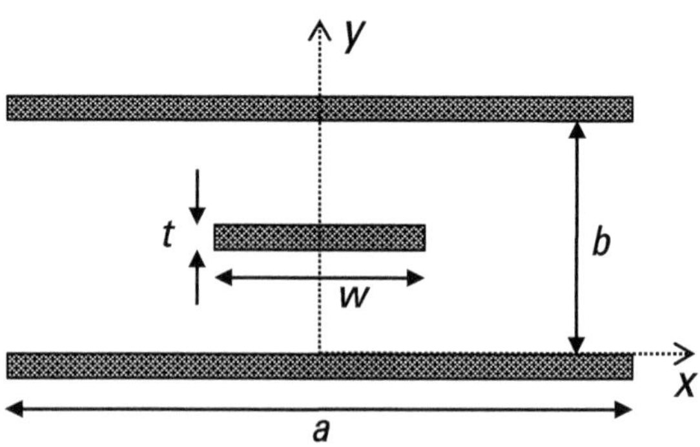

Figure 5.17 Dimensions of a stripline transmission line.

Equation (5.36) can be used to design a stripline system for material measurements by ensuring that the intrinsic impedance is 50Ω so that it matches the input impedance of the microwave analyzer. For example, a stripline with dimensions of $t = 6.35$ mm, $b = 57.15$ mm, and $w = 65.0875$ mm has an intrinsic impedance of 50.26Ω when calculated by (5.36). This intrinsic impedance can be verified by using a computational code to model the stripline and solving the Laplace equation for electric potential with a standard relaxation method. This computational method calculates an intrinsic impedance of 50.17Ω, which is within a fraction of a percent of the estimate from (5.36). Either of these values of impedance is close enough to the standard 50-Ω input impedance to provide an excellent match to a microwave analyzer.

Another factor that goes into design of a stripline material measurement fixture is the upper frequency of use. This is driven by when higher-order modes can propagate, which complicates the interactions with material specimens. The first higher-order mode, called TE_{10}, occurs at frequencies above where the separation between the two outer conductors is a half-wavelength apart [30]. For the example design described above ($t = 6.25$, $b = 57.15$, $w = 65.0875$ mm), this mode should start at a frequency just above 2.6 GHz.

A third aspect to consider for a stripline fixture design is the width of the outer conductors. The outer conductors need to be long enough so that there is no longer much E-field at the edges of the transmission line (or inserted material specimen). Collin conducted a variational analysis of the stripline geometry and points out that the E-field decays exponentially in the x-direction (see Figure 5.17) with a dependence of $e^{-\pi x/b}$ [31]. At $x \approx 0.5w + 1.5b$ the field is down to about 1% of its strength at the origin, which should minimize effects from the edge such as field truncation and radiation to structures outside of the stripline. In other words, a reasonable rule of thumb is that the outside conductors should extend beyond the center conductor by at least 1.5 times the separation between the two outer conductor plates.

Since the outer conductors of a stripline are not physically connected to each other, a significant disadvantage of the stripline is that it is not easily constructed in multiple finite length sections. This contrasts with waveguides, which have flanged ends for bolting sections together, or coaxial airlines, which have standard connectors or flanges for connecting sections together. Instead, a stripline fixture consists of transitions and a straight section that are made of a single piece of metal for each conductor (two outer conductors and one inner conductor). This makes calibrations such as TRL impractical since the stripline cannot be lengthened by insertion of a finite length section.

Instead, a stripline can be successfully calibrated with different offset electrical shorts, which are just metal specimens clamped into different locations along the transmission line. These offset shorts combined with a thru measurement provide the necessary independent parameters for accurately determining the propagation constant and correcting for errors from the transitions and the microwave network analyzer [32, 33].

Once calibrated, the stripline can then invert material properties much the same way as in a coaxial airline. The fundamental mode is nondispersive with a cutoff frequency of zero. So, the various inversion methods described in Chapter 2 for free-space techniques also apply to scattering parameters measured with a specimen in a stripline fixture.

One of the advantages of the stripline method over other guided-wave fixtures is in how it can be used to manage air gaps. Air gaps are the dominant source of measurement errors in waveguide and coaxial airlines because the E-field lines within cross from the material specimen into the air gap. Even small airgaps can have a strong influence on the measured permittivity for moderate- to high-dielectric materials. The stripline significantly reduces this error because of its ability to be slightly deformed by clamping so that the outer conductors squeeze the sample. This is shown schematically in Figure 5.18. The specimens are made in two pieces that fit within the fixture between the inner and outer conductors. They can be made just slightly undersized so that when clamped, the only air gap is on either side of the inner conductor, where the E-field strength is relatively low.

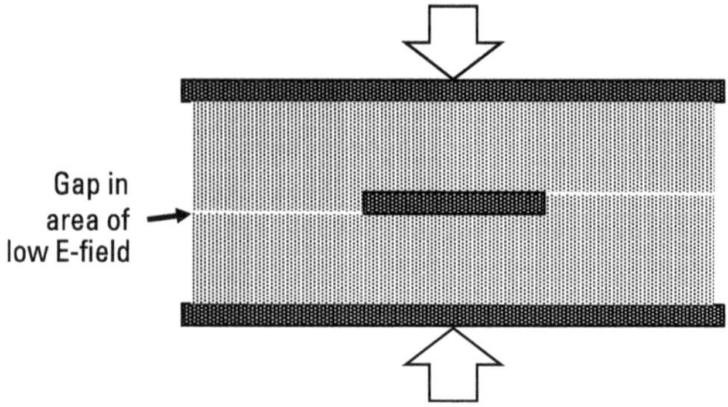

Figure 5.18 Drawing of sample within a stripline cross-section, clamped so that the only airgaps are at a low-E-field region on the sides of the center conductor.

5.4 Other Transmission Line Methods

A fundamental truth in RF material measurements is that no one measurement technique is the solution for all situations. There are a variety of physical properties that make some materials difficult to machine or form into shape. There can also be complications from inhomogeneity or anisotropy that restrict what devices a material can be effectively measured in. Other factors that influence the choice of measurement fixture include frequency range, bandwidth, ease of calibration, accuracy, cost, and size, all of which determine what method is the best for a given situation. The methods described in Sections 5.1–5.3, while the most common, may not always be the best choice, driving a need to develop new methods for new situations. This section describes a sample of alternative transmission line methods that have been documented in the literature.

Circular waveguides have been discussed in the literature as an alternative to rectangular waveguides for transmission/reflection measurements of material coupons. However circular waveguides are not as good when there is material anisotropy, and they have even more limited bandwidths than rectangular waveguide. On the other hand, there is a growing need for measuring specimens at higher and higher frequencies ranging from millimeter-wave to terahertz. Moreover, the air gap is especially hard to control as the waveguide dimensions shrink into the millimeter size range.

An alternative that has been introduced recently is the use of a corrugated waveguide. While a material is not easily placed inside a corrugated waveguide, it is possible to sandwich an oversized specimen between two open-ended corrugated waveguides [35]. At these higher frequencies, the fractional bandwidths are less of a disadvantage, and the corrugated waveguide method shows promise as an alternative to conventional rectangular waveguide measurements. A potential disadvantage, however, is that because the specimen is sandwiched between two open waveguides, this creates a gap though which energy can couple, leading to additional evanescent modes excited at the specimen. The corrugated waveguide transmits a hybrid mode, HE_{11}, that simulates a Gaussian beam with the E-field linearly polarized and peaked in the middle of the transmission line. So, the performance is comparable to the focused-beam methods discussed in Chapter 4. However, published data show some ripple in higher-permittivity materials, as well as slowly rising permittivity versus frequency [35].

Another alternative waveguide that has been explored for material measurement is the ridged waveguide. While corrugated waveguide have ridges oriented around the inside of the pipe, perpendicular to the propagation direction, ridged waveguides have one or two ridges that are in the center of the waveguide, parallel to the propagation direction. Like in a rectangular waveguide, material measurements must be made above the cutoff frequency for the fundamental mode and below the cutoff for the next higher-order model. The advantage of a ridged waveguide over a rectangular one is that the bandwidth where there is only the fundamental mode is about three-times greater in a ridged waveguide. The trick is to calculate what the fundamental mode cutoff is, since the internal shape of a ridged waveguide is much more complicated than the simple rectangle. Two methods have been derived for determining this cutoff frequency and demonstrated on material measurements of a single specimen from 6 to 18 GHz [36]. This increased bandwidth has the advantage of providing the same data as three separate rectangular waveguide measurements. However, air gap uncertainties are still a potential problem with this method, and the specimens can be more difficult to machine or cut since they are H-shaped to fit within the ridged line. As an alternative, this ridged waveguide idea can also clamp larger-sized specimens in between two open-ended ridged waveguides [37]. This avoids the issues associated with precision cutting of specimens or air gaps but requires a significantly more complicated analysis to extract material properties.

Finally, while not within the scope of wideband materials characterization discussed in this book, the idea of using waveguide and transmission lines as resonant cavities is also a way to measure dielectric or magnetic properties of materials at specific frequencies (narrowband). The general principle is that when a section of transmission line is truncated, then resonant modes can exist within that structure. A material specimen can then be introduced that perturbs the resonance so that it shifts in frequency, and the quality of the resonance decreases. Characterization of the frequency shift and resonance quality change can then be used to calculate real and imaginary permittivity or real and imaginary permeability, depending on the specifics of the cavity. Cavities include fixtures made from rectangular waveguide [38], stripline transmission lines [39], cylindrical cavities [40], and many others.

References

[1] Packard, K. S., "The Origin of Waveguides: A Case of Multiple Rediscovery," *IEEE Trans. Microwave Theory and Techniques*, Vol. MTT-32, No. 9, September 1984, pp. 961–969.

[2] Branin, F. H., "A Microwave Method of Determining Dielectric Constant and Loss by Phase Comparison," *1949 Conference on Electrical Insulation*, Pocono Manor, PA, November 7–9, 1949, pp. 57–59.

[3] Redheffer, R. M., "The Measurement of Dielectric Constants," Chapter 10 in *Technique of Microwave Measurements*, C.G. Montgomery (ed.), New York, NY: McGraw-Hill, 1947, pp. 561–76.

[4] Harrington, R. F., *Time-Harmonic Electromagnetic Fields*, New York, NY: McGraw-Hill, 1961.

[5] Ramo, S., J. R. Whinnery, and A. Van Duzer, *Fields and Waves in Communication Electronics* (Third Edition), New York, NY: John Wiley & Sons, 1994.

[6] Rytting, D. K., "Network Analyzer Accuracy Overview," *58th ARFTG Conference Digest*, San Diego, CA, 29–30 November 2001.

[7] Shoaib, N., *Vector Network Analyzer (VNA) Measurements and Uncertainty Assessment*, Cham, Switzerland: Springer International Publishing, 2017.

[8] Engen, G. F., and C. A. Hoer, "Thru-Reflect-Line: An Improved Technique for Calibrating the Dual 6-Port Automatic Network Analyzer," *IEEE Trans. on Microwave Theory and Techniques*, Vol. MTT-27-12, December 1979, pp. 98–987.

[9] Baker-Jarvis, J., "Transmission/Reflection and Short-Circuit Line Permittivity Measurements," NIST Technical Note 1341, July 1990.

[10] Nicolson, A. M., and G. F. Ross, "Measurement of the Intrinsic Properties of Materials by Time-Domain Techniques," *IEEE Trans. Instrumentation and Measurement*, Vol. 19, No. 4, 1970, pp. 377–382.

[11] Bryant, J. H., "Coaxial Transmission Lines, Related Two-Conductor Transmission Lines, Connectors, and Components: A U.S. Historical Perspective," *IEEE Trans Microwave Theory and Techniques*, Vol. MTT-32, No. 2, September 1984, pp. 970–983.

[12] Weir, W. B., "Automatic Measurement of Complex Dielectric Constant and Permeability at Microwave Frequencies," *Proceedings of the IEEE*, Vol. 62, No. 1, Jan. 1974, pp. 33–36.

[13] Barr, J. T., and M. J. Pervere, "A Generalized Vector Network Analyzer Calibration Technique," *34th ARFTG Conference Digest*, 1989, pp. 51–60.

[14] Champlin, K. S., and G. H. Glover, "Gap Effect in Measurement of Large Permittivities," *IEEE Trans Microwave Theory and Techniques*, Vol. 14, No. 8, 1966, pp. 397–398.

[15] Westphal, W. B., "Techniques of Measuring the Permittivity and Permeability of Liquids and Solids in the Frequency Range 3 c/s to 50 kmc/s," M.I.T. Laboratory for Insulation, Cambridge, MA., Res. Tech. Report. No. 36, 1950.

[16] Marcuvitz, N., *Waveguide Handbook*, New York, NY: McGraw-Hill, 1951.

[17] Baker-Jarvis, J., et al., "Transmission/Reflection and Short-Circuit Line Methods for Measuring Permittivity and Permeability," NIST Technical Note 1355-R, December 1993.

[18] Schultz, J. W., "Numerical Analysis of Transmission Line Techniques for RF Material Measurements," *Antenna Measurement Techniques Association (AMTA) Symposium Proceedings*, Irvine CA, October 19–24, 2003.

[19] Mattar, K. E., D. G. Watters, and M. E. Brodwin, "Influence of Wall Contacts on Measured Complex Permittivity Spectra at Coaxial Line Frequencies," *IEEE Trans. Microwave Theory and Techniques*, Vol. 39, No. 3, March 1991, pp. 532–537.

[20] Schultz, J. W., G. Mohler, and J. G. Maloney, "Numerical Analysis of the Coaxial Airline Fixture for RF Material Measurements," *Antenna Measurement Techniques Association (AMTA) Symposium Proceedings*, Austin TX, October 22–27, 2006.

[21] Jones, C. A., J. H. Grosvenor, and C. M. Weil, "RF Material Characterization Using a Large-Diameter (76.8 mm) Coaxial Air Line," NIST Technical Note 1517, February 2000.

[22] Archer, O., et al, "Permeability Measurement on Ferromagnetic Thin Films from 50 MHz up to 18 GHz," *J. Magnetism and Magnetic Materials*, Vol. 136, 1994, pp. 269–278.

[23] Adenot, A.-L., et al., "Broadband Permeability Measurement of Ferromagnetic Thin Films, or Microwires by a Coaxial Line Perturbation Method," *J. Appl. Phys*, Vol. 87, 2000, pp. 5965–5967.

[24] Tekeya, S., and K. Shimada, "New Measurement Method of RF Absorber Characteristics by Large Square Coaxial Line," *1993 International Symposium on Electromagnetic Compatibility*, Dallas, TX, August 9–13, 1993, pp. 397–402.

[25] Wadell, B. C., *Transmission Line Design Handbook*, Norwood, MA: Artech House, 1991.

[26] IEEE Std 1128-1998, "IEEE Recommended Practice for Radio-Frequency (RF) Absorber Evaluation in the Range of 30 MHz to 5 GHz," reaffirmed June 8, 2012.

[27] Damaskos, N., B. J. Kelsall, and J. E. Powell, Jr., "Square Coaxial Lines and Materials Measurements," *Microwave Journal*, Vol. 55, No. 2, February 2012, pp. 104–108.

[28] Steer, M., *Microwave and RF Design Transmission Lines*, Volume 2, Chapel Hill, NC: UNC Press, 2019.

[29] Chen, T.-S., "Determination of the Capacitance, Inductance, and Characteristic Impedance of Rectangular Lines," *IRE Trans. On Microwave Theory and Techniques*, September 1960, pp. 510–519.

[30] Weil, C. M., et al., "On RF Material Characterization in the Stripline Cavity," *IEEE Trans. Microwave Theory and Techniques*, Vol. 48, No. 2, February 2000.

[31] Collin, R. E., *Field Theory of Guided Waves* (Second Edition), New York, NY: Wiley-Interscience, reprinted by IEEE, 1991.

[32] Lewandowski, A., et al., "A Multireflect-Thru Method of Vector Network Analyzer Calibration," *IEEE Trans. Microwave Theory and Techniques*, Vol. 65, No. 3, March 2017, pp. 905–915.

[33] Hoffmann, J. P., P. Leuchtmann, and R. Valdieck, "Over-Determined Offset Short Calibration of a VNA," *2008 71st ARFTG Microwave Measurement Conference*, Atlanta GA, 2008.

[34] Wang, Y., et al., "Characterization of Dielectric Materials at WR-15 Band (50–75 GHz) Using VNA-Based Technique," *IEEE Trans. Instrumentation and Measurement*, Vol. 69, No. 7, July 2020, pp. 4930–4939.

[35] Wang, Y., "Material Measurements Using VNA-Based Material Characterization Kits Subject to Thru-Reflect-Line Calibration," *IEEE Trans. Terahertz Science and Technology*, Vol. 10, No. 5, September 2020, pp. 466–473.

[36] Hyde, M. W., "Broadband Characterization of Materials Using a Dual-Ridged Waveguide," *IEEE Trans. Instrumentation and Measurement*, Vol. 62, No. 12, December 2013, pp. 3168–3176.

[37] Hyde, M. W., and M. J. Havrilla, "A Broadband, Nondestructive Microwave Sensor for Characterizing Magnetic Sheet Materials," *IEEE Sensors Journal*, Vol. 16, No. 12, June 2016, pp. 4740–4748.

[38] ASTM D2520, "Standard Test Methods for Complex Permittivity (Dielectric Constant) of Solid Electrical Insulating Materials at Microwave Frequencies and Temperatures to 1650°C."

[39] Waldron, R. A., "Theory of a Strip-Line Cavity for Measurement of Dielectric Constants and Gyromagnetic Resonance Line-Widths," *IEEE Trans. Microwave Theory and Techniques*, 1964, pp. 123–131.

[40] Vanzura, E. J., R. G. Geyer, and M. D. Janezic, "The NIST 60-Millimeter Diameter Cylindrical Cavity Resonator: Performance Evaluation for Permittivity Measurements," NIST Technical Note 1354, 1993.

6

Scatter and Surface Waves

6.1 Diffuse Scatter

So far, this book has discussed aspects of measuring intrinsic properties of materials with free-space and transmission-line systems. Accurate determination of these intrinsic properties (e.g., permittivity and permeability) relies on measurement of specular scatter from a well-defined, homogeneous slab of material. Specular scatter is described in part by Snell's law,

$$n_1 \sin\theta_1 = n_2 \sin\theta_2 \tag{6.1}$$

where $n = \sqrt{\mu\varepsilon}$ is the index of refraction on each side of an interface between two media, such as air and the material slab, as shown in Figure 6.1. Thus, a wave traveling in air and then entering a slab at some angle other than normal incidence will refract or bend according to the difference in the intrinsic properties between air and the material. Some of the power from that same wave will also be reflected so that it stays in the medium from which it originated. In this case Snell's law simplifies to $\theta^{\text{incident}} = \theta^{\text{reflected}}$.

More generally, the Fresnel equations [1] describe the behavior of electromagnetic waves at an interface in terms of complex reflection and refraction. These Fresnel equations describe the reflection coefficients used in Chapter 2 ((2.21) and (2.22)) to derive inversion algorithms for determining μ and

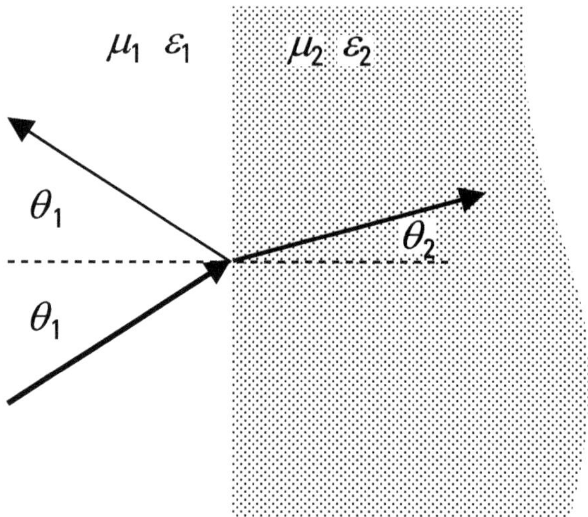

Figure 6.1 Specular refraction and reflection at an interface between two media.

ε. The assumption of (2.21) and (2.22) is that the interface is smooth and homogeneous. On the other hand, the discussion in this chapter is about characterizing the *diffuse* scatter that occurs in inhomogeneous materials or structures. When electromagnetic energy encounters specimens with a rough surface, a portion of the incident energy may be scattered diffusely. The difference between specular reflection and diffuse reflection at microwave frequencies is similar to the difference between a wall with gloss or flat paint at optical frequencies.

An example of microwave scatter from a rough surface, including diffuse and specular components, is shown in Figure 6.2. The data shown was calculated with a two-dimensional FDTD simulation of a 60-cm-wide rough surface on an infinite ground plane. It was illuminated with a 10-GHz collimated beam at 45-degrees incidence, and the roughness was randomly generated on the conductive ground plane with a peak-to-trough height of 3 mm ($\sim\lambda/10$). The relative amplitude of the scatter in logarithmic units is shown in a polar plot and is sometimes colloquially called fuzzball scatter. The radial scale is 10 dB per division, so the diffuse scatter varies from 10 to 30 dB below the specular lobe in this case. In typical engineered composites, the surface roughness is typically not this large. However, in some materials, there can still be enough roughness to result in a small but still measurable amount of diffuse scatter.

Besides surface roughness, there are also other sources for diffuse scatter. For example, a material surface may be smooth, but if it is a composite

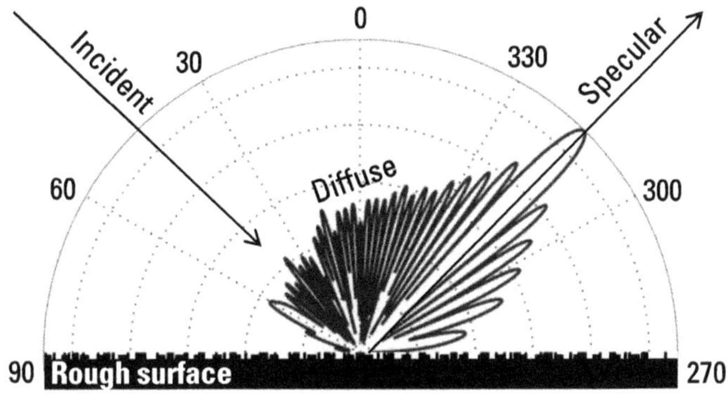

Figure 6.2 Calculated diffuse scatter from a 60-cm-wide rough surface at 10 GHz on an infinite ground plane.

of two or more constituents, then the local dielectric or magnetic properties may vary spatially. Such an effect can be caused by variations in a composite weave or by changes in local volume loading of a pigment-loaded polymer. Discontinuities such as an edge or a gap in the middle of a material specimen can also diffract energy, causing diffuse scatter, and, finally, periodic arrays, whether in the form of a composite material like honeycomb or as an engineered surface like a frequency-selective surface or a metamaterial [2, 3], can have geometric variations and imperfections. Periodic arrays can also exhibit coherent addition of the scatter from the array elements resulting in another type of scatter called grating lobes or Floquet scatter [3]. With the different sources of electromagnetic scatter, there are different parameters for characterizing this scatter, some of which are summarized in Table 6.1. Sections 6.1.1–6.1.6 also discuss these different parameters.

Table 6.1
Summary of Different Free-Space Scatter Parameters

Measured Parameter	Brief Description	Units
RCS	Total scatter from a 3D object illuminated by a plane wave	dBsm (dB square meters)
Scattering coefficient	RCS per unit area of a surface or 1D RCS	dBsm/sm
Echo width	RCS per unit length of a linear discontinuity or edge, or 2D RCS	dB-meter or dB-knife edge
Cross-polarized scatter	RCS (1D, 2D, or 3D) where input plane wave and output are orthogonal polarizations	dBsm, dBsm/sm, or dB-meter

6.1.1 RCS

Free-space methods such as the focused-beam system provide a means to measure some of these nonspecular forms of scatter and are convenient for determining the monostatic backscatter portion of the diffuse fuzzball. The term monostatic means that the transmitter and receiver are in the same location, so that the only scatter measured is energy reflected back in the direction of the source or backscatter. In this mode, a focused-beam system can act as an alternative to a compact RCS measurement range [4]. Measuring microwave scatter or RCS from an object is usually a difficult and expensive prospect involving construction of complex models or test bodies and subsequent measurement in RCS range facilities. RCS ranges are typically large facilities that require significant levels of supporting manpower and equipment to obtain scatter data from a target under test. This is in contrast to a focused-beam system, which is a smaller, laboratory-scale apparatus that can operate at a significantly lower cost than an RCS range.

Before discussing the use of a focused beam to probe the scatter from an object, we must first quantify the concept of RCS. The RCS of a complex object is a function of its size, shape, and material properties and is expressed in terms of an effective scattering cross-sectional area in units such as square meters. Specifically, it is a measure of the reflective intensity of a target under radar illumination. When discussing the RCS at arbitrary directions, the term bistatic is used to denote that the incident and measurement angles are not necessarily the same. Bistatic RCS is plotted in the polar plot of Figure 6.2. However, when the direction of scatter under study is restricted to the direction back toward the illuminating radar, then the terms monostatic RCS or backscatter are used.

The RCS, σ, is defined as an effective cross-sectional area for an equivalent isotropic scatterer when illuminated by a plane wave. It is calculated by equating the captured incident power, to the power reradiated by an isotropic scatterer. The captured incident power is equal to the effective cross-sectional area of the scatterer, σ, multiplied by the incident power per unit area, W_i. The power reradiated by an isotropic scatterer is the scattered power density, W_s, at a distance R from the scatterer, multiplied by the surface area of a sphere with that same radius: $4\pi R^2$. Equating these terms and rearranging to solve for RCS gives

$$\sigma = 4\pi R^2 \frac{W_s}{W_i} \tag{6.2}$$

Because power density is proportional to the square of the electric or magnetic field, we can rewrite the ratio of the scattered and incident power densities in terms of fields. Furthermore, because of the dependence of (6.2) on *R*, we can standardize by taking the limit as *R* goes to infinity. Applying this to (6.2), gives the formal definition of RCS,

$$\sigma = \lim_{R \to \infty} 4\pi R^2 \left| \frac{E_s}{E_i} \right|^2 \tag{6.3}$$

where E_i and E_s are the incident and scattered E-fields, respectively. Note that (6.3) could alternately be written in terms of the scattered magnetic fields. Furthermore, the usual convention is to express this cross-section in a logarithmic scale: decibel-square meters (dBsm), which is equal to $10\log_{10}\sigma$, where σ is the RCS in square meters.

Actual measurement of RCS requires an RCS measurement range or facility to create a near–plane wave condition over some volume of space, called a quiet zone. A target placed inside that quiet zone then scatters energy as if it were experiencing a far-field plane wave. The energy scattered from that target then determines the RCS. Since the measurement range has a finite size and *R* is not infinite, a calibration measurement of a known target is used to determine the quantitative RCS. The return signal from an RCS measurement includes scattering from a variety of phenomena. These phenomena include scatter from rough surfaces, material inhomogeneities, edges, and other discontinuities within the object, as well as from the overall shape, size, and orientation of the object. The plurality of scattering mechanisms from a given object makes it difficult to distinguish the individual contributions caused by each phenomenon. The focus of this book is characterization of electromagnetic materials, so we are specifically concerned with determining diffuse scatter caused by material roughness or inhomogeneity.

Similar to an RCS range, a focused-beam system, such as that discussed in Chapter 4, also simulates a plane wave at the focal point of the beam. However, in contrast to an RCS range, the focused beam usually only illuminates a small portion of a larger specimen or target. So, a focused-beam system can have the advantage of isolating the effects of individual discontinuities or material inhomogeneities from the scatter that comes from the target shape. In essence, it can focus on a particular portion of an object to distinguish between different scattering phenomenology and between individual scattering sources on a larger body.

The idea of a focused beam has also been applied to enable numerical electromagnetic simulations of rough surface scatter. Collin showed that Gaussian beam illumination could be used to approximate the scattering coefficient of an "infinite" rough surface as long as a planar phase front is maintained, and the correlation lengths of the rough surface are smaller than the illuminating beam diameter [5]. While Collin emphasized the use of a tapered beam for numerical analysis, the first few sections of this chapter demonstrate the use of a Gaussian beam for experimental backscatter characterization.

When the scatter from a rough surface or inhomogeneous material is measured, it is quantified in terms of the RCS per unit area or dBsm/sm, which is designated by a subscript zero, σ_0. This σ_0 is a commonly measured parameter in remote-sensing applications [6], where radar backscatter of the planet's surface is mapped for environmental, agricultural, and other applications. σ_0 is sometimes called the scattering coefficient. In addition, when characterizing the scatter from a linear discontinuity, such as a gap or an edge, the relevant scattering parameter is RCS per unit length [7]. This is also known as the echo width and is expressed in units of dBsm per meter, or more compactly as a two-dimensional RCS: decibel-meters (dB-m).

6.1.2 Scattering Coefficient Measurement

As noted, a free-space focused-beam system illuminates only a portion of a target. A representative geometry of this is shown in Figure 6.3, where a single horn and lens are used to illuminate a specimen. When that specimen is rotated so that the surface normal is not pointing back toward the lens, the diffuse backscatter is then received, while the specular reflection is directed away from the lens and horn. Calibration of a focused beam is also different than in a traditional RCS range, which uses conductive spheres or squat cylinders. In a focused beam, a flat metal plate that extends outside the beam is used to establish a reflection reference so that it looks to be infinitely large.

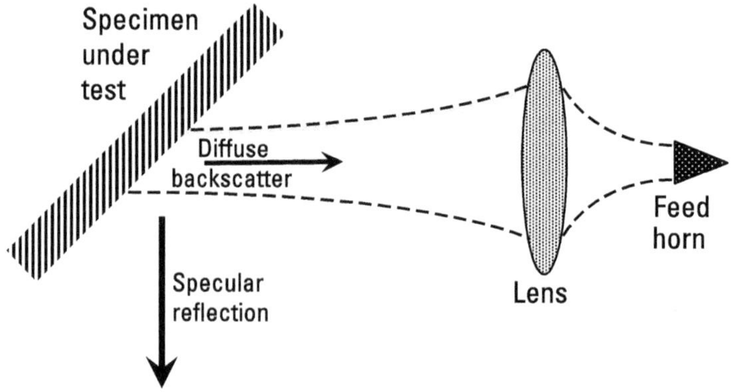

Figure 6.3 Geometry for using a focused-beam system to measure diffuse scatter from a material specimen.

Plane wave–spectrum decomposition is a way to analyze the focused beam scatter and derive the RCS of focused-beam illuminated surfaces. Assuming the case of a linearly polarized field centered on the z-axis and oriented in the x-direction, the plane wave spectrum can be represented with a Fourier transform pair [8],

$$\frac{E_x(x,y,z)}{E_0}=\frac{1}{(2\pi)^2}\int_{-\infty}^{\infty}\int_{-\infty}^{\infty}F_x(k_x,k_y)e^{-i(k_x x+k_y y+k_z z)}\,dk_x\,dk_y \tag{6.4}$$

$$F_x(k_x,k_y)=\int_{-\infty}^{\infty}\int_{-\infty}^{\infty}\frac{E_x(x,y,z=0)}{E_0}e^{i(k_x x+k_y y)}\,dx\,dy \tag{6.5}$$

Additionally, the scatter generated by a target within the illuminating beam has a spectrum defined by

$$F_x^{\text{scatter}}(k_x,k_y)=\int_{-\infty}^{\infty}\int_{-\infty}^{\infty}F_0(k'_x,k'_y)R_s(k'_x,k'_y,k_x,k_y)\,dk'_x\,dk'_y \tag{6.6}$$

where the spectrum of the incident beam is F_0, and R_s is the appropriate component of the plane-wave scattering matrix [9]. The plane-wave scattering matrix relates the amplitude of the scattered plane wave in the direction (k_x,k_y) to a plane wave incident from the direction (k'_x,k'_y). This scatter is then collected by a lens, which also imposes a spectral-weight function, $F_x^{\text{lens}}(k_x,k_y)$. In a focused-beam system, the scatter is measured by an equivalent lens and horn to that which generates the incident field. So reciprocity dictates that $F_x^{\text{lens}} = F_0$. The received signal is obtained by multiplying the scattered spectrum with the spectral-weight function of the lens,

$$\frac{V_r}{V_0}=\int_{-\infty}^{\infty}\int_{-\infty}^{\infty}F_x^{\text{scatter}}(k_x,k_y)F_0(k_x,k_y)\,dk_x\,dk_y \tag{6.7}$$

As noted, the reflection calibration is referenced to a conductive plate that extends well beyond the extent of the beam and that is centered at the beam waist, $z = 0$. Because it is an ideal reflector and effectively infinite in extent relative to the beam, the plane-wave scattering matrix is described by $R_s(k'_x,k'_y,k_x,k_y) = \delta(k_x - k'_x)\delta(k_y - k'_y)$, where δ is the Dirac delta function. Applying this to (6.6), and combining it with (6.7), the received signal for the metal reference plate is then

$$\frac{V_{\text{metal}}}{V_0} = \int_{-\infty}^{\infty}\int_{-\infty}^{\infty}\left[F_0\left(k_x,k_y\right)\right]^2 dk_x\, dk_y \tag{6.8}$$

Assuming that the focused beam closely approximates a plane wave, and the phase taper is negligible, then $F_0(k_x,k_y)$ is approximately real. Furthermore, Parseval's theorem states that the integral of the square of a function is equal to the integral of the square of its Fourier transform. Using this, we can rewrite (6.8) as

$$\frac{V_{\text{metal}}}{V_0} = (2\pi)^2 \int_{-\infty}^{\infty}\int_{-\infty}^{\infty}\left|\frac{E_x\left(x,y,z=0\right)}{E_0}\right|^2 dx\, dy = (2\pi)^2 A_{\text{eff}} \tag{6.9}$$

where we note that the integral of the power is equivalent to the effective illumination area of the focused beam. When the beam follows a Gaussian amplitude taper, this is equivalent to the area in which the power density is within $1/e$ of the peak.

$$A_{\text{eff}} = \int_{-\infty}^{\infty}\int_{-\infty}^{\infty} e^{-x^2/w_x^2} e^{-y^2/w_y^2}\, dx\, dy = \pi w_x w_y \tag{6.10}$$

This corresponds to the area of an ellipse defined by the horizontal and vertical beam waists.

RCS is defined in terms of an equivalent isotropic scatterer. With a similar plane wave analysis, we can also compute the scatter from a small isotropic scatterer within the focused beam. A small target located at a position $(x_0,y_0,z_0,)$ within the beam will have a scattering matrix that can be approximated by

$$R_s\left(k_x',k_y',k_x,k_y\right) \approx R_s(0)e^{-i\left(k_x'x_0+k_y'y_0+k_z'z_0\right)}e^{-i\left(k_xx_0+k_yy_0+k_zz_0\right)} \tag{6.11}$$

As above, the k-vectors represent the scattered plane waves, and the primed k-vectors represent the incident plane waves. Unlike the scattering matrix for the infinite plate reflector, the plane waves emanating from a small isotropic scatterer show an exponential dependence with distance from that scatterer. As in the case of the infinite metal plate reflector shown above, we can combine the scattering matrix of the small isotropic scatterer with (6.6) and (6.7) to obtain

$$\frac{V_s}{V_0} = R_s(0)\left[\int_{-\infty}^{\infty}\int_{-\infty}^{\infty} F_0\left(k_x,k_y\right)e^{-i\left(k_xx_0+k_yy_0+k_zz_0\right)}\, dk_x\, dk_y\right]^2 \tag{6.12}$$

From (6.4), (6.12) can be rewritten in terms of the incident electric field,

$$\frac{V_s}{V_0} = R_s(0)\left[(2\pi)^2 \frac{E_i(x_0, y_0, z_0)}{E_0}\right]^2 \tag{6.13}$$

From (6.9) and (6.13), and assuming that the scatterer is centered at the origin, we can now calculate the scattered signal relative to the metal plate,

$$\frac{V_s}{V_{\text{metal}}} = \frac{(2\pi)^2 R_s(0)}{A_{\text{eff}}} \tag{6.14}$$

We can also write general expressions for the incident and scattered fields. Since we have placed the scatterer at the origin, $x = y = z = 0$, the following incident field is derived from (6.4)

$$\frac{E_i}{E_0} = \frac{1}{(2\pi)^2} \int_{-\infty}^{\infty} \int_{-\infty}^{\infty} F_0(k_x, k_y)\, dk_x\, dk_y \tag{6.15}$$

The scattered field can be written in terms of the spectral representations of three-dimensional far-field radiated fields [8].

$$\frac{E_s}{E_0} = \frac{ik_0}{2\pi r} F_x^{\text{scatter}}(k_x = 0, k_y = 0)\, e^{-ik_0 r} \tag{6.16}$$

where (6.6) tells us that

$$F_x^{\text{scatter}}(0,0) = \int_{-\infty}^{\infty} \int_{-\infty}^{\infty} F_0(k_x, k_y) R_s(k_x, k_y, 0, 0)\, dk_x\, dk_y \tag{6.17}$$

These equations can then be applied to (6.3) to get the RCS,

$$\sigma = 4\pi r^2 \left|\frac{E_s}{E_i}\right|^2 = 4\pi \frac{\frac{k_0^2}{(2\pi)^2} \left|F_x^{\text{scatter}}(0,0)\right|^2}{\frac{1}{(2\pi)^4} \left|\int_{-\infty}^{\infty} \int_{-\infty}^{\infty} F_0(k_x, k_y)\, dk_x\, dk_y\right|^2} \tag{6.18}$$

If we approximate the incident field as a single plane wave, the spectral weight function is given by $F_0(k_x, k_y) = \delta(k_x)\delta(k_y)$ and $F_x^{\text{scatter}}(0,0) = R_s(0)$. Then the RCS simplifies to $\sigma = 2(2\pi)^3 k_0^2 |R_s(0)|^2$. Combining this with (6.14) then results in

$$\sigma = 4\pi \frac{A_{\text{eff}}^2}{\lambda^2} \left| \frac{V_s}{V_{\text{metal}}} \right|^2 \tag{6.19}$$

Thus, this measured and calibrated voltage-reflection coefficient (e.g., $S_{11} = V_s/V_{\text{metal}}$) depends not only on the scattering properties of the material under test, but also on the illumination area of the incident beam.

The relationship between the voltage-reflection coefficient and RCS can also be derived more intuitively by assuming that the reflection from the response calibration (metal plate) is equivalent to the RCS of a flat metal plate at normal incidence to the receiver. In this assumption, the metal plate has an effective surface area equal to the illumination area of the beam (in the high-frequency limit), $|V_{\text{metal}}|^2 \propto 4\pi(A_{\text{eff}}^2/\lambda^2)$. Then the calibrated scatter from the sample, $|V_{\text{cal}}|^2$, is equal to the ratio of the sample RCS to the metal plate RCS. With algebraic rearrangement, the sample RCS is expressed by

$$\sigma = 4\pi \frac{A_{\text{eff}}^2}{\lambda^2} \left| V_{\text{cal}} \right|^2 \tag{6.20}$$

which is equivalent to (6.19). In traditional compact-range measurements, calibration is accomplished by comparing the scatter of the target under test to that of a known standard shape, such as a metal sphere or squat cylinder. The advantage of (6.20) is that it is referenced to a metal plate, which is the usual calibration standard for focused-beam reflection measurements. It doesn't require any additional calibration targets, and cylinders or spheres can be used instead as verification standards. That said, it does require that the beam waist has been characterized so that A_{eff} is known.

For an inhomogeneous surface, the quantity of interest is not RCS, but scattering coefficient. Instead of a target illuminated by a larger plane wave in an RCS range, a tapered plane wave is used to illuminate a larger specimen in the focused-beam system. Because the target in this case is effectively infinite in size relative to the beam, the scattering coefficient (σ_0) is defined as RCS per unit physical area, where the illuminated (physical) area is the projected area divided by $\cos\theta$,

$$\sigma_0 = \frac{\sigma}{A_{\text{eff}}/\cos\theta} = 4\pi\cos\theta \frac{A_{\text{eff}}}{\lambda^2} \left| V_{\text{cal}} \right|^2 \tag{6.21}$$

where θ is the angle between the surface normal and the direction of the illuminating beam.

While the abovementioned equations were derived relative to a flat metal plate at normal incidence, this simple response calibration still leaves considerable noise in the signal due to other undesired reflections within the system. This is akin to the effects of background clutter in a traditional RCS measurement. Typical RCS calibration methods also include a background subtraction step to minimize the effects of other scatterers within the measurement range. In focused-beam measurements, this is equivalent to the aforementioned response-and-isolation calibration. In particular, an isolation measurement, made with no metal plate or specimen present, is vector subtracted from both the response signal and from the specimen signal.

$$V_{\text{cal}} = \frac{S_{11}^{\text{specimen}} - S_{11}^{\text{background}}}{S_{11}^{\text{metal plate}} - S_{11}^{\text{background}}} \tag{6.22}$$

where S_{11} designates the monostatic reflection scattering parameter measured for the specimen, background, or metal plate. As in the case of most RCS measurement ranges as well as in the focused-beam material measurements described previously in Section 2.3 of this book, time-domain processing (windowing or gating) is also used to further minimize errors from background scatter and multipath.

6.1.3 Examples of Scattering-Coefficient Measurement

When a focused-beam system is first configured for scattering-coefficient measurement, a validation measurement of an ideal specimen should be performed to establish the minimum measurable backscatter levels. This validation also ensures that the system is operating within normal specifications. An ideal specimen in this case is a flat metal plate that should have negligible diffuse backscatter. Figure 6.4 shows example measurement data from a metal plate. This data was from an aluminum plate, 61 cm on a side, corresponding to approximately 20 wavelengths at 10 GHz. The plate was rotated azimuthally so that monostatic reflection data was taken as a function of both frequency and angle. In this geometry, Figure 6.3 represents a bird's-eye view of the measurement fixture. Thus, the plate was much larger than the illuminating beam, minimizing edge-diffraction effects except at the most extreme angles. In addition, an absorber fence constructed from pyramidal foam absorber and placed around the measurement system minimized secondary reflections from walls or objects in the laboratory.

Near zero degrees in azimuth, a reflection lobe is evident in Figure 6.4 due to the specular reflection of the beam at normal incidence intersecting the

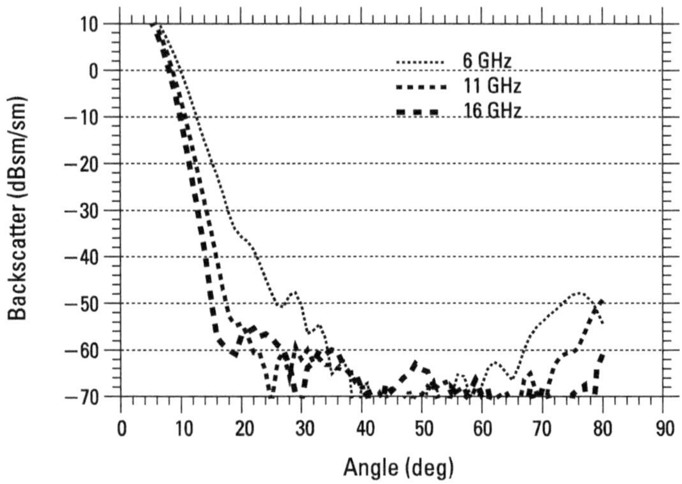

Figure 6.4 Diffuse scatter measurement of a flat metal plate versus angle at 6, 11, and 16 GHz.

lens. The calibration and data-processing method derived above was used to convert the measured S-parameter data into dBsm per square meter. Therefore, the peak of the specular lobe is above 0 dB due to this RCS calibration. In other words, the amplitude and width of this specular lobe is a function of the illumination beamwidth, and meaningful nonspecular (i.e., diffuse) backscatter is only measurable outside this specular lobe (beyond approximately 20 degrees). The lens used in this measurement was designed to provide a normalized beam waist of $k_0 w_0 \approx 12$. Lenses can be designed for wider beamwidths in which the plane-wave distribution narrows, and this lobe also narrows. The nonspecular scatter levels are as low as −65 dBsm/sm at some angles, indicating the sensitivity limit of the measurement apparatus. The residual signal measured below this level is due to internal network analyzer noise and the limits of the calibration methodology for eliminating multipath reflections. At near-grazing angles, the edges of the finite-size plate begin to interact with the main part of the beam, causing increased scatter.

With a reasonable noise floor thus established, the backscatter of various specimens of interest can then be characterized. Figure 6.5 is an example of such a measurement [4]. The specimen was 61 cm by 122 cm in dimension and consisted of a random arrangement of metal spheres in a two-dimensional plane. The spheres consisted of metal BBs (shot pellets) and were 4.4 mm in diameter. A total of approximately 6,000 spheres were used to make the random surface. To contain the spheres, they were sandwiched between two

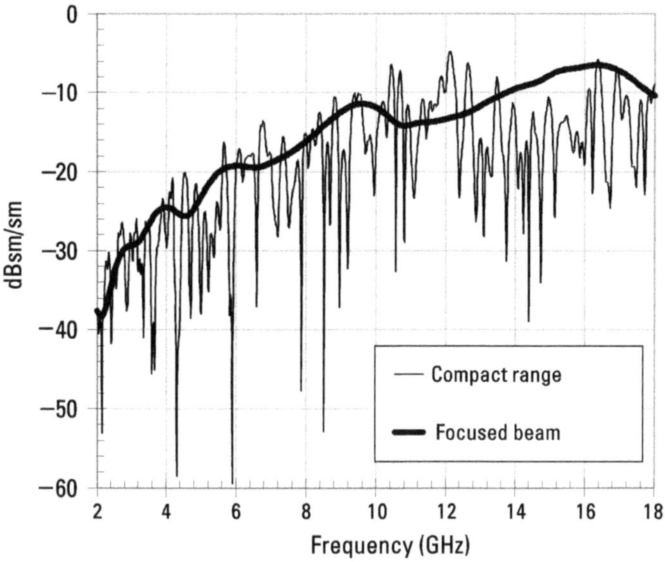

Figure 6.5 Comparison of the RCS per unit area of an inhomogeneous 61 x 122 cm specimen of metal spheres randomly arranged in a plane.

pieces of low-dielectric foam and anchored with adhesive. Spray adhesive was used on each of the Styrofoam substrates to ensure that the spheres were anchored. The random pattern was generated by sprinkling the spheres onto the adhesive covered foam. Then the second adhesive-covered foam substrate was pressed on top of the spheres, and the edges were taped to hold the entire assembly together. An image of the arrangement of the spheres within the foam sandwich is shown in Figure 6.6.

This model rough surface was measured in both a compact RCS range and in the focused-beam system with the rough scattering methodology developed above. The data in Figure 6.5 compares results from these two measurements at a monostatic incidence angle of 45 degrees. Because of the limited beamwidth relative to the size of the specimen, the focused-beam data shown is an average of measurements made at five different locations across the panel. The compact range data follows the same trend as the focused-beam data; however, it contains more fine structure. This effect is due to interference effects from edge illumination in the compact range, which does not occur as strongly with the focused-beam methodology. In other words, the compact-range measurement illuminated the whole specimen including the edges, while the focused beam was more localized and avoided edge illumination.

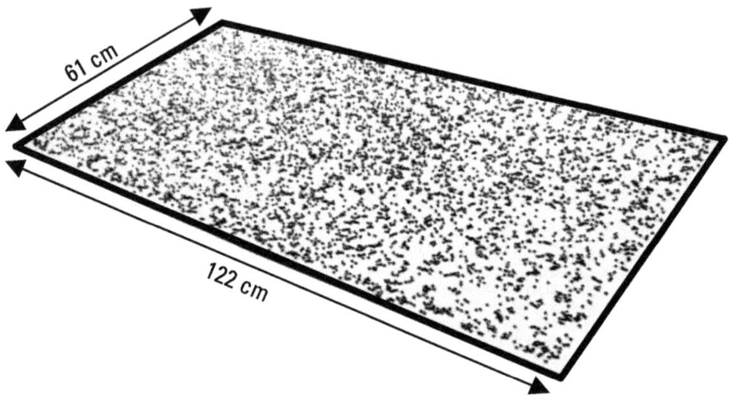

Figure 6.6 Image of an inhomogeneous panel, in which a random arrangement of metal spheres was glued onto a foam substrate.

Other than the edge-diffraction effects, the data in Figure 6.5 shows general agreement between the two measurement methods.

The focused-beam method can also measure nonspecular scatter of periodic materials such as a frequency-selective surface (FSS). FSS materials are engineered composites that typically consist of a periodic pattern etched in a conductive metal sheet and supported on a low-loss dielectric substrate [3]. They are resonant structures optimized to be microwave transparent at some frequencies and reflective at others. The periodic patterning of FSS materials also leads to additional nonspecular scatter due to diffraction phenomena. For example, Figure 6.7 shows a contour plot of 26.5–40-GHz backscatter measured from a rectangular array of Jerusalem-cross–shaped slots etched in a thin sheet of copper on a fiberglass substrate. The polarization of the incident beam was with the magnetic field transverse to the plane defined by the surface normal and the incident beam direction (TM). The periodicity of the Jerusalem-cross patterning was 6.1 mm.

There are two main features evident in Figure 6.7. First is the specular reflection at the angle of 0 degrees. As noted, the width of this specular lobe is determined by the plane wave spectrum width of the incident beam, which is inversely related to the beam waist at the focus. The second feature is the grating lobe that extends from approximately 69 degrees at 26 GHz to 38 degrees at 40 GHz. The angular direction of this grating lobe is described by Bragg diffraction theory and a simple diffraction equation can be derived,

$$2\pi\left(\sin\theta_r + \sin\theta_i\right) = \pm 2m\pi\frac{\lambda}{d} \tag{6.23}$$

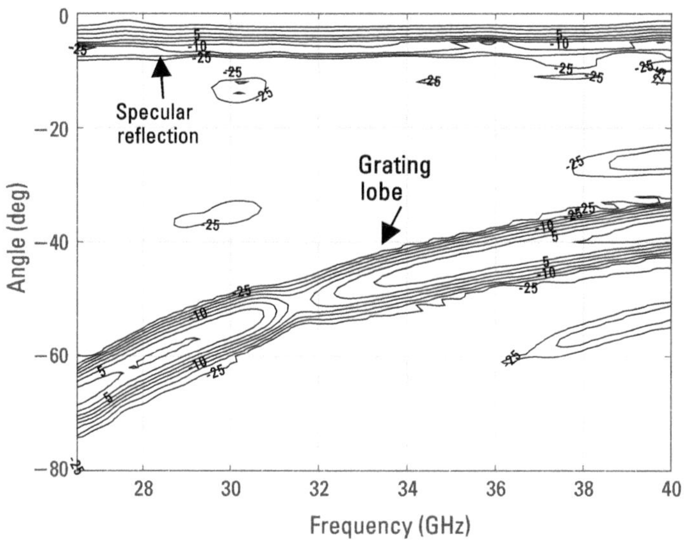

Figure 6.7 Diffuse backscatter measured from a Jerusalem-cross slot FSS versus frequency and angle. The data shows a strong grating lobe that spans from 69 to 38 degrees in the measured frequency band.

where $\theta_r = \theta_i$ for monostatic backscatter, $m = 1$, and d is the lattice constant of 6.1 mm. Equation (6.23) applies to rectangular arrays and shows that grating lobes first appear when periodicity of the array corresponds to one-half of the incident wavelength. While not highlighted in the range of the contour plot of Figure 6.7, other sources of diffuse scatter also exist in this data at lower amplitudes. These other diffuse scatter effects include inhomogeneity in the underlying substrate, which is a fiberglass-reinforced dielectric composite.

6.1.4 Echo-Width Measurement

A focused-beam backscatter methodology can also be applied to linear discontinuities such as edges, cylinders, joints between materials, or slots. Instead of RCS per unit area, the scattering figure of merit is called echo width and is measured in decibel-meters. In terms of incident and scattered fields, echo-width or two-dimensional RCS is defined by

$$\sigma_{2D} = \lim_{\rho \to \infty} 2\pi\rho \left| \frac{E_{\text{scattered}}}{E_{\text{incident}}} \right|^2 \tag{6.24}$$

where ρ is the distance from the scattering discontinuity. For example, a typical measurement geometry for a linear discontinuity such as a groove in a specimen slab is shown in Figure 6.8.

Referring to the coordinates in Figure 6.8, a two-dimensional scatterer oriented along the y-direction and positioned at x and $z = 0$, we assume that the illuminating beam propagates parallel to the z-direction with the beam waist centered at x, z, and $y = 0$. The receiver (lens and horn) is also positioned at x and $y = 0$, and the target is rotated about the y-axis, so there is symmetry with respect to the x-z plane. The contribution to the y-component of the received field from the $y > 0$ side of the target is canceled by the $y < 0$ contribution. Thus, the measured backscatter from these linear targets can be described as two-dimensional. This dictates that the measurement can only be made when the axis of the discontinuity and the propagation vector are orthogonal, which is at 0-degrees elevation.

Similar to the RCS per unit area measurement, the echo-width measurement is referenced to a normal-incidence metal plate. In contrast, however, echo width is calculated by assuming that the normal incidence metal plate is equivalent to a two-dimensional strip with its width corresponding to the 1/e width of the beam. This 1/e width is a direct result of the assumption that the incident beam follows a Gaussian amplitude profile. Along similar lines to the derivation provided previously for the RCS per unit area, the echo width of a metal strip at normal incidence is given by $2\pi(W_{\text{eff}}^2/\lambda)$, where W_{eff} is the illuminating beam width. Then the echo width of the sample is given by

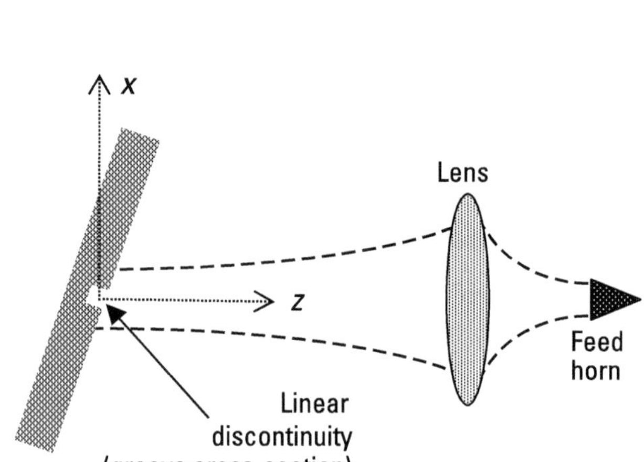

Figure 6.8 Example measurement geometry of a linear discontinuity (groove) in a specimen slab, where the groove is centered in the illuminating beam, and its axis is in/out of the page.

$$\sigma_{2D} = \frac{2\pi W_{\text{eff}}^2}{\lambda} \left| V_{\text{cal}} \right|^2 \tag{6.25}$$

Of course, this method requires accurate characterization of the beam profile, and if the lens is reconfigured or a different feed horn is used, a new beam profile must be measured. An alternative method may also be used to calibrate, which instead uses known calibration standards, much like traditional RCS calibration in a compact range. In particular, long cylindrical rods can be used as standards as long as the rods extend well beyond the illumination area of the beam. In this way, the rod appears to be effectively infinite in length, and an analytical expression for the echo width of an infinite conductive (metal) cylinder can be applied. In logarithmic units, the calibrated echo width is computed from the return loss by

$$\sigma_{2D}^{\text{target}} = V_{\text{cal}}^{\text{target}} + \sigma_{2D}^{\text{cylinder}} - V_{\text{cal}}^{\text{cylinder}} \tag{6.26}$$

where V_{cal} is the return loss data, calibrated by the method described in (6.22). The value of $\sigma_{2D}^{\text{cylinder}}$ depends on the diameter of the metal cylinder and can be calculated analytically [7, 10].

6.1.5 Examples of Echo-Width Measurement

An example application of this calibrated echo-width method is provided in Figure 6.9, which schematically shows the measurement geometry of nonspecular backscatter caused by diffraction from a simple conductive wedge. The scatter from an infinite edge can be derived analytically [11]. For the TE polarization (where the E-field is parallel to edge axis), and ignoring the specularly reflected components, the diffraction component of the E-field is given by

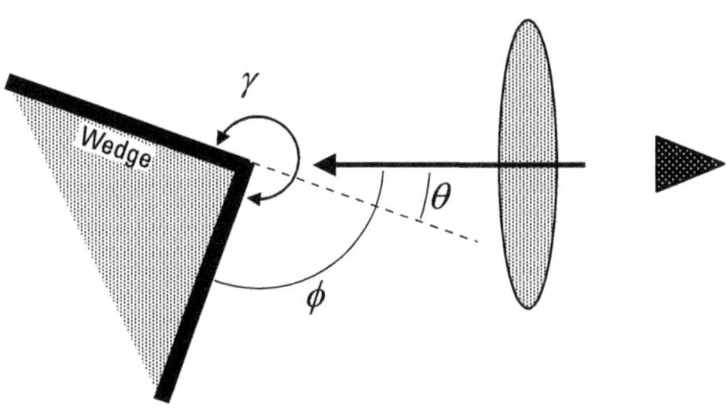

Figure 6.9 Schematic of measurement geometry for echo width of a wedge.

$$E_z^{\mathrm{TE}} = \frac{\pi}{\gamma}\sin\left(\frac{\pi^2}{\gamma}\right)\left[\frac{1}{\cos\left(\frac{\pi^2}{\gamma}\right)-1} - \frac{1}{\cos\left(\frac{\pi^2}{\gamma}\right)-\cos\left(2\phi\frac{\pi}{\gamma}\right)}\right]\frac{e^{-i\rho k_0}e^{i\pi/4}}{\sqrt{2\pi\rho k_0}} \quad (6.27)$$

where $k_0 = 2\pi/\lambda_0$ is the wave number, γ is the exterior angle of the wedge, and ϕ is the incident angle, as shown in Figure 6.9. Similarly for the TM polarization (where the E-field is perpendicular to edge axis), the magnetic field due to diffraction is given by

$$H_z^{\mathrm{TM}} = \frac{\pi}{\gamma}\sin\left(\frac{\pi^2}{\gamma}\right)\left[\frac{1}{\cos\left(\frac{\pi^2}{\gamma}\right)-1} + \frac{1}{\cos\left(\frac{\pi^2}{\gamma}\right)-\cos\left(2\phi\frac{\pi}{\gamma}\right)}\right]\frac{e^{-i\rho k_0}e^{i\pi/4}}{\sqrt{2\pi\rho k_0}} \quad (6.28)$$

Using the definition for two-dimensional RCS, the TE and TM echo widths are then

$$\sigma_{2D}^{\mathrm{TE}} = \frac{1}{k_0}\left\{\frac{\pi}{\gamma}\sin\left(\frac{\pi^2}{\gamma}\right)\left[\frac{1}{\cos\left(\frac{\pi^2}{\gamma}\right)-1} - \frac{1}{\cos\left(\frac{\pi^2}{\gamma}\right)-\cos\left(2\phi\frac{\pi}{\gamma}\right)}\right]\right\}^2 \quad (6.29)$$

and

$$\sigma_{2D}^{\mathrm{TM}} = \frac{1}{k_0}\left\{\frac{\pi}{\gamma}\sin\left(\frac{\pi^2}{\gamma}\right)\left[\frac{1}{\cos\left(\frac{\pi^2}{\gamma}\right)-1} + \frac{1}{\cos\left(\frac{\pi^2}{\gamma}\right)-\cos\left(2\phi\frac{\pi}{\gamma}\right)}\right]\right\}^2 \quad (6.30)$$

Figure 6.10 compares the diffraction echo width predicted by this classical diffraction theory to a focused beam measured echo width at 10 GHz for the 90-degree wedge. The measurement data is from a 90-degree metal wedge, with 45-cm-wide sides centered at the focus of the beam. The wedge was mounted on a turntable, and backscatter was collected as a function of frequency and azimuth angle. The data at 10 GHz is plotted as a function of angle, θ, where $\theta = \phi - 90$ is defined in Figure 6.9. This data shows quantitative agreement for most of the angles measured. Poor agreement exists near $\theta = 0$, where the measured scatter includes a specular component due to the normal incidence of one side of the wedge, and where the classical diffraction theory no longer applies. At angles near this 0-degree specular angle, the effects of finite beamwidth are also evident.

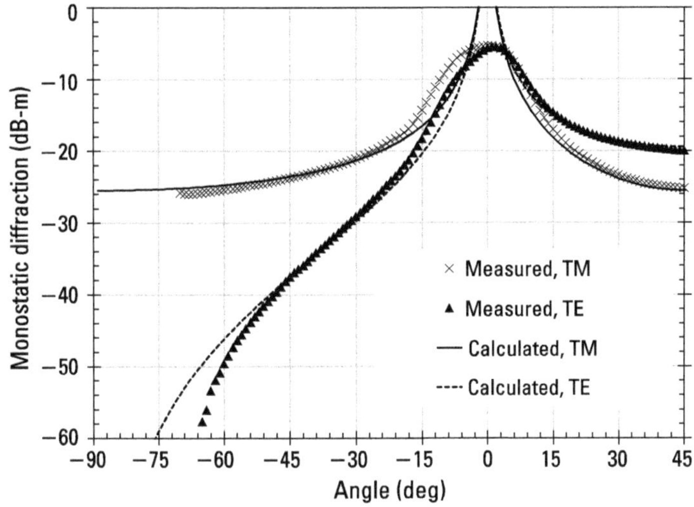

Figure 6.10 Comparison of measured and calculated diffraction from a 90-degree wedge at 10 GHz.

While the wedge is a simple analytical shape, sometimes edges are designed with a variety of absorbing materials to minimize scatter. The abovementioned method provides a relatively low-cost laboratory-scale method for experimenting with various material treatments for edge-scatter reduction. Echo-width measurements are also useful when trying to measure the diffuse-scatter effects in structures that have access doors. In this case, a gap may exist between two panels, and the abovementioned method can quantify how much scatter that gap creates. It can also verify different treatment strategies for reducing the scatter from gaps.

6.1.6 Cross-Polarized Scatter

Cross-polarized scatter is when the polarization of the received energy is orthogonal to the polarization of the incident wave. Anisotropic materials and inhomogeneities can contribute a cross-polarization component to a scattered signal, particularly when the inhomogeneity is asymmetrical. Measuring this component with a free-space system requires correcting for the cross-polarized contributions of all error sources, including the transmit and receive coupling errors of the feed antenna. It is possible to utilize a response-and-isolation calibration methodology for calibrating cross-polarized measurements based on a wire-grid polarizer. In particular, the analytical expressions for reflection of parallel and perpendicularly polarized energy from a wire grid polarizer are as follows [12]

$$R_{\text{power}}^{\|} = \frac{1}{1-\left(\frac{2d}{\lambda}\ln\left(\frac{d}{\pi a}\right)\right)^{2}} \tag{6.31}$$

and

$$R_{\text{power}}^{\perp} = \frac{\left(\frac{\pi^{2}a^{2}}{2\lambda d}\right)^{2}}{1-\left(\frac{\pi^{2}a^{2}}{2\lambda d}\right)^{2}} \tag{6.32}$$

where the polarizer is made up of a periodic array of wires of radius, *a*, that are spaced with a periodicity of *d*. When the polarizer is rotated within the measurement plane to an angle of 45 degrees, the reflection of the polarizer in both polarizations is approximately –6 dB. The exact reflection coefficient is calculated from (6.31) and (6.32) and used as a relative standard for normalizing the raw reflection loss. In a typical dual-polarized horn antenna, the cross-polarization isolation is only 20–30 dB below the copolarized signal, and improved isolation may be desired for some measurements.

A more accurate, full-polarimetric calibration methodology, based on a dihedral calibration standard is a preferred alternative to the polarizer standard. This cross-polarization measurement methodology depends on a dual-polarized antenna to enable simultaneous detection of two orthogonal polarizations. Measurements are performed with a VNA with one of the ports (e.g., port 1) connected to the horizontal polarization feed of a dual-pol feed antenna and the other port (port 2) connected to the vertical polarization feed as shown in Figure 6.11. With this configuration the antenna is simultaneously used in both receive and transmit mode. For the remainder of this section, the S-parameter designations are transformed from the port-centric designation to the corresponding co- and cross-polarization designations: 11 → *hh*, 12 → *hv*, 21 → *vh*, and 22 → *vv*.

A dihedral corner reflector for calibration can be made from two adjacent metal plates at right angles to each other so that incoming energy is reflected directly (or specularly) back to a transmitting antenna. A dihedral is shown schematically in Figure 6.12. For the focused-beam system, if the dihedral is built large enough to capture the entire quasi-plane wave beam focused onto

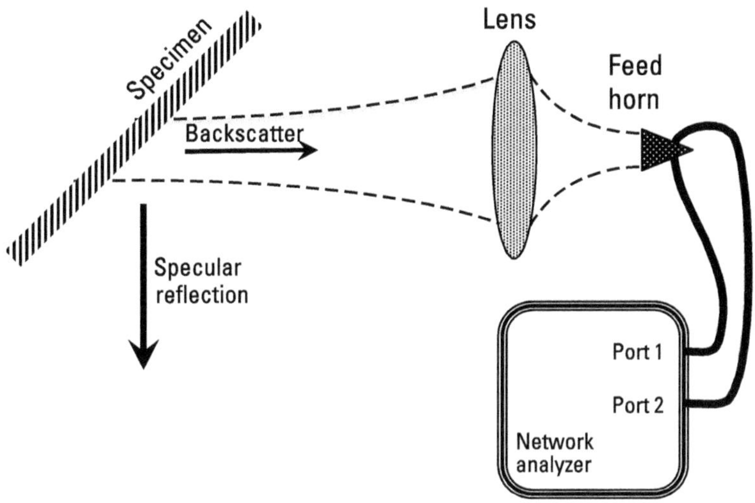

Figure 6.11 Focused-beam measurement configuration for cross-polarization measurements.

it, all the energy should be reflected back as an identical quasi-plane wave. When the junction of the plates is oriented vertically, a right-angle dihedral has a reflection characteristic such that a horizontally polarized incident wave experiences a voltage reflection coefficient of $\Gamma = -1$, while a vertically polarized wave sees a reflection coefficient of $\Gamma = 1$. The opposite signs of the reflection coefficient are due to a 180° phase shift difference (or rotation) from the double bounce a wave encounters during the roundtrip inside the dihedral depending upon its polarization. This characteristic provides a polarization-dependent reflection coefficient, whereas the reflection coefficient of a normal-incidence flat-metal plate is independent of orientation or polarization.

Because of the polarization-dependent reflection coefficient, rotation of the dihedral about its center axis as shown in Figure 6.12 provides a strong cross-polarization reference signal, useful for a full cross-polarization calibration algorithm. In particular, the E-field of the signal incident upon a vertically oriented dihedral is rotated 180° for horizontal polarization and is not rotated for vertical polarization. Thus, the scattering of a rotated dihedral is determined by separating the incident signal into its horizontal and vertical components; then the horizontal component is rotated 180° and recombined to obtain the total received signal. The vector components of the received signal are analyzed with respect to the cross- and co-polarization axes. Measurement of complex voltage by network analyzers allows a representation of the scattering from complex targets as a polarization scattering matrix (PSM),

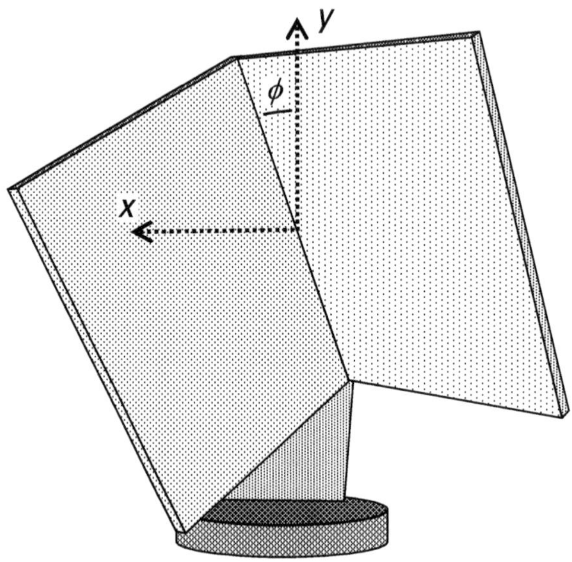

Figure 6.12 Schematic of a conductive dihedral on a stand, rotated about its center by an angle ϕ relative to vertical.

$$\begin{bmatrix} a^- \\ b^+ \end{bmatrix} = \begin{bmatrix} S_{hh} & S_{hv} \\ S_{vh} & S_{vv} \end{bmatrix} \begin{bmatrix} a^+ \\ b^- \end{bmatrix} \tag{6.33}$$

where the a wave corresponds to horizontal polarization, and the b wave corresponds to vertical polarization. The + and − designations are defined in Figure 6.13, and the geometry of the dihedral and the orientation (ϕ) of its axis relative to the measurement system is shown in Figure 6.12. The PSM at each frequency is specified by eight scalar quantities, four amplitudes, and four phases where one phase angle is arbitrary and is used as a reference for the other three.

Based on concepts originally developed for remote sensing, Chen, Chu, and Chen [13] developed a calibration technique that uses three calibration standards in an anechoic chamber to determine the PSM of an unknown target. The relationship between the actual target PSM and the measured PSM can be described as $\mathbf{S}^m = \mathbf{X} + \mathbf{RST}$ where $\mathbf{X}$ is an isolation error matrix resulting from residual reflections and coupling between transmitting and receiving channels when no target is present. $\mathbf{T}$ and $\mathbf{R}$ are transfer matrixes that account for frequency response, mismatches, and cross-polarization coupling, $\mathbf{S}$ is the target scattering matrix, and $\mathbf{S}^m$ is the measurement. Barnes introduced this RST model in 1986 [14], and in expanded form the measured signal is

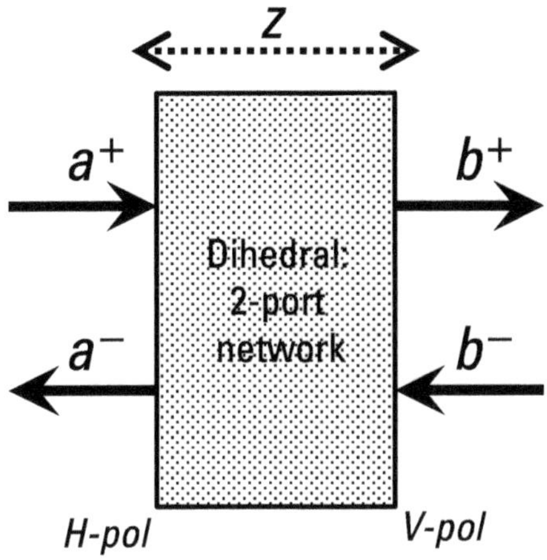

Figure 6.13 Two-port network definitions for a PSM of a dihedral calibration standard.

$$\begin{bmatrix} S_{hh}^m & S_{hv}^m \\ S_{vh}^m & S_{vv}^m \end{bmatrix} = \begin{bmatrix} X_{hh} & X_{hv} \\ X_{vh} & X_{vv} \end{bmatrix} + \begin{bmatrix} R_{hh} & R_{hv} \\ R_{vh} & R_{vv} \end{bmatrix} \begin{bmatrix} S_{hh} & S_{hv} \\ S_{vh} & S_{vv} \end{bmatrix} \begin{bmatrix} T_{hh} & T_{hv} \\ T_{vh} & T_{vv} \end{bmatrix} \tag{6.34}$$

which can be rearranged when solving for a measured target to

$$\mathbf{S} = \mathbf{R}^{-1}\left(\mathbf{S}^m - \mathbf{X}\right)\mathbf{T}^{-1} \tag{6.35}$$

This expression represents a set of coupled nonlinear equations. Thus, calibration involves finding **X**, **R**, and **T**. **X** is simply the isolation measurement (matched load) where $\mathbf{S} = 0$. Yueh proposed a solution that obtains normalized quantities of the **R** and **T** matrixes [15]. The solution requires three different calibration standards: flat-metal plate, dihedral, and rotated dihedral. The PSM of a flat metal plate is

$$\mathbf{S}_1 = \begin{bmatrix} -1 & 0 \\ 0 & -1 \end{bmatrix} \tag{6.36}$$

Inversion of this calibration standard is possible because the diagonal elements of the flat plate PSM are nonzero. The metal plate standard establishes a relation between **R** and **T** of $\mathbf{R} = \mathbf{S}_1^m\mathbf{T}^{-1}$; therefore derivation of **T** will also give **R**. The normalized **T** matrix can be defined as

$$\mathbf{T} = \begin{bmatrix} T_{hh} & T_{hv} \\ T_{vh} & T_{vv} \end{bmatrix} = T_{vv}\begin{bmatrix} w & w/u \\ v & 1 \end{bmatrix} = T_{vv}\mathbf{T}' \tag{6.37}$$

Solutions of u and v are obtained using the measurement of the second independent calibration standard, with the requirement that the diagonal terms of the standard are independent of each other. The dihedral corner reflector fills this requirement with a PSM of

$$\mathbf{S}_2 = \begin{bmatrix} -1 & 0 \\ 0 & 1 \end{bmatrix} \tag{6.38}$$

where one polarization acts as an electrical equivalent of a short, and the orthogonal polarization acts as an electrical open. A solution for w is obtained using the third calibration standard, the rotated ($\phi = 22.5°$) dihedral. This rotated dihedral provides independent cross-polarization information on the diagonals necessary for solving w. The PSM of a rotated dihedral is

$$\mathbf{S}_3 = \begin{bmatrix} -\cos 2\phi & \sin 2\phi \\ \sin 2\phi & \cos 2\phi \end{bmatrix} \tag{6.39}$$

R and **T** are then solved by evaluating (6.35) with these PSM expressions of the various calibration standards. Noting that T_{vv} is canceled out, the result is normalized to the S_{vv} component of the flat-plate calibration.

6.2 Near-Field Probe Measurements

The aforementioned far-field scattering measurements can provide information on radiated fields from an object but supply little information on evanescent fields such as surface or local cavity modes. These nonradiating fields can eventually contribute to radiated fields in the presence of local perturbations, such as geometric discontinuities or material property variations. Therefore, near-field measurements of scattering bodies and materials provide additional insight into these scattering mechanisms by measuring both radiated and nonradiated fields. Near-field measurements, which can also detect covered or buried inhomogeneities in dielectric or magnetic media, are used to measure evanescent fields in antennas and microwave circuits, providing information about their electromagnetic behavior. In addition to a near-field probe, such measurements also require a microwave source to excite the object under investigation. In this case, a focused beam is beneficial since it can provide a controlled and localized illumination of an object or a portion of an object. This section describes a method for using a focused-beam system combined with a near-field probe to investigate evanescent microwave scatter phenomena associated with discontinuities such as edges [16].

When investigating scatter, analyses are simplified when the illuminating energy can be approximated as a plane wave. Additionally, accurate measurement of the scatter from a feature or inhomogeneity requires that the extent of the illuminating beam be greater than the physical dimension of the scattering feature. For example, a rounded edge with radius of curvature r, must be encompassed within the illuminating beam, including both the width and depth. This is analogous to compact-range RCS measurements where the target under test must be contained within the quiet zone. Recall from Chapter 4 that a measure of the depth of focus is the Rayleigh length, defined as the distance from the beam waist location to where the spot size has increased to $w(Z_R) = \sqrt{2}w_0$,

$$Z_R = \frac{k}{2}w_0^2 \tag{6.40}$$

For a focused beam where $w_0 \sim 2\lambda_0$, and assuming the physical size of the scattering feature must be less than $2Z_R$, then $2Z_R < 8\pi\lambda_0$, or approximately 25 wavelengths in the direction of the beam propagation. A more restrictive dimension is in the directions orthogonal to the incident propagation direction—the width of the illuminating beam. In analogy with the compact range quiet zone, we can define a similar quiet zone within the focused beam where the amplitude taper decreases by no more than 3 dB. Note that even if these rules of thumb are followed, there may still be finite size effects [17], so data interpretation should always consider the presence of these effects.

The rest of this section illustrates near-field probe measurements combined with focused-beam illumination by describing measurements of the fields around a simple, linear edge. To facilitate this measurement, one port of a two-port network analyzer is connected to an antenna and lens to illuminate an area under test, and the second analyzer port is connected to an electrically small probe that measures the local electric or magnetic field as it is moved around the region of interest. The total transmission measured is therefore proportional to the intensity of the fields measured by the probe. Probes can be either electric dipoles or their complement, a small magnetic loop [18]. An example of a loop probe is shown schematically in Figure 6.14. In this case, an electrically small loop is made from a semi-rigid coaxial cable, and two gaps are cut in the outer conductor. The two ends of the semi-rigid cable are connected to a 180-degree hybrid junction and the network analyzer measures the difference signal, which minimizes pickup of E-field and maximizes H-field sensitivity.

Now we consider a focused beam illuminating the edge of a metal sheet with dimensions, 122-cm-wide × 91-cm-tall × 0.04-cm-thick. The

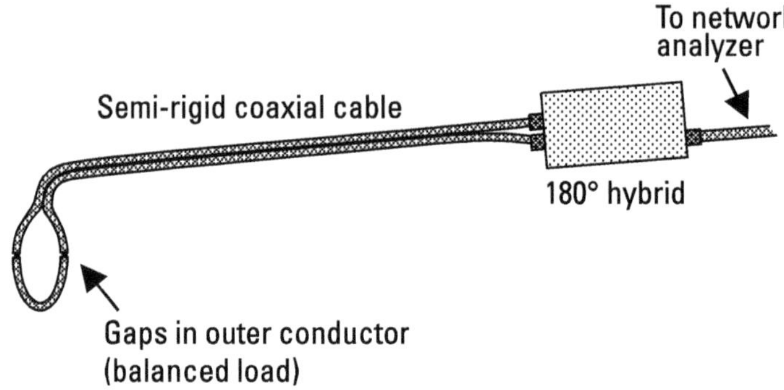

Figure 6.14 Drawing of a loop probe for measuring H-fields constructed from semi-rigid coaxial cable.

measurement geometry is shown in Figure 6.15. The edge is positioned at the beam's focus in the region of approximately constant phase. A small loop probe is spatially scanned either linearly or in a raster pattern in the near-field region immediately in front of the sample with a separation of approximately 4 mm between the sample surface and the probe center. Thus, the probe is $\leq \lambda/9$ from the sample surface within the measured frequency range. The induced signal in the probe is proportional to the total local *H*-field, which depends on spatial position and frequency. This assumes that the probe does not significantly shadow the incident beam onto the specimen under test. This is helped primarily by orienting the feed cables to the probe to be orthogonal to the incident E-field polarization to minimize probe scatter.

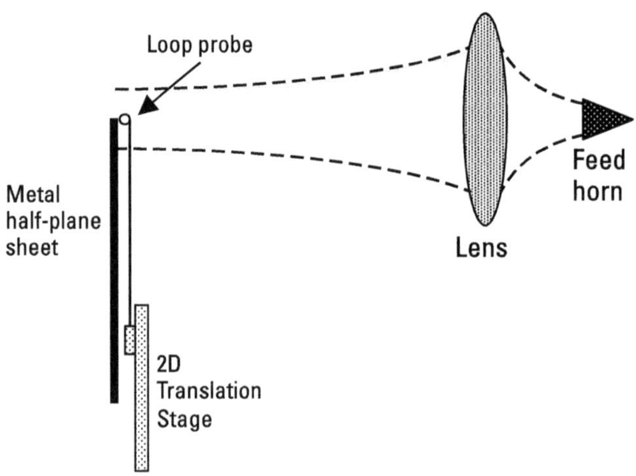

Figure 6.15 Geometry of focused-beam and near-field probe measurements of a conductive half-plane edge.

In the measurement procedure, a scan is made with the specimen in place, and a second scan is made with no specimen. The second, or clear site, scan is proportional to the incident field, while the first scan includes both incident and scattered fields. The measured field data is normalized to the incident (clear sight) field at the center of the beam, and the data can be expressed as near-field amplitude or phase. For the data shown, the probe was scanned linearly across the edge and parallel to the plane of the sheet, as indicated in Figure 6.15. Figure 6.16 shows the total field measured with the conductive half-sheet present and the polarization of the incident illumination such that the H-field is parallel to the edge and E-field perpendicular to the edge. The data is plotted as a function of position and frequency. The edge of the metal sheet is at position = 0 and extends to position > 0. Thus, the higher field amplitude apparent at position > 0, is due to the sum of the incident and scattered energy, while at position < 0, there is no metal sheet and therefore no specular reflection. Vector-subtracting the incident field (clear site) from the total field results in the scattered field, which is shown at 6 GHz in Figure 6.17, along with the total field. The scattered field shows that there is little energy at positions < 0, which are away from the metal sheet. Furthermore,

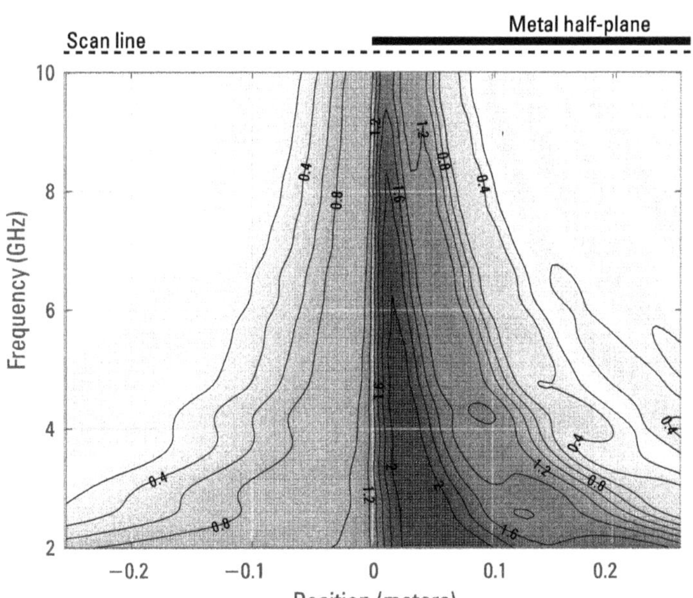

Figure 6.16 Amplitude of total H-field (incident + scattered) from half-plane illuminated at normal incidence, TM polarization (H-field parallel to the edge), normalized to incident field at center of beam.

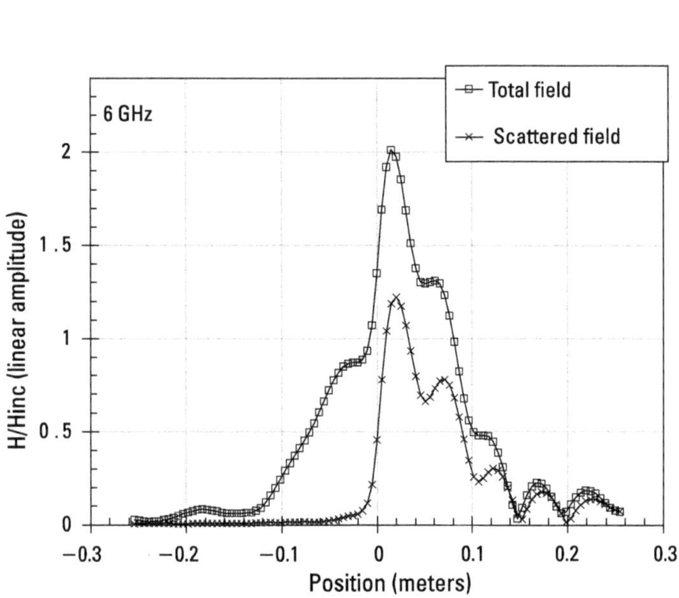

Figure 6.17 Amplitude of total and scattered H-field at 6 GHz from half-plane illuminated at normal incidence, TM polarization, normalized to incident field at center of beam.

an interference pattern is evident in both scattered and total fields at positions > 0, with regular peaks and nulls versus position. This interference is due to superposition of diffraction scatter from the edge with the specular reflection.

An alternative way to view the measured probe data is in a $k-\omega$ diagram of the plane wave spectrum. Since data is acquired as a function of position on the plate, the plane wave spectrum can be calculated via Fourier transform with respect to spatial position [8],

$$\frac{F_y(k_x)}{F_0} = \int_{-\infty}^{\infty} H_x(x) e^{ik_x x}\, dx \tag{6.41}$$

This transform requires that a scan of the entire width of the illumination be conducted to avoid truncation errors. Figure 6.18 shows the plane-wave spectra of the scattered H-fields for TM illumination of the edge (H-field parallel to edge) at normal incidence. With this polarization, the incident electric field is perpendicular to the edge. $k_x = \pm k_0$ light lines are indicated as thin black lines on the graph. These light lines define the borders beyond which only evanescent fields can exist. In other words, plane-wave energy outside of these lines does not propagate to the far field.

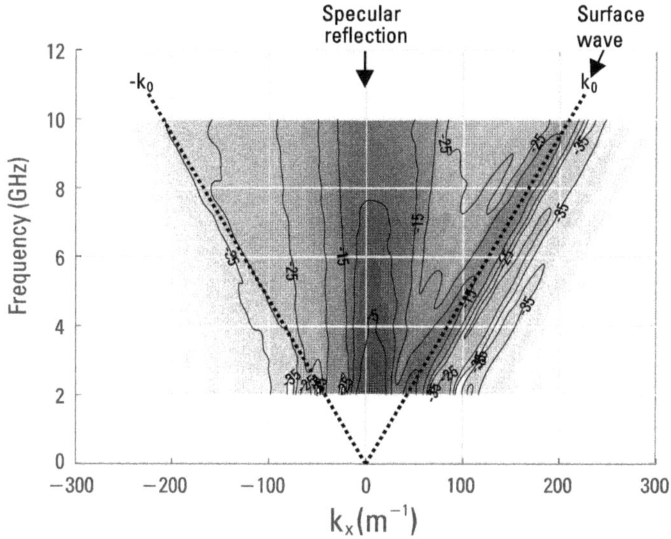

Figure 6.18 Plane-wave spectra of scattered H-fields from half-plane edge, illuminated at normal incidence, TM polarization.

The data in Figure 6.18 shows both specular reflection from the metal sheet and edge wave scatter. Because of the normal incidence illumination, the spectra show a primary lobe centered at $k_x = 0$. The width of this k-vector lobe depends on the width of the illuminating beam and reflects the plane-wave distribution of the incident beam. Superimposed with this specular lobe is a broader distribution of plane waves that span the k-space between the light lines at approximately 20–30 dB below the specular lobe. These broader spectra are due to diffracted energy from the edge. In this data the strongest component of diffracted propagation is in the $k_x = k_0$ direction (light line on the right), which is the direction exactly parallel to the surface of the metal plane. This $k_x = k_0$ energy is a surface-traveling wave, which is discussed in more detail in Section 6.3.

This plane wave spectra data illustrates the power of combining a near-field probe measurement with a focused-beam system. In this case, we have a method for experimentally characterizing the physical phenomena responsible for scatter behavior from edge discontinuities. Similarly, this method can be used to understand the electromagnetic behavior for applications ranging from component scatter to FSSs and radomes. In structures such as FSSs, there can be substantial evanescent energy due to the resonant nature of these materials, and these modes would be indicated by having plane-wave energy outside of the light lines. Similarly, surfaces with material layers may trap energy, which would also be indicated by plane-wave spectrum energy outside the light lines.

6.3 Surface-Traveling Wave

Surface-traveling waves consist of electromagnetic energy where the propagation of that energy is exactly parallel or grazing to the plane of a surface. This contrasts with the phenomena of specular reflection, where incident energy is at some other angle besides grazing, and the surface than reflects the incident energy into a complimentary angle determined by Snell's law. Surface-wave energy is a phenomenon that occurs when a radar system illuminates a target and is important because the surface-traveling waves propagate along the surface of the illuminated body until they encounter discontinuities such as gaps or edges. When surface waves interact with discontinuities, they then scatter and radiate out. Surface-wave energy is also an issue in vehicles and structures that have multiple antennas for wireless communication and sensing. Two or more antennas located on a surface may experience cosite interference when energy radiated from one of the antennas is received by another antenna and interferes with the RF electronics that the antenna is connected to.

Traveling waves on a conductive body occur when there is a component of the E-field normal to the surface. More precisely, the exact surface normal and the propagation-direction vector define a plane, and surface-traveling waves can form when the E-field is oriented parallel to that plane. This is also known as a TM polarization, because when the E-field is parallel to that plane, the magnetic field is transverse to it. For the case of a body illuminated by a faraway radar, it may experience a buildup of current on its conductive surfaces from radar illumination. When current reaches a discontinuity, such as an edge, it then reradiates because of reflections at the discontinuity. For the case of cosite interference, antennas built on or in a surface launch currents that travel along that surface until they encounter another antenna. One method for reducing the negative consequences of surface waves is to include a coating that absorbs the current and associated fields so that it doesn't build up, which for cosite interference, reduces the direct coupling between two interfering antennas [19]. Similarly, a radar target that has material appropriately applied reduces the backscattered energy and lowers its RCS [7].

Finally, a surface-traveling wave can also be attached to the outside of a wire. In fact, this idea was developed in earnest by Goubau [20] who added a dielectric layer around a wire to more tightly bind the wave to the wire so that it could be used as a transmission line. In most situations, however, the surface wave attached to the outside of a wire is undesired and can cause interference problems between electronic components. Ferrite chokes are often clamped around the outside of a wire to absorb RF energy and prevent these surface-traveling waves from propagating.

6.3.1 Surface-Wave Attenuation

Surface-traveling waves are sometimes described in terms of surface currents and a space wave. However, using these two terms can be confusing, since in reality the surface currents are coupled to the space wave, and attenuation of one also attenuates the other. In a magnetic-absorber coating, the magnetic loss factor interacts strongly with the magnetic field near the surface, which is coupled to both the surface currents and the space wave. Extracting energy out of these magnetic fields reduces the surface current and simultaneously draws energy out of the coupled space wave. Traveling-wave attenuation can be evaluated in terms of the resultant scatter from an illuminated body. See Figure 6.19 for a simple illustration of this, showing a flat conductive plate that has been coated on both sides with magnetic absorber. The traveling wave is excited on the surface and travels back and forth, shedding scattered energy at the discontinuities (edges of the plate).

Modeling this simple geometry in an FDTD solver provides the calculated RCS data shown in Figures 6.20 and 6.21. Each of these polar plots shows two sets of bistatic RCS data—one for the metal plate without any treatment (solid lines) and a second one in which the metal plate was coated on both sides with the magnetic absorber (dashed-dot). The dielectric and magnetic properties of the applied coating were based on a commercially available carbonyl iron–filled polyurethane, and a 1.5-mm-thick coating was used. The plotted data show that the dominant scatter is in the forward and

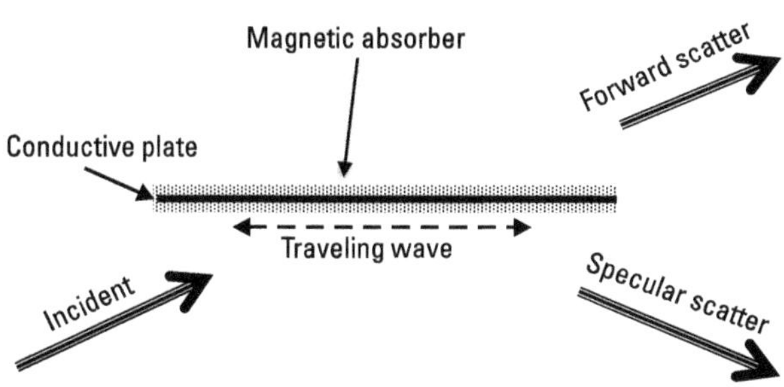

Figure 6.19 Geometry of a flat metal plate illuminated by an incident plane wave, showing forward and specular scatter directions as well as traveling-wave propagation directions.

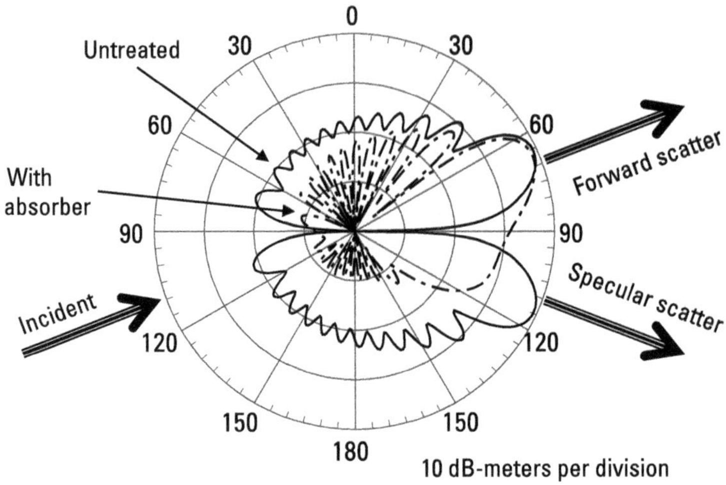

Figure 6.20 Bistatic scatter from a 20-cm flat metal plate with and without magnetic absorber coating at 9 GHz.

specular scatter directions. The incident plane wave was at an angle of 20 degrees from grazing, which is why the forward and specular scatter are at complementary angles. Both Figures 6.20 and 6.21 are on common scales, and the division markings in the radial direction correspond to 10-dB increments. The simulations were two-dimensional, so the plotted scatter levels are in units of two-dimensional RCS, decibel-meters.

The only difference between Figures 6.20 and 6.21 is the size of the metal plate, 20-cm-long in Figure 6.20 and 40-cm-long in Figure 6.21. As a result, the scattered RCS is somewhat different between the two plate sizes. The first lobe that is near the backscatter direction (i.e., at an angle of approximately 110–115 degrees) is sometimes called the Peters lobe after the researcher who first studied traveling-wave scatter from ogives [21, 22]. It is a strong indication of the strength of the traveling wave on the body. In Figure 6.20, the difference between the treated and untreated scatter at these angles is approximately 13 dB. Since the length of the plate is 20 cm, we can therefore say that the absorption effectiveness of the magnetic absorber treatment is about 0.65 dB per cm. For the longer plate in Figure 6.21, the difference between the treated and untreated RCS is smaller, only about 8 dB, and the total body is twice as long. Thus, the magnetic absorber treatment for this larger body had an effectiveness of only 0.2 dB/cm. In other words, the traveling-wave absorption of a given absorber depends on the overall geometry of the body on which it

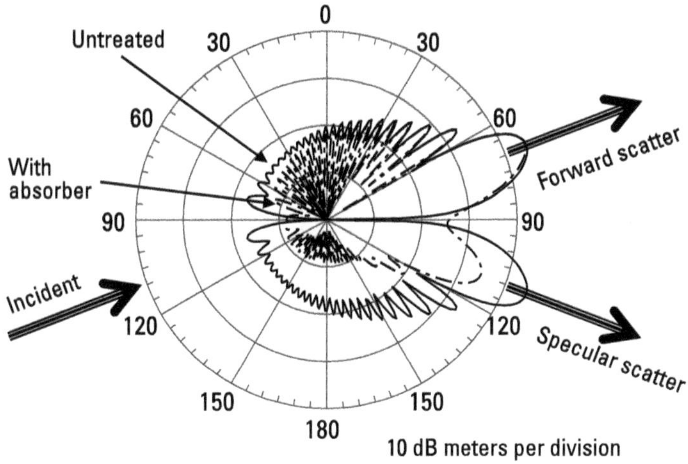

Figure 6.21 Bistatic scatter from a 40-cm flat metal plate with and without magnetic absorber coating at 9 GHz.

is applied, including length and curvature. There is no universal "decibel per unit length" metric, and it is only possible to measure relative effectiveness of a coating specific to the conditions of that measurement. In other words, any measurement device that can be conceived for measuring traveling-wave absorption can only measure a qualitative value, since there is no well-defined quantitative parameter.

6.3.2 Surface-Wave Attenuation Measurement

The RCS calculation in the previous section is easily done with a computational electromagnetic simulation tool, but more difficult to perform experimentally. Experimental measurement of RCS is typically done in a range facility that includes an antenna and radar system for illumination, a target mounting column or pylon along with fixturing, and a significant amount of distance so that the wave interacting with the target can be approximated as a far-field plane wave [23]. A compact RCS range shortens the required distance by including a parabolic reflector that transforms the energy emanating from the antenna into a collimated beam, similar to the lens on a lens-based focused-beam system. In addition, depending on the signal sensitivity needed, a microwave network analyzer can sometimes be used as the radar system.

As for a target, simple shapes like the flat plate in Section 6.3.1 can be used as test targets. To further simplify analyses, the leading edge of the test target can be tapered so that it has a low scatter level. This leads to only the trailing edge as a significant discontinuity for scattering the traveling wave. The measurement sequence then measures the target both bare and with traveling-wave absorber material applied so that the difference can be determined. This method measures backscatter only, as opposed to the bistatic RCS calculated earlier.

While RCS range measurements characterize traveling-wave absorbers under semirealistic conditions, they are expensive and time-consuming. There are also some smaller-scale laboratory methods that can determine relative absorption performance of traveling-wave absorbers with a smaller, less expensive setup. Probably the most common is the traveling-wave table, pictured schematically in Figure 6.22. This fixture has two feeds, one that transmits and a second that receives. A material specimen is placed between the feeds, and the relative attenuation of the material is determined by measuring the insertion loss between the two feeds with a microwave network analyzer. Calibration consists of measuring the traveling-wave transmission on this table without any treatment, and the traveling-wave performance is measured as the insertion loss when a specimen is placed on the table, divided by the calibration measurement. This insertion loss can be expressed in decibels per unit length by also dividing by the width of the specimen under test.

A potential disadvantage of a traveling-wave table is that when the feeds are electrically small, the radiation from them is point-source-like. When a material is placed in between, the material can act as a waveguide or lens and artificially focus the traveling wave so that the measured insertion loss due to the material is less than that for a more realistic plane-wave source. Ideally the feeds for a traveling-wave table should consist of linear antenna arrays with a

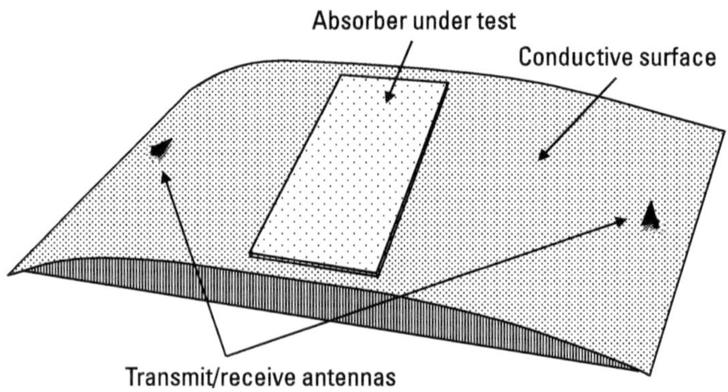

Figure 6.22 Drawing of a commonly used table fixture for measuring attenuation by traveling-wave materials on conductive surfaces.

wider physical aperture, so that the traveling wave is more plane wave–like, similar to a target being illuminated by a distant radar. The drawing in Figure 6.22 also shows a curved surface, which is commonly done to ensure that the transmit-and-receive antennas are shadowed from each other. However, this is not necessary since, as discussed previously, the space wave and surface currents are coupled, and attenuation of one also attenuates the other.

Another laboratory method, inspired by the idea of surface waves attaching to wires, is shown in Figure 6.23. This is a cylindrical surface-wave measurement fixture that uses a cylindrical diameter large enough to wrap a surface-wave absorber around the circumference. It behaves somewhere between a wire and an infinite plane. I used a fixture like this previously to measure magnetic surface wave absorbers down to 100 MHz without requiring an anechoic chamber. It consists of a conductive cylinder with coaxial cables attached at each end. The transition from the center conductor of the coaxial cable at the transmit end launches a surface-traveling wave that attaches to the long cylinder. A material specimen is placed at the center of the cylinder, wrapped around the circumference, and then another transition at the other end of the cylinder receives the traveling wave after it has traversed through the specimen. The attenuation per unit length is determined by dividing the insertion loss by the length of the specimen under test. Like the traveling wave table, the specimen data is first calibrated by dividing it by a clear-site measurement. The novelty of this cylindrical geometry is that since the attached traveling wave is oriented radially around the cylinder, it behaves similarly to an infinite plane wave—minimizing the finite size effects from point source illumination. It also minimizes the required specimen size since the specimen need only be big enough to wrap around the circumference. That said, it is not exactly the same as a flat surface, and the cylindrical curvature tends to more tightly bind the surface wave than a flat planar surface.

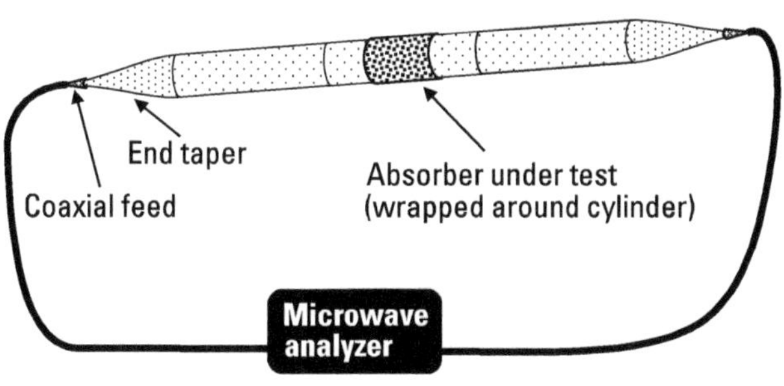

Figure 6.23 Cylindrical surface-traveling wave fixture for measuring attenuation by absorber specimens at UHF and VHF frequencies.

A fourth method for characterizing surface-traveling waves is shown schematically in Figure 6.24. This fixture can be small enough to be handheld and is a portable device for measuring surface-wave attenuation of absorber coatings in-situ on a surface [24]. The fixture acts similarly to the surface-wave table shown in Figure 6.22, but with a few important differences. The transmit-and-receive antennas are linear-array antennas that launch a wave from a distributed aperture instead of a point radiator. Microwave combiners can be used as feed networks to split a single input into multiple channels that feed each element of the array antennas. This provides a more collimated wave of energy going across the measurement area so that focusing effects are reduced. The spot-probe configuration also includes a structure for physically connecting the two antennas so that they are held a fixed distance from each other with a known opening between them. The spot probe is then placed on a surface to be tested, and the transmission scattering parameter (S_{21}) is measured to characterize the surface-wave attenuation of a coating. Calibration is done by placing the same spot probe on a flat conducting surface, which provides the nonattenuating reference, and the sample measurement is divided by the calibration S_{21} to obtain a relative insertion loss by the coating. The decibel-per-unit-length attenuation is calculated by dividing the relative insertion loss by the separation between the transmit-and-receive antenna arrays.

An example of surface-wave attenuation measured by a spot probe device is shown in Figure 6.25 for three different thicknesses of a commercial magnetic absorber. The commercial absorber consists of iron powder mixed into a polyurethane rubber, and the three thicknesses are 0.76, 1.6, and 2.4 mm. The absorber was measured on a flat aluminum surface. Since the volume fraction

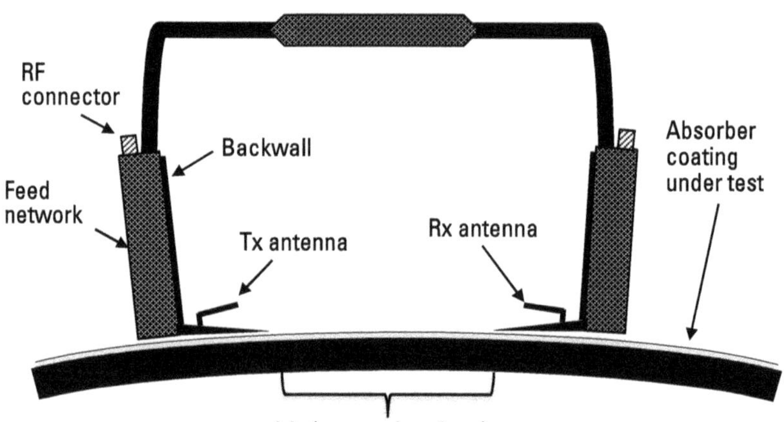

Figure 6.24 A traveling-wave spot probe for measuring absorber performance.

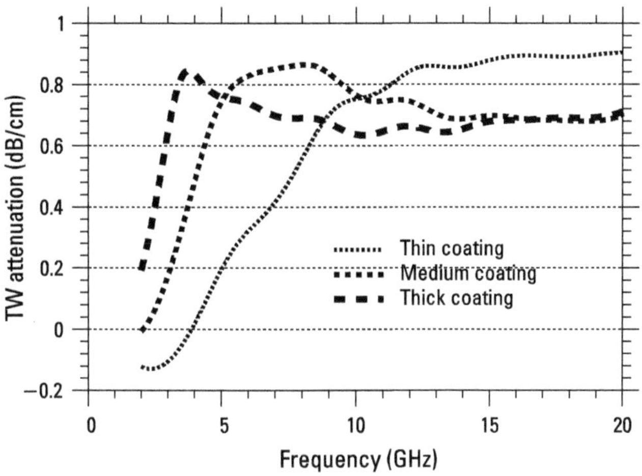

Figure 6.25 Measured surface-wave attenuation of three different magnetic absorber coatings of different thicknesses.

of iron powder was the same in all three specimens, the variation shown in the plotted data is due to absorber thickness. The trend is that thicker coatings show better performance at lower frequencies, which is consistent with them also being electrically thicker.

Another interesting aspect to the absorber performance occurs at the lowest frequencies, where the attenuation performance not only approaches 0 dB/cm but can even dip into the negative. Negative attenuation is equivalent to gain, but energy conservation dictates that negative attenuation should be nonphysical in materials such as this. To understand this apparent contradiction, it is first important to know that a commercial iron-based absorber such as this will typically have an approximately constant real permittivity with not much dielectric loss and a dispersive magnetic permeability with significant magnetic loss in the 2–18-GHz range. For the material measured in Figure 6.25, the permittivity was approximately 17 across the frequency band, and the magnetic permeability ranged from 2.8–1.8i at 4 GHz to 0.8–1.4i at 18 GHz. In other words, the magnetic loss tangent (imaginary divided by real μ) is higher at the upper frequencies and decreases as frequency is decreased.

To put this in context, recall that the refractive index of a material is given by, $n = \sqrt{\mu\varepsilon}$, so that the absorber slab has both decreasing magnetic loss tangent and increasing refractive index at the lower frequencies. For this reason, the surface-traveling wave, which extends into free space away from the surface, is more strongly drawn into the magneto-dielectric layer on the surface while also experiencing a lower relative magnetic loss. In other words, the absorber slab acts as a magneto-dielectric waveguide at the lowest frequencies, which concentrates the surface wave's power closer to the surface. Since the measurement was calibrated with no absorber, the magneto-dielectric waveguide draws power out of the part of the wave in free space and increases the power that goes into the receive antenna array. The negative attenuation measurement is the net result of this surface-layer waveguiding effect.

6.3.3 Surface-Wave Backscatter

Surface waves can be of interest because they interact with discontinuities on a surface or body, which results in scatter. Therefore, measurement of traveling-wave backscatter is a way to explore different strategies for controlling surface wave–related backscatter. The surface-wave measurement methods described in Section 6.3.2 can be used not only to measure transmission, but also reflection or S_{11}. The calibration method for backscatter is somewhat different than for traveling-wave attenuation. Backscatter requires both a reflection standard and a matched load (or isolation) standard, since the reflection from the probe itself must be subtracted to obtain reasonable signal to noise levels. A useful reflection standard is a flat metal sheet placed vertically on the ground plane and in front of the transmit/receive antenna. The isolation standard is simply the response of the probe on an ideal flat metal plate with no defect. The calibrated backscatter is then calculated with (6.22), which is repeated here for convenience,

$$\text{Backscatter} = \frac{S_{11}^{\text{specimen}} - S_{11}^{\text{isolation}}}{S_{11}^{\text{response}} - S_{11}^{\text{isolation}}} \tag{6.42}$$

The vertical metal plate response standard, while experimentally convenient, provides only a relative backscatter signal. If a more quantitative backscatter is desired, it is possible to use a trailing conductive edge as the response calibration standard. This will then enable the backscatter to be expressed in decibels to knife edge (dBke), a unit sometimes used in the scatter community. For this calibration to be accurate, it must be located at approximately the same position as the scatterer of interest. The backscatter wave, while approximately collimated, is not truly a plane wave and will diverge as a function of distance from the transmit/receive antenna.

An experimental demonstration of surface-wave backscatter measurement is shown in Figure 6.26. This data was measured with the traveling-wave spot probe described in Section 6.3.2. A penny on a conductive ground plane is a small scatterer, and the data in Figure 6.26 shows the relative scatter of a single penny and two stacked pennies. Also shown in Figure 6.26 is a measurement of the metal ground plane with no penny, which provides an indication of the measurement sensitivity or noise floor. The noise floor is determined by a number of potential measurement errors, such as microwave network analyzer drift, thermal expansion–induced changes within the antenna array and feed network, and scatter from nearby objects. For this reason, time-domain gating is also used to minimize the noise from these unwanted scatter sources. For the data in Figure 6.26, a 0.75-ns time-domain gate was used.

The frequency dependence of the penny backscatter has characteristic nulls near 16 GHz and at a frequency approximately half that, which are independent of the height (one penny or two). Nulls such as these are indicative of multiple scattering centers interfering with each other. Knowing that a penny is a bit over 19 mm in diameter, we can predict the locations of these nulls, which is where that diameter is approximately a half wavelength and one wavelength. In other words, there are reflections from both the leading and trailing edges of the penny that destructively interfere at these corresponding frequencies.

Traveling-wave backscatter can also be used to find small defects on a surface, even if those defects are underneath a coating or otherwise difficult to detect visually. To do this, backscatter data must be acquired over a sufficiently wide bandwidth so that it can be transformed into time domain with enough resolution to locate when a scattering center returns to the receiver. An example of this time-domain measurement is provided in Figure 6.27, which

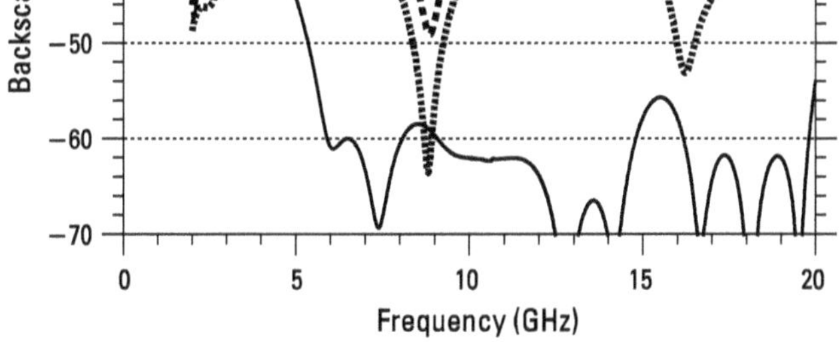

Figure 6.26 Measured surface-wave backscatter from one penny and two stacked pennies on a metal ground plane.

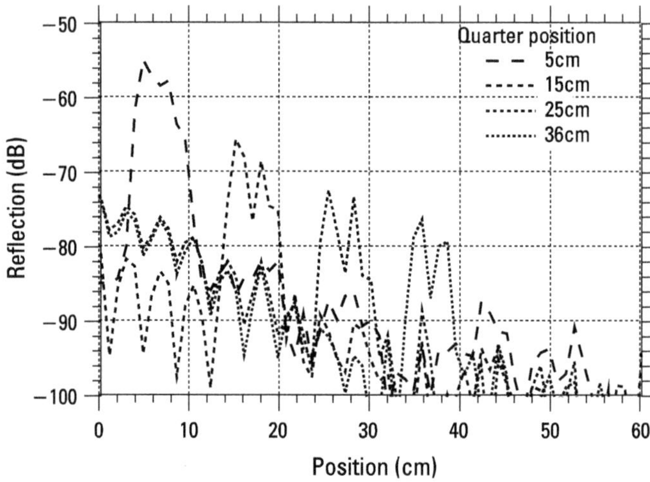

Figure 6.27 Surface-wave backscatter versus down-range distance of a quarter on a resistive sheet measured at four different antenna-to-quarter distances.

shows the measured signal from a quarter positioned on a resistive sheet at several different distances from the transmit/receive antenna. The data in this plot was acquired over the 2–18-GHz frequency band and then transformed into time domain. The time-domain axis was converted into down-range distance by assuming the speed of light of the surface wave to be 30 cm/ns, and the time of the signal was assumed to correspond to roundtrip time for the surface-traveling wave. Another interesting feature of the reflections from the quarter is that they have double peaks, corresponding to the reflections from the leading and trailing edges of the quarter, consistent with the frequency-domain interference pattern observed in Figure 6.26 for the penny backscatter.

References

[1] Jackson, J. D., *Classical Electrodynamics* (Third Edition), Hoboken, NJ: Wiley, 1998.

[2] *Introduction to Complex Mediums for Optics and Electromagnetics*, W.S. Weiglhofer and A. Lakhtakia (eds.), Bellingham, WA: SPIE Press, 2003.

[3] Munk, B. A., *Frequency Selective Surfaces*, Hoboken, NJ: Wiley, 2000.

[4] Schultz, J. W., et al., "A Focused-Beam Methodology for Measuring Microwave Backscatter," *Microwave and Optical Technology Letters*, Vol. 42, No. 3, August 5, 2004, pp. 201–205.

[5] Collin, R. E., "Scattering of an Incident Gaussian Beam by a Perfectly Conducting Rough Surface," *IEEE Trans. Antennas and Propagation*, Vol. 42, 1994, pp. 70–74.

[6] Long, M. W., *Radar Reflectivity of Land and Sea* (Third Edition), Norwood, MA: Artech House, 2001.

[7] Knott, E. F., J. F. Schaeffer, and M. T. Tulley, *RCS* (Second Edition), Norwood, MA: Artech House, 1993.

[8] Smith, G. S., *An Introduction to Classical Electromagnetic Radiation*, Cambridge, United Kingdom: Cambridge University Press, 1997.

[9] Kerns, D. M., *Plane-Wave Scattering-Matrix Theory of Antenna-Antenna Interactions*, National Bureau of Standards, Monograph 162, U.S. Gov Printing Office, 1981.

[10] Senior, T. B. A., and P. L. E. Uslenghi, "The Circular Cylinder," in *Electromagnetic and Acoustic Scattering by Simple Shapes*, Amsterdam, Netherlands: North-Holland Publishing Company, J. J. Bowman, et al., (eds.), 1969.

[11] Ruck, G. T., *RCS Handbook*, Volume 1, New York, NY: Plenum Press, 1970.

[12] Larson, T., "A Survey of the Theory of Wire Grids," *IRE Trans. Microwave Theory and Techniques*, Vol. 10, No. 3, May 1962, pp. 191–201.

[13] Chen, T.-J., T.-H. Chu, and F.-C. Chen, "A New Calibration Algorithm of Wide-Band Polarimetric Measurement System," *IEEE Trans. Antennas and Propagation*, Vol. 39, No. 8, August 1991, pp. 1188–192.

[14] Barnes, R. M., "Antenna Polarization Calibration Using In-Scene Reflectors," *Proceedings of the Tenth DARPA/Tri-Service Millimeter Wave Symposium*, U.S. Army Harry Diamond Lab., Adelphi, MD, April 8–10, 1986.

[15] Yueh, S. H., et al., "Calibration of Polarimetric Radars Using In-Scene Reflectors," *Progress in Electromagnetics Research (PIER 3–Polarimetric Remote Sensing)*, J. A. Kong (ed.), New York, NY: Elsevier Publishing, 1990.

[16] Schultz, J. W., E. J. Hopkins, and E. J. Kuster, "Near-Field Probe Measurements of Microwave Scattering from Discontinuities in Planar Surfaces," *IEEE Trans. Antennas and Propagation*, Vol 51, No. 9, September 2003.

[17] Petersson, L. E. R., and G. S. Smith, "On the Use of a Gaussian Beam to Isolate the Edge Scattering From a Plate of Finite Size," *IEEE Trans. Antennas and Propagation*, Vol. 52, No. 2, 2004, pp. 505–512.

[18] Harms, P. H., et al., "A System for Unobtrusive Measurement of Surface Currents," *IEEE Trans. Antennas and Propagation*, Vol. 49, No. 2, Feb. 2001, pp. 174–184.

[19] Hamid, S., and D. Heberling, "Experimental Demonstration of Antenna Isolation Improvement using Planar Resonant Absorbers," *Proc. of the 2019 Int. Symposium on Electromagnetic Compatibility (EMC Europe 2019)*, Barcelona, Spain, Sep. 2–6, 2019, pp. 351–354.

[20] Goubau, G., "Open Wire Lines," *IRE Trans. On Microwave Theory and Techniques*, Vol. 4, No. 4, October 1956, pp. 197–00.

[21] Peters, L., "End-Fire Echo Area of Long, Thin Bodies," *IRE Trans on Antennas and Propagation*, Vol. 6, No. 1, January 1958, pp. 133–139.

[22] Stoyanov, Y. J., C. R. Schumacher, and A. J. Stoyanov, "RCS Calculations of Traveling Surface Waves," Report DTRC-90/014, David Taylor Research Center, Bethesda, MD, May 1990, AD-A227032.

[23] Knott, E. F., *RCS Measurements*, Raleigh, NC: SciTech Publishing Inc., 2006.

[24] Schultz, J. W., et al., "Traveling Wave Spot Probe," U.S. Patent 9995694B2, 2014.

7

CEM-Based Methods

7.1 CEM

CEM modeling is a method of applying Maxwell's equations to model electromagnetic behavior in situations where analytical approximations are not sufficient or feasible. In some of the previous chapters, we discussed the concepts of network-analysis methods to solve boundary-value problems. For example, inverting material properties from free-space measurements approximates the measurement as plane-wave interaction with an ideal slab specimen to derive an equation relating intrinsic properties to extrinsic S-parameters. While this process does use a computer to iteratively solve the equation(s), it is still restricted to certain simple geometries that can be described with just one or two equations. On the other hand, there are many problems in areas such as electromagnetic scatter or antenna design, that have complex geometries not describable with a simple equation. These more complex geometries require a different approach to solve. Similarly, expanding material measurements into new and more complex fixture designs also requires a different approach.

CEM modeling capability has grown with computer technology, though many of the CEM methods in use today were originally envisioned in the early days of the microprocessor [1, 2]. Since then, there have been advances in efficiency and applicability of these methods [3]. In general, CEM methods apply Maxwell's equations to a finite simulation space to solve for how it interacts with incident electromagnetic energy. In cases such as geometrical or physical optics, CEM methods are approximate and use a high-frequency assumption where a wave is treated as a vector or ray propagating through the simulation space. In other cases, a full-wave solution to Maxwell's equations is used by dividing the simulation space into facets or cells [4].

Full-wave CEM methods are considered exact in that the only approximation is from discretization error due to the size of the facet or cell used to divide the simulation space. Different conditions are placed on these units depending on the presence of conductor or material or free space. Boundary conditions are also assigned to the perimeters of the simulation space, and they can include constructs such as perfectly matched layers (PMLs) where the fields are treated as if they can continue to infinity, or electrical or magnetic conductive boundaries where the fields are reflected. They can even be periodic boundary conditions where the fields are folded from one side of the simulation space to the other. One of a variety of different methods is then used to iteratively apply Maxwell's equations to each cell or facet [4]. Different CEM methods have different strengths and weaknesses, and the choice of method depends on the type of problem and considerations such as bandwidth, speed, simulation size, and desired outputs.

The prevalence and relative speed of modern processors also enables easy use of CEM tools even on everyday computer resources. Numerous open-source codes exist that can be downloaded, compiled, and used. These resources provide good starting points for conducing CEM simulations at low cost. However, using open-source CEM software can require a significant investment in time or training to understand its use or to adapt it for a particular application. Another option for quickly obtaining a CEM simulation capability is to purchase a commercial code. There are numerous commercial CEM codes available, which are actively maintained and have most of the latest techniques for enhancing simulation speed and efficiency [5]. They also include sophisticated visualization tools to create model inputs and analyze outputs. Of course, these sophisticated capabilities come at a financial cost since complex simulation tools require care and maintenance.

The use of CEM in materials measurement problems, or in any application, is usually one of two roles: (1) design or (2) analysis. In the design role, CEM calculations optimize a device's geometry for a set of requirements. This is done iteratively where simulations are run over and over to evaluate different design configurations. Needing many iterations requires a code to be fast so that those many iterations can be evaluated in a reasonable time with reasonable computing resources. To keep simulations fast, high accuracy is not necessarily required; rather the need is to know if one design configuration is better or worse than another. In this role, design simulations use a coarser grid to minimize the number of calculations and memory requirements of the simulation array. A coarse simulation may increase the numerical error and lead to other systematic errors such as anomalous dispersion, but that is usually an acceptable trade-off for being able to more quickly design an electromagnetic device.

On the other hand, the second typical role of CEM simulations is analysis, where a given geometry is modeled with higher fidelity to get a more quantitative assessment of its behavior. In this case, fewer configurations need to be modeled, and so each model calculation can take more computer resources or time to compute. Since quantitative accuracy is more important, a finer grid with smaller cells or facets can be used, which reduces the various computational errors and increases the accuracy of the result. In the analysis role, CEM calculations can also help understand physical phenomena. For example, CEM methods provide a mechanism for visualizing fields within a geometry and obtaining insight on how features within that geometry affect the overall performance.

The measurement methods described in this book have the common theme of wide-bandwidth, nonresonant techniques. For this reason, the CEM method that is heavily used throughout this book is the FDTD method, in which a simulation space is usually divided into cubic or rectangular cells, and each cell is assigned to be a conductor, free space, or material. FDTD works in time domain by launching fields at the beginning of a simulation and then iteratively marching through time, updating the fields throughout the simulation space at each time increment. The introduced fields have a time-dependent amplitude, and the exact pulse shape used depends on the desired bandwidth within the simulation. The simulation marches through a series of time steps until the introduced fields dissipate. The time dependencies of the fields are monitored when they cross a bounding surface, and these figures are used to calculate near-field outputs or are transformed into far-field equivalents [5]. The time-dependent outputs can be transformed into frequency domain via a Fourier transformation, providing a calculation of frequency-dependent behavior from a single simulation. This ability to determine behavior over a wide frequency range with a single simulation is what makes FDTD an advantageous method for modeling broadband material measurement devices.

As discussed in Chapter 1, electromagnetic materials are dispersive with frequency-dependent intrinsic properties. A disadvantage of the FDTD method is that it cannot model arbitrary frequency dependencies of materials. Instead, it is restricted to modeling dispersive dependencies that have corresponding Fourier-transform pairs in frequency and time domain. The Debye, Lorentz, and Drude dispersion models described in Chapter 1 can be reformulated in time domain by taking their inverse Fourier transform [6]. Thus, it is possible to model dispersive material behavior by fitting it to these dispersion models for representation in FDTD simulations. If these single-pole dispersion models are insufficient for capturing material dispersion over a wide bandwidth, additional terms or poles can be added to better fit actual frequency-dependent behavior. Fortunately, additional poles can also be represented in time domain via Fourier transform.

7.2 CEM Inversion of Broadband Materials

Section 7.1 described the two typical CEM modeling roles: design and analysis. There is a third role for CEM modeling: CEM-model material-inversion. CEM inversion methods for material property extraction are relatively new, since the ability of common desktop or laptop computers to sufficiently run full-wave computational electromagnetic tools is also recent. An early example of this method utilized partially filled waveguides with an arbitrarily shaped specimen, which could be modeled with a finite-element method [7]. Since then, others have also explored the use of full-wave CEM methods for material property inversion [8, 9]. These approaches have generally used an optimization scheme to run the full-wave CEM solver iteratively until the best model/measurement match is achieved.

The advantage of this CEM inversion approach is that a measurement fixture is no longer restricted to something that can be described with analytical equations. In other words, material measurements in waveguides or coaxial airlines or free space all can be described with a simple equation or small set of equations that relate the intrinsic material properties to the reflection and transmission within the fixture. With CEM inversion, new material measurement fixture designs can be used even if they cannot be characterized with a set of analytical equations. As long as the fixture can be described accurately with a CEM model, it is viable as a material measurement device.

Another advantage of CEM inversion over conventional methods is the potential for simplified calibration procedures. Imperfections in measurement fixtures lead to evanescent coupling, additional reflections, or multipath effects, which are not included in the analytical equations describing the fixture. Conventional methods must measure a plurality of calibration standards to account for these imperfections. For example, Chapter 5 described some of the multistandard calibration methods used in waveguides and coaxial airlines. In contrast, fixtures designed to use CEM inversion can have much simpler calibration procedures. The example fixtures described in Sections 7.3.1–7.3.3 use just a single calibration method instead of the typical three (e.g., short, open, and load) needed for determining reflection in a conventional transmission line setup.

Simplified calibration in CEM methods is possible since fixture imperfections can be explicitly modeled with CEM tools. For example, a waveguide or coaxial measurement fixture typically has a transition to couple an RF cable into the transmission line. While these transitions are optimized to minimize mismatch reflections, the conversion from an RF cable or other input into the transmission line geometry is never ideal, and some level of reflection still exists. The purpose of a conventional calibration is to de-embed that small but otherwise unknown reflection so that it can be separated from the desired signal. With CEM inversion, the paradigm can shift to include the transition as part of the inversion model. In other words, the details of the transition are explicitly included in the fixture model so that they do not need to be calibrated out.

A potential disadvantage of the CEM inversion method is that the operation of a CEM code requires significant expertise not normally found in a materials-measurement laboratory. The CEM inversion method has been shown to be accurate [7], but it was originally restricted to advanced research laboratories because of the knowledge needed to apply a CEM solver to measurement data. Furthermore, depending on the complexity of the model, a CEM solver can also take time to converge to a solution, preventing rapid characterization of multiple material samples. In QA situations, many measurements are needed to verify materials being applied or manufactured. Scaling iterative CEM inversion to handle a large measurement quantity is impractical, so a variation of CEM inversion was developed to avoid requiring users to have both measurement and CEM modeling expertise. This modified CEM method instead precalculates solutions across a range of different expected specimen properties before any measurements are made [10, 11]. Modeling different combinations of properties across the range of expected values results in a database of CEM results that are later compared to measurement data. A measurement then uses this precomputed data to find the closest match to the current measurement. Interpolation refines the best match into a result with sufficient accuracy. In other words, the CEM modeling part of the process is done ahead of time so that the actual inversion is fast and easily applied in industrial situations.

Sections 7.3.1–7.3.3 describe examples of material measurement fixtures using precomputed CEM inversion models. These measurement fixtures, which span a variety of applications, are summarized in Table 7.1. With traditional methods, there is a limited palate of choices, limiting applicability. In contrast, fixtures that use CEM inversion can be designed for a much wider range of applicability. Some of the examples listed in Table 7.1 are appropriate for nondestructive testing in industrial environments, and these are just a small sampling of the possibilities.

7.3 CEM Inversion Example: RF Capacitor

A first example of CEM inversion is inspired by the quest for measurement methods of inhomogeneous and anisotropic materials at VHF and UHF frequencies. This has long been one of the primary stretch goals of the advanced RF materials measurement community. Recent interest in anisotropic metamaterials and devices made from these materials have also increased the need for advanced RF material characterization methods. Until recently, the most practical method for these types of materials was VHF waveguide, which is large, expensive, and cumbersome to operate. While there are other methods that work at these frequencies, such as coaxial airlines or stripline waveguides, their nonuniform fields are poorly suited for materials that are inhomogeneous and anisotropic.

Table 7.1
EM-Inversion Fixtures Described in This Chapter

Fixture	Description
RF Capacitor	Measures bulk materials in cube-shaped specimens 60–800 MHz; Real ε: 1 to > 100; Imag ε: 0.01 to > 100 Measures all three tensor directions in a single specimen
Epsilon measurement probe	Measures bulk materials with arbitrary thickness 30–500 MHz; Real ε: 1 to > 2,000; Imag ε: 0.1 to > 100 Nondestructive (e.g., in-line measurement)
Mu measurement probe	Measures bulk materials with arbitrary thickness 60–500 MHz; Real μ: 1 to > 100; Imag μ: 0.1 to > 100 Nondestructive, handheld sensor
Slotted R-coax	Measures thin sheet materials 60–700 MHz; Impedance: 100–70,000+ Ω/square; Real ε: 1 to > 100, Imag ε: 0.1 to > 10 Nondestructive (e.g., in-line measurement)

One method that has successfully characterized homogenous dielectric materials is the impedance analyzer, which forms a capacitor out of the specimen under test. This idea of creating a capacitor out of a material specimen has been in use for many decades. It forms the basis of the discipline of dielectric spectroscopy, which measures thin, flat specimens sandwiched between two parallel electrodes. The key assumption of dielectric spectroscopy is that the capacitor formed by the specimen is small, so that it can be modeled as a lumped-circuit element. This assumption works when the specimen is electrically much smaller than a wavelength in its largest dimension and when the specimen is much thinner than it is wide so that it can be made into a parallel-plate capacitor with minimal fringing fields. These analytical models restrict the available geometries and frequency ranges that a measurement fixture can have so that fringe field effects do not dominate. That said, such dielectric spectroscopy methods have been extended to UHF and even microwave frequencies [12, 13]; however, the specimen size is necessarily small for these methods, and parasitic impedances can still dominate [14]. While there is no definitive rule for relating the maximum electrode dimension to when the lumped-circuit approximation becomes invalid, the approximation has been applied to capacitive fixtures with electrode diameters as high as a thirtieth of a wavelength. Metamaterials, on the other hand, have interior structures with periodicities that are a significant fraction of a wavelength in dimension. Even some more traditional composites, such as honeycomb, are inhomogeneous, with length scales that are too large to be accurately represented by conventional dielectric spectroscopy geometries, especially at higher frequencies.

A more recent attempt to address materials such as dielectrically lossy honeycomb core in a capacitor-like fixture was done by Choi [15]. Choi measured cubed-shaped specimens at different orientations to obtain anisotropic material properties. The specimens are large enough, 76-mm (3-in) cubes, to encompass the characteristic inhomogeneity length scales of honeycomb core, but this is too thick for measurement with conventional dielectric spectroscopy fixtures. Nevertheless, Choi's method makes assumptions based on the lumped-circuit model. Errors induced by this simplified model are dealt with by applying multiple calibration standards. However, the inherent disadvantage of this method is the lack of representative calibration standards with accurately known dielectric loss. Moreover, the 76-mm cube is large enough to have significant fringe fields and radiation from the specimen, which are attenuated by the use of absorber around the fixture. This potentially makes it difficult to account fully for the dielectric loss induced by the unknown specimen, since not all the power is accounted for during the measurement. Leveraging the idea of CEM inversion, an alternative approach to measuring inhomogeneous anisotropic materials combines the concepts behind low-frequency capacitance and high-frequency coaxial airline devices to make an RF capacitor fixture that works at VHF and UHF frequencies. This approach uses a CEM solver to model the measurement geometry exactly, accounting for all fringe fields as well as parasitic capacitances and inductances that plague conventional impedance analysis methods [10].

7.3.1 RF Capacitor Design

With the idea that CEM inversion enables more complex fixture design, the principles underlying the RF capacitor assume that a specimen cannot be simply modeled as a lumped-circuit element. Instead, the material under test is also assumed to be part of a high-frequency transmission line. Furthermore, cube-shaped specimens are desired to enable independent measurement of the principal tensor directions when the material is anisotropic, which leads to a square coaxial airline as a basis for the fixture. The conceptual design of the RF capacitor is shown in Figure 7.1. The fixture consists of a square-coaxial airline section with both inner and outer conductors, that is transitioned to an RF cable on one side and terminated to an electrical short on the other. The coaxial airline is sized so that the specimen replaces a section of the center conductor, which makes this setup go beyond something that can be modeled with a simple equation or two. A metal plate is positioned adjacent to the material specimen to form the shorted end. Since this fixture is a one-port microwave network, it can be used with one of the many low-cost vector reflectometers that are commercially available [16], or alternately with any existing multiport laboratory network analyzer [17].

Because the fixture is a coaxial transmission line, it is straightforward to impedance-match it to a typical 50-Ω RF cable by ensuring that the cross-sectional dimensions of the square coax also correspond to a 50-Ω line impedance. Measurement of a specimen consists of measuring the reflection coefficient of the fixture with the specimen inserted. An important feature of this fixture is that it is a closed system. In other words, the outer conductor prevents radiation from the specimen so that all energy is accounted for—either absorbed by the material specimen or reflected back to the microwave analyzer. Another feature is that this geometry creates an electric field within the specimen that is oriented predominantly in one direction as shown by Figure 7.1. As will be shown in Section 7.3.2, there is little E-field in the other directions so that the dielectric properties in each tensor direction are determined by only a single measurement.

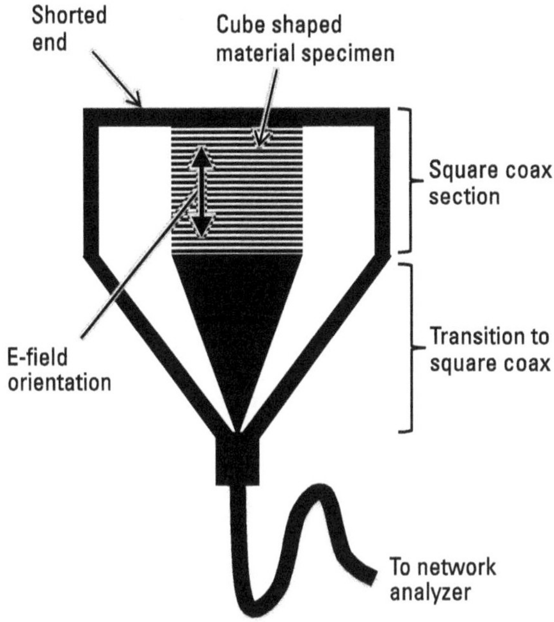

Figure 7.1 Cross-section of RF capacitor measurement fixture.

As Figure 7.1 shows, the center conductor on one side and the short on the other side are in direct contact with the material specimen forming the RF equivalent of a parallel plate capacitor. Normally, air gaps or electrode-blocking effects may be of concern in such a geometry. To prevent this, the shorted end of the fixture is made into a removable plate that can slide in and out of the fixture. When a specimen is inserted into the fixture, the metal plate sits on top of the specimen. The plate is weighted so that intimate contact is ensured with no air gaps. However, the plate is not so heavy as to distort the specimen's shape.

Calibration of this fixture consists of one measurement: the RF capacitor fixture with a low-dielectric foam spacer inserted as a test specimen. When subsequent unknown specimens are measured, the fully calibrated specimen data is then calculated as a simple ratio to the calibration data, also known as a response calibration,

$$S_{11}^{\text{calibrated}} = \frac{S_{11}^{\text{specimen}}}{S_{11}^{\text{response}}} \tag{7.1}$$

where S_{11} is the measured or calculated reflection coefficient. With this calibration method, no extra signal processing or time-domain gating is necessary. This calibration is considerably simpler than conventional high-frequency impedance analysis methods, which require three calibration standards to properly perform the fixture compensation: short, open, and load. Similarly, Choi's method requires even more calibration standards [15]: short, open, and two loads.

Once calibration is complete, specimens are inserted and measured. The complex dielectric permittivity is inverted from the calibrated reflection coefficient. Both the amplitude and phase of the reflection are measured as a function of frequency, so the real and imaginary permittivity can be determined on a frequency-by-frequency basis. Because coaxial airlines are broadband, a wide range of frequency-dependent data can be obtained from a single measurement. For the RF capacitor fixture, material properties have been successfully obtained from 60 to 800 MHz with a single measurement [10].

Inversion of the complex permittivity is done by a table-lookup algorithm where the measured reflection coefficient is compared to precomputed reflection coefficients from a variety of virtual specimens. The exact geometry of the RF capacitor was modeled with a full-wave FDTD solver with the results used to build a table correlating dielectric properties to calibrated S_{11}. In the FDTD method, dielectric materials are most easily modeled by a dielectric constant (i.e., real permittivity) and a bulk conductivity, which is related to imaginary permittivity. Therefore, the inversion table is constructed from a series of simulations that span the expected range of dielectric properties to be measured with this fixture.

In the data of Figure 7.2, a sampling of the phase and amplitudes are shown for some combinations of real permittivity and conductivity (sigma) used to invert the intrinsic properties from the complex S_{11}. Like the calibration procedure for the measurement, this data is normalized to the case of an air cube. The top plot of Figure 7.2 shows a series of different permittivities for when conductivity is zero; the bottom plot shows a series of different conductivities for when the real permittivity is three. Figure 7.2 is just a small sample of the simulations, and the full database contains more than 1,200 combinations of permittivities and bulk conductivities. With a wideband pulse, the FDTD computational method has the convenient ability of creating broadband data. Therefore, each single simulation provides data spanning over a decade of frequency. Once a data table is constructed for a given specimen shape, there is no need to run these simulations again.

For this RF capacitor design, the behavior of the reflection coefficient as a function of the specimen dielectric properties is monotonic over most

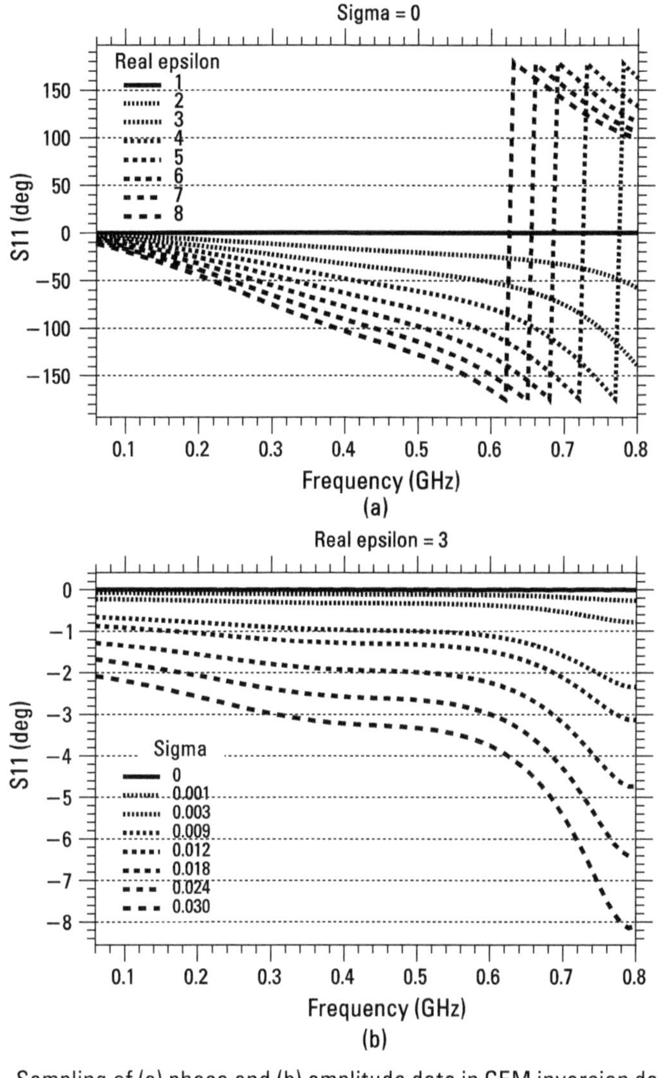

Figure 7.2 Sampling of (a) phase and (b) amplitude data in CEM inversion database for RF capacitor fixture.

of the frequency range of interest. So, simple interpolation is used to obtain arbitrarily fine resolution of the inverted properties. The spacing of permittivity and conductivity combinations are chosen so that interpolation errors are minimal. Usually, a good test of this is to compare linear interpolation to a higher-order method such as spline and check that the interpolated values do not change significantly. At higher frequencies, multiple solutions are occasionally possible, but this is easily dealt with by limiting the lookup table search at each subsequent frequency to be within the neighborhood of solutions obtained at the previous frequency. So, an inversion is done in series starting at the lowest frequency where there are not multiple solutions, and then it progresses up in frequency with bounds to prevent solution hopping. Table lookups and interpolations are fast so this precomputed inversion method can take just seconds to convert measured S_{11} data into complex dielectric permittivity, depending on the size of the lookup table. Once an inversion table is constructed, no special computational electromagnetics expertise or high-power computers are needed to use it. This makes the precomputed CEM inversion method especially appropriate for use not just in advanced laboratories, but also in automated manufacturing settings.

7.3.2 RF Capacitor Uncertainty

While the geometry of the RF capacitor fixture is modeled exactly, there are still a few simplifying assumptions made in building the inversion lookup table. To keep the lookup table tractable, the material under test is modeled as isotropic. This is usually a reasonable assumption because the E-field within the specimen is predominantly axial, and there is not much cross-component of the E-field interacting with the material. In other words, even though the specimen under test may be anisotropic, the E-field is predominantly only in one direction—parallel to the axis of the square-coaxial geometry. Therefore, it should not matter what values of dielectric permittivity and conductivity are used in the other two directions, as they should have a negligible effect on the RF capacitor response.

To test this assumption, a set of full anisotropic simulations were made of a material specimen that had a real relative permittivity of $\varepsilon = 4$, and a microwave conductivity, σ, of 0.3 S/m in the measurement direction. These dielectric parameters are representative of some commercially available absorbing honeycomb core materials. In the two directions orthogonal to the measurement direction, the permittivity and microwave conductivity varied by multiplicative factors of 0.25X, 0.5X, 2X, and 4X relative to the measurement direction. The resulting S_{11} from these simulations were then inverted using the lookup table that was generated from isotropic simulations. The resulting error of this anisotropy is shown in Figure 7.3. As this data shows, the isotropic assumption induces very low errors for low frequencies. As the frequency increases, the error increases, but it is still at manageable levels for most cases.

Another potential source of error for the RF capacitor fixture is due to the interface between the specimen and the metal-short and square-coaxial

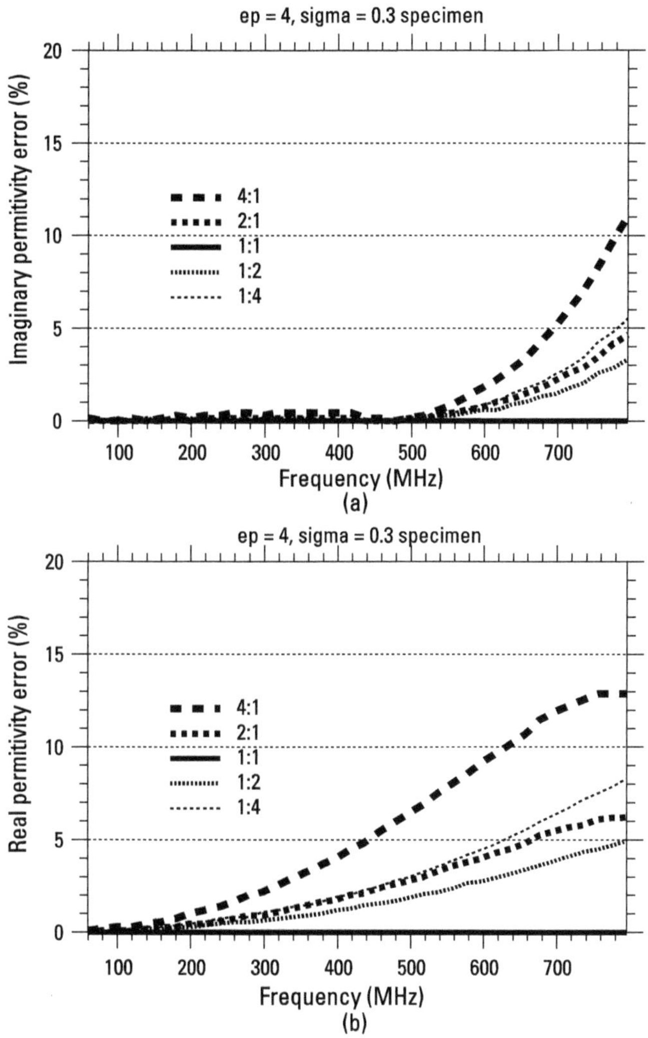

Figure 7.3 Percent error of (a) imaginary and (b) real permittivity for different in-plane/out-of-plane permittivity ratios.

center-conductor. Because the short is a metal plate that slides, gravity will always ensure that (1) it is in intimate contact with the specimen and (2) the specimen is in intimate contact with the coaxial center-conductor on its other side. However, when a material such as honeycomb core is machined into a cube specimen, there may be damage to the surface of that specimen that slightly alters the material properties near the cut. In other words, while there will never be a full air gap between the specimen and the metal, there may be a partial air gap due to this cutting damage. Additionally, contact with the metal may be incomplete if the specimen surface is not perfectly flat.

Figure 7.4 shows the computed errors in the inverted real permittivity (bottom) and imaginary permittivity (top) of a lossy dielectric specimen (real permittivity of $\varepsilon = 4$ and microwave conductivity of 0.4 S/m). These errors are for a damaged surface region of 1-mm thickness, and the damage was simulated by assuming that the dielectric properties of that 1-mm surface region were a percentage of the bulk specimen properties. The error is then the difference between an ideal specimen and one with the 1-mm damaged surface region. In Figure 7.4, 100% corresponds to air, and 0% corresponds to an undamaged, perfectly flat surface. This data shows that even if the surface region has properties that are only 50% of that in the bulk of the material, the resulting uncertainties are only a few percent or less.

Other potential errors that may exist in a measurement with a fixture such as the RF capacitor include general network analyzer uncertainties such as noise and drift. This error is not unique to this fixture, and estimates based on repeatability measurements are usually the best way to determine the impact of these errors. Finally, uncertainties from the specimen dimensions are potentially a significant error source as well. The CEM-inversion database assumes a specific size (76.2-mm cube), and deviations from these dimensions will bias the measurement results. This error can be minimized by specifying that specimens are within a certain maximum tolerance (e.g., 76.2 mm +/− 1%). Determining the actual error then requires model simulations of under- and oversized specimens to gauge the impact on inverted permittivity with the ideal size assumed.

7.3.3 Example Measurements

This section demonstrates the use of CEM inversion with the RF capacitor to determine dielectric properties on several 76.2-mm cube specimens. The first two examples are simple isotropic materials: acrylic and Delrin®. Acrylic is the trade name for poly(methyl methacrylate) and is well-known to have a relative dielectric constant just above 2.6 across all microwave frequencies. It is also known to have a small dielectric loss factor [18]. Delrin is the trade name for polyoxymethylene (POM), which is one of the few polymers that has significant dielectric loss at microwave frequencies. POM also has a slightly higher real permittivity than acrylic. Figure 7.5 shows the measured amplitude and phase of specimens of these simple polymers.

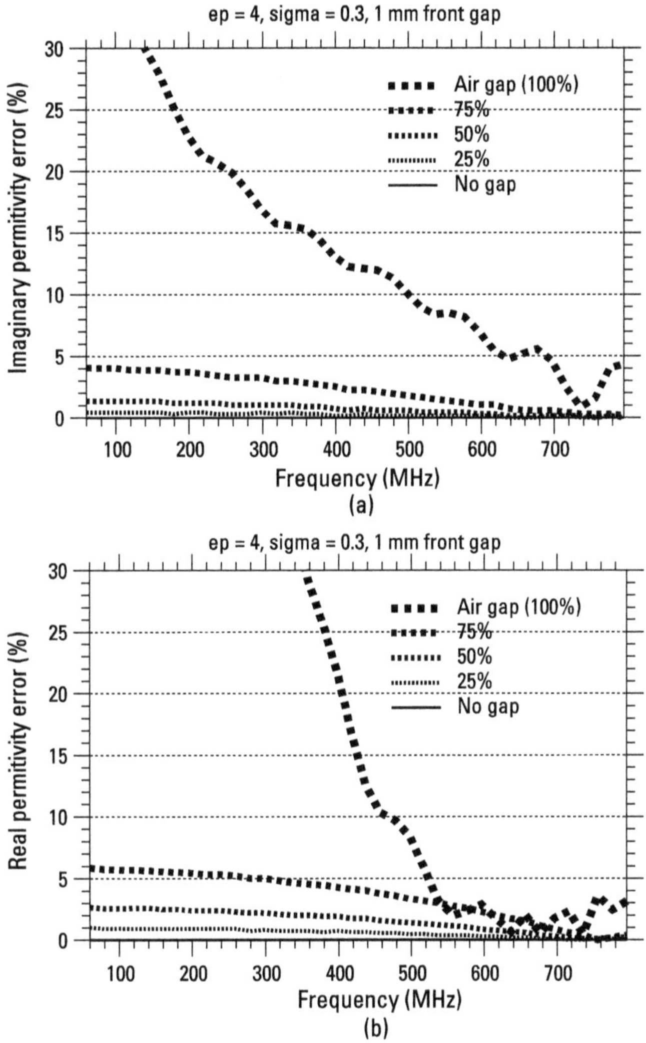

Figure 7.4 Percent error of (a) imaginary and (b) real permittivity for a 1-mm gap of damaged material with varying degrees of lower conductivity.

The reflection loss amplitude for acrylic is close to 0 dB except at the highest frequencies. Since the amplitude is mostly related to the energy absorbed by the specimen and because acrylic has a low dielectric loss factor, it is expected to be close to zero for acrylic. On the other hand, POM exhibits an insertion loss of a few tenths of a decibel because its dielectric loss factor is higher than

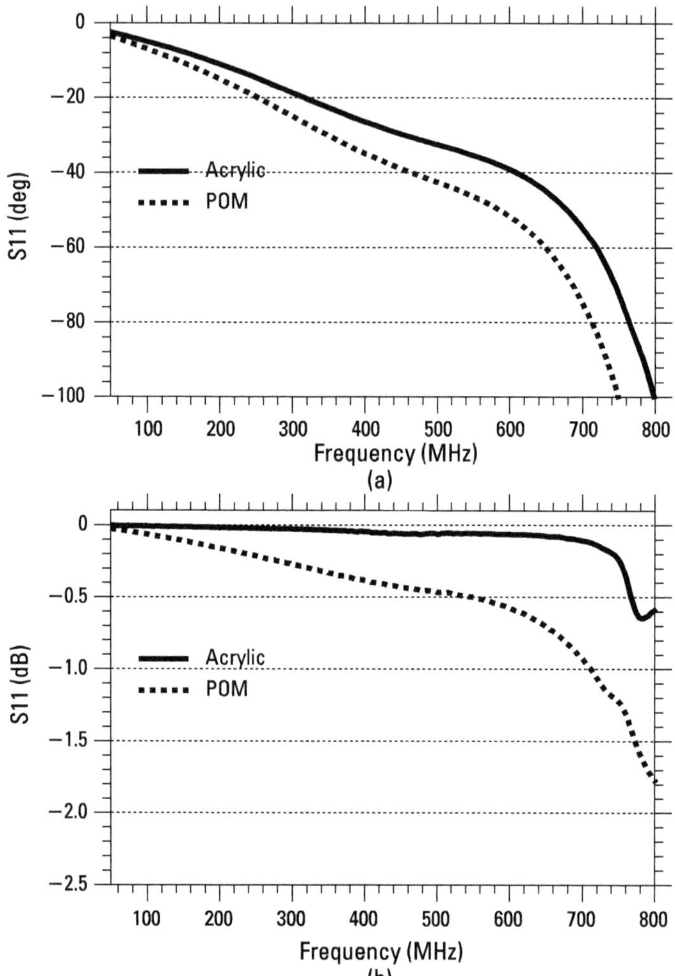

Figure 7.5 Calibrated S_{11} phase (a) and amplitude (b) from acrylic and POM specimens measured in the RF capacitor.

acrylic at these frequencies. The phase of the reflection coefficient represents the delay in the energy propagating through the specimen, corresponding to the speed of light within the material and therefore its real dielectric permittivity. As shown in Figure 7.5, POM shows a slightly increased phase delay compared to the acrylic, indicating a higher real dielectric permittivity.

Based on the calibrated reflection loss data of Figure 7.5, the inverted real dielectric permittivity (a) and imaginary permittivity (b) are shown in Figure 7.6. In addition to the inverted specimen data, these plots also include

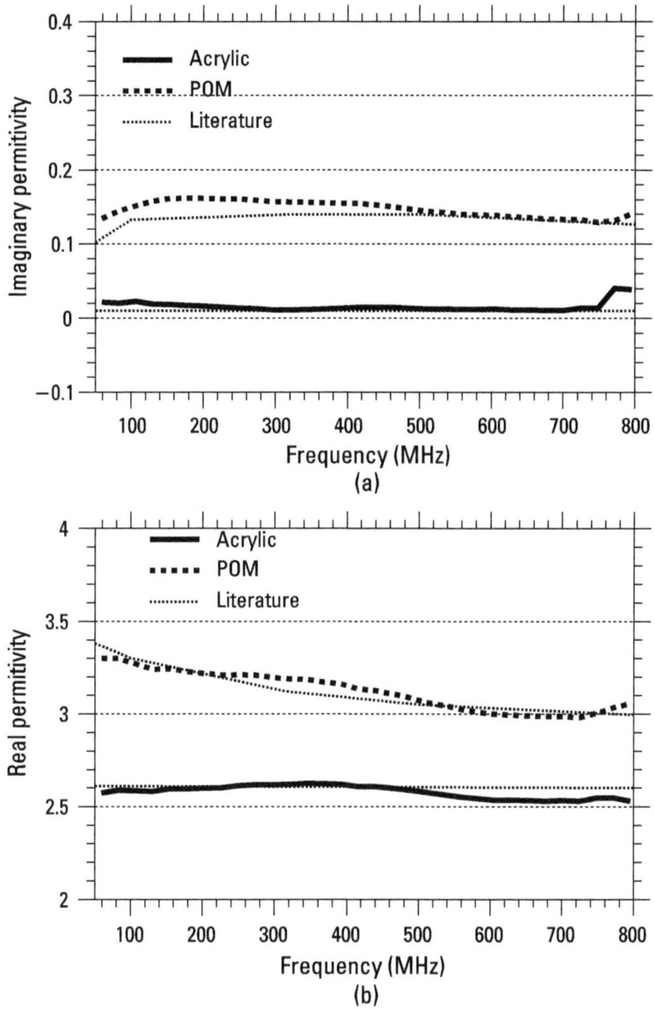

Figure 7.6 (a) Imaginary and (b) inverted real permittivity of acrylic and POM, along with characteristic data from literature showing agreement.

the known (literature-published) dielectric properties of acrylic, which agrees well with the RF capacitor–measured results. Similarly, the plots also include measured properties of low-moisture POM from the literature using a conventional impedance-analyzer method [19]. The literature properties agree well with the inverted POM properties from the RF capacitor.

Another, more interesting, example is measurement of an anisotropic artificial dielectric material constructed from interspaced layers of conductive carbon-loaded foam between layers of low-dielectric foam. The carbon foam layers were 3.175-mm-thick, while the low-dielectric layers were 6.35-mm-thick. Such a layered material will have high loss in the directions parallel to the plane of the layers and low loss in the direction perpendicular to the layers. This is because the parallel direction has fully connected carbon foam in which current can flow, while the perpendicular direction has low-loss dielectric layers interrupting current flow. This anisotropy in the dielectric properties is shown in the measured and inverted permittivity data of Figure 7.7. The thicker lines are the real (solid) and imaginary (dashed) permittivity with the E-field parallel to the layer orientations, while the thinner lines are the real and imaginary permittivity perpendicular to the layers. The effect of current interruption is evident in the perpendicular direction where the imaginary part is close to zero, and the real permittivity is relatively low.

Figure 7.7 also compares the RF capacitor measured data of the artificial dielectric to high-frequency, free-space measurements. The lower-frequency curves are from the RF capacitor. The higher-frequency curves are from a free-space method used to measure both the carbon foam and low dielectric foam constituents of the artificial dielectric [20]. The high-frequency composite permittivity is calculated using a simple series or parallel-circuit effective medium model. As this data shows, the RF capacitor results are in line with

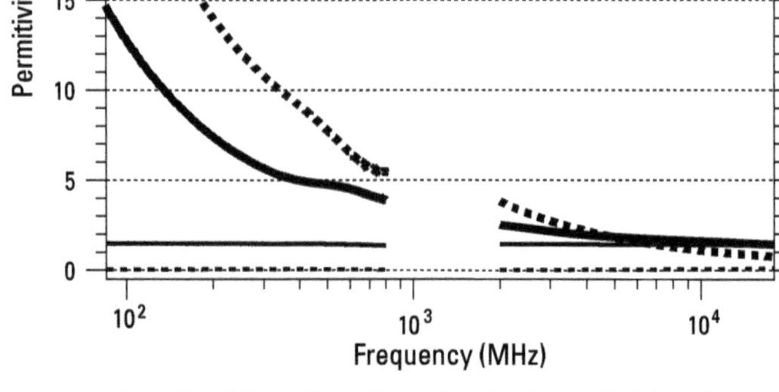

Figure 7.7 Inverted real (solid) and imaginary (dashed) permittivity of an artificial dielectric made of carbon foam and low-loss foam layers, measured parallel (thick lines) and perpendicular (thin lines) to the layer orientation.

the high-frequency properties of the specimen. These results demonstrate the power of CEM inversion–based fixtures over older, conventional methods. In this example, the RF capacitor can obtain full anisotropic properties of a dielectric material from a single, moderately sized specimen. In non-CEM-based material measurements, such a material can only be characterized with multiple specimens measured in a fixture such as a large UHF or VHF waveguide. Having to use multiple specimens not only increases time and cost, but also adds measurement uncertainty, since differences may occur both from anisotropy as well as material inhomogeneity.

7.4 CEM-Inversion Example: Nondestructive Measurement Probes

The CEM-inversion method opens new possibilities for material measurements not possible with traditional methods, particularly at UHF and VHF frequencies where wavelength is relatively large. Additionally, there is always a need for nondestructive measurements, particularly in material-manufacturing situations. To obtain intrinsic properties at UHF and VHF frequencies usually requires materials to be cut into waveguide-shaped specimens, which can still be very large, or into toroids for coaxial airlines, both of which are considered destructive testing. Another method that has been used before for nondestructive testing at low frequencies is the open-ended transmission line probe. The idea of an open-ended transmission line for measuring dielectric properties has been around for many decades [12]. A known drawback to traditional coaxial probe methods is their sensitivity to air gaps between the open end of the probe and the specimen under test. This can be a significant source of measurement error for these conventional devices since the air gap also depends on the specimen being ideally flat and smooth.

Another complication for open-ended transmission line probes is the complexity of inverting material properties. For traditional open-ended coaxial probes, analytical expressions based on lumped-circuit models can sometimes be used to invert permittivity. This is possible under ideal conditions and where the wavelength is sufficiently large relative to the probe electrical dimensions. Calibration in this case is based on conventional transmission-line calibration standards. Conventional open-ended probes have also been used in less ideal situations, where empirical calibration based on known dielectric specimens is required. Empirical calibration is complicated by the need to accurately know the dielectric properties of the calibration standards, which were presumably measured by other methods. Sections 7.4.1–7.4.2 describe a couple of applications for one-port, open-ended transmission-line probes that leverage the CEM inversion method to get around some of these difficulties and accomplish what has not been possible with traditional open-ended probes using conventional inversions.

7.4.1 Epsilon Measurement Probe

When a transmission line terminates with an open circuit, there is a strong reflection at that open circuit. This reflection occurs because of the mismatch between the end of the transmission line, which is typically a standard 50-Ω, and the impedance of free space beyond it which is about 377Ω. For example, the coaxial probe used in [12] was both an open circuit and literally an open-ended transmission line, in which the center conductor and outer conductor are simply terminated without any electrical connection between them. When such a probe is placed in contact with a planar specimen, the mismatch is now between the 50-Ω coaxial line and the impedance of the adjacent material, which is no longer 377Ω and depends on the permittivity and permeability of the material. This change in the mismatch affects both the amplitude and phase of the reflection.

Leveraging the increased fixture-design options provided by CEM inversion, an alternative to the conventional open-ended coaxial line probe is shown in Figure 7.8. This is also an open-ended transmission line, but it has a few design features that are different than the simple coaxial airline probe. One of these differences is the bent-over transmission line. The sensor has a conventional RF connector on one end, which is where the microwave signal is brought into the sensor from the network analyzer. The inner conductor of the RF connector attaches to a conductive section that runs a short length and then ends. The outer conductor of the RF connector is electrically connected to a surrounding conductor of the sensor, which helps to shield the sensor from external interference. These two conductors within the sensor behave as a two-conductor transmission line, similar to a microstrip line. A material under test is placed in contact with the sensor as shown in Figure 7.8, so that it interacts with the fields within the sensor. Because this open-ended transmission line runs for a short distance along the material under test, this configuration can have better sensitivity than a conventional open-ended coaxial probe, which simply terminates at the test sample and doesn't run along it. On the other hand, this bent-over configuration of the sensor is not a simple geometry that can be modeled analytically. Only the CEM-inversion method as described in this chapter can possibly relate the measured reflection behavior to intrinsic properties of the test specimen.

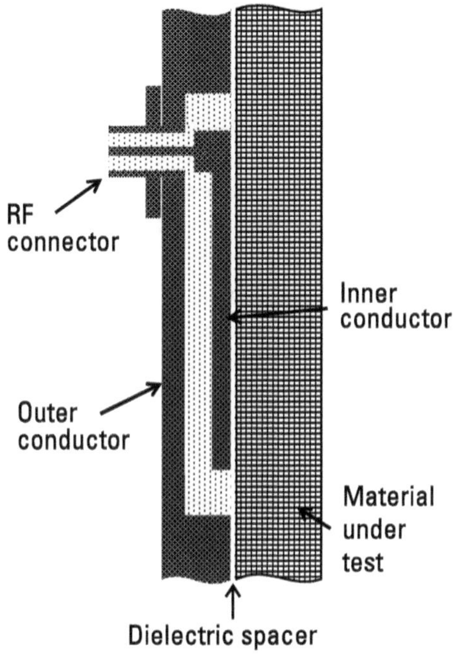

Figure 7.8 Cross-sectional drawing of an epsilon-measurement probe, showing a transmission line ending in an electrical open circuit.

Another differentiating feature of this sensor design is the dielectric spacer that is on the open face of the sensor and separates it from the material under test. Recall that air gaps between a probe and the specimen under test have a strong effect on the measured signal. Since it can sometimes be difficult to have test specimens perfectly flat and smooth, this air gap is a significant source of measurement error. With the sensor design of Figure 7.8, the dielectric spacer is designed to insert a known gap between the specimen and the sensor face. A known gap can then be included in the CEM model of the sensor so that it is included in the material inversion rather than being an unknown variable. Since this type of sensor relies on evanescent interaction with the material under test, the inclusion of a dielectric spacer does reduce the overall sensitivity. However, the increased sensitivity from the bent transmission line design makes up for loss in sensitivity from the liftoff distance.

Much like the RF capacitor described in Section 7.3.1, calibration of a sensor like this uses just a single measurement of a known standard. In this case that standard is no material or free space. Subsequent measurement of a material under test is then ratioed to the clear-site sensor measurement for comparison to the CEM model. In the RF capacitor fixture, the size of the specimen is always the same. The epsilon measurement probe, however, may be used to measure specimens of different thicknesses. This requires an expanded set of CEM simulations to create a database not only for different combinations of permittivity and loss, but also for different specimen thicknesses. The inversion process uses the known thickness of the specimen to first interpolate from two closest thicknesses in the CEM database. Then a second stage of interpolation obtains the exact permittivity and loss from the closest points in the thickness-interpolated reflection data.

Figure 7.9 shows example inverted properties using this simple calibration combined with the CEM inversion. The real and imaginary permittivities are shown by the solid and dashed lines, respectively. The thicker lines are for a fiberglass specimen, while the thinner lines are for a polyoxymethylene polymer material. Also provided in Figure 7.9 are measured permittivity of these same specimens from a higher-frequency free-space focused-beam measurement, showing agreement with the probe-measured results. This measurement probe can also be used to determine the dielectric permittivity of magnetic materials. Figure 7.10 shows example data of two different commercial magnetic-absorbing materials made from carbonyl iron mixed into urethane rubber. The primary difference between these two materials is the amount of iron powder mixed in. As expected, a greater amount of iron results in a higher dielectric permittivity. Also shown in Figure 7.10 are free-space measurements

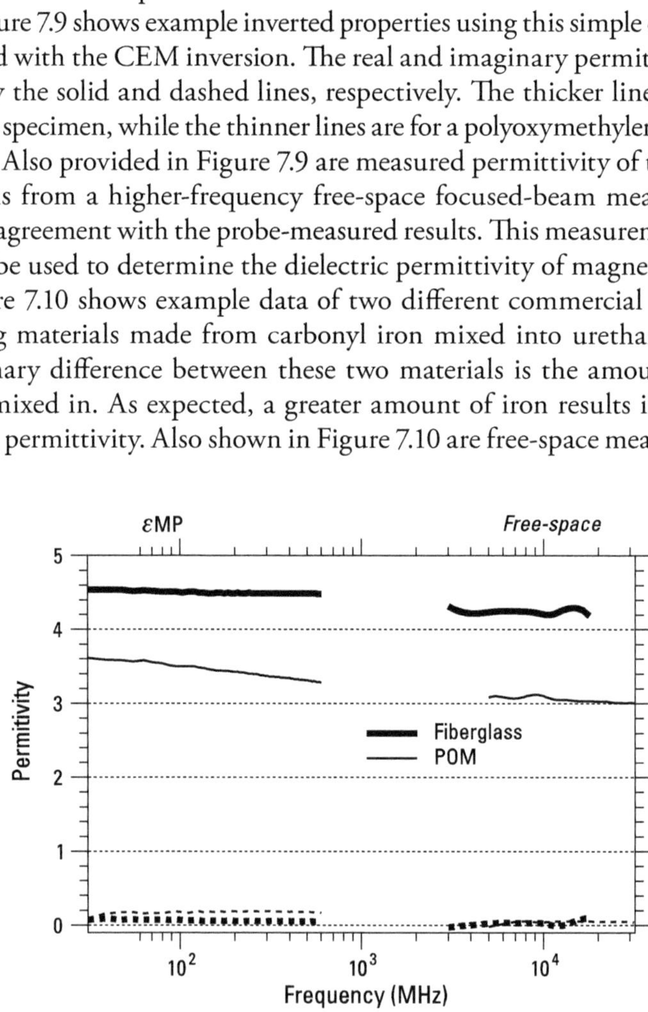

Figure 7.9 Inverted real (solid lines) and imaginary (dashed) of simple dielectric materials using the epsilon measurement probe, compared to free-space results.

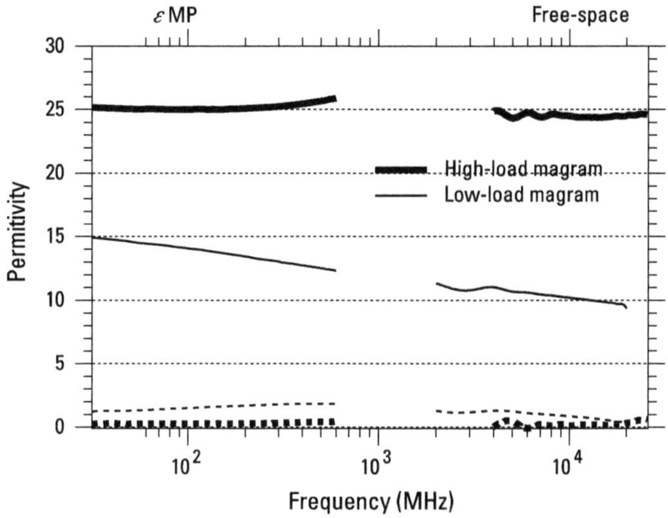

Figure 7.10 Inverted real (solid lines) and imaginary (dashed) of magnetic absorber materials using the epsilon measurement probe, compared to free-space results.

of these same specimens showing the relative agreement. The epsilon measurement probe is not particularly sensitive to magnetic permeability as these inversions were based on a database constructed from permittivity and loss combinations that assume the relative magnetic permeability is 1. That said, the higher-loaded absorber shows a gentle rise in the real part of the inverted permeability. This anomalous increase in the apparent permittivity can be attributed to the higher permeability of the increased iron loading. Improved accuracy can therefore be obtained by also creating additional databases for nontrivial magnetic permeability and including an additional interpolation step to account for permeability. Of course, this requires that the permeability will have been measured by another fixture.

7.4.2 Mu Measurement Probe

Using a similar idea as the epsilon measurement probe in Section 7.4.1, a probe can be designed to obtain magnetic permeability. The electrical open that terminates the epsilon measurement probe creates a situation where the dominant fields at the end of the probe are E-fields. To create dominant magnetic fields, the probe should be terminated with an electrical short instead. This works by a condition predicted from Maxwell's equations, which say that the tangential E-field should go to zero at a conductive boundary. Since power is conserved, the energy that was in the E-fields is converted into a stronger magnetic field making that the dominant interaction with an adjacent material specimen.

Using this idea, a mu measurement probe can be designed from a shorted sensor that generates a magnetic field in the specimen under test. The sensor measures the response to the imposed H-field through the amplitude and phase of the reflected signal. Such a device is shown in Figure 7.11. Like the epsilon measurement probe, this device has a two-conductor transmission line that runs along the surface of the active sensor area. Instead of an electrical open, the transmission line is terminated by connecting the two conductors so there is an electrical short, and it is suspended above a nearby specimen material. Additionally, the specimen under test is backed by a conductive ground plane so that the magnetic field lines are concentrated into the specimen material. In Figure 7.11, the magnetic field is predominantly in and out of the plane of the image. So, if there is anisotropy in the material under test, the probe can be rotated to measure the two orthogonal orientations.

The electrical short termination suppresses much of the E-field. However, the E-field is not exactly zero, so there is some interaction between the probe

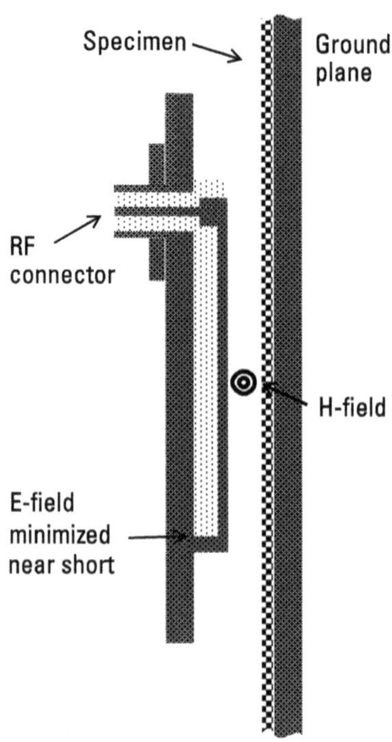

Figure 7.11 Schematic cross-section of a mu measurement probe based on a transmission line terminated by an electrical short.

and the dielectric properties. Furthermore, at microwave frequencies, most magnetic materials have a dielectric permittivity that is significantly larger than the relative magnetic permeability. For this reason, the CEM inversion requires input of an estimated dielectric permittivity to improve measurement accuracy. This interaction is strongest as frequency increases and wavelength decreases. In the case of a varying permittivity versus frequency, the value at the highest frequency in the measurement band should be used since that is where the influence is strongest. Based on these input variables, CEM simulations must be made for different combinations of (1) magnetic permeability and loss, (2) permittivity, and (3) thickness. These various tables are interpolated in multiple stages to obtain a complex permeability for a given specimen measurement. The general algorithm flow is shown in Figure 7.12.

The mu measurement probe is calibrated by simply measuring no specimen—that is, the sensor placed over the empty conductive ground plane. The calibrated reflection is the ratio of the specimen measurement to this no-sample measurement. No further signal conditioning or time-domain process is required, and this signal is used in the algorithm flow in Figure 7.12. An example measurement result from this probe provided in Figure 7.13 is compared to a 7-mm coaxial-airline measurement. The coaxial airline, of course, is a destructive method and only samples a small 7-mm area, whereas the measurement probe has a sensing area of a few inches across and is nondestructive. Thus, the difference in the results between the two methods may

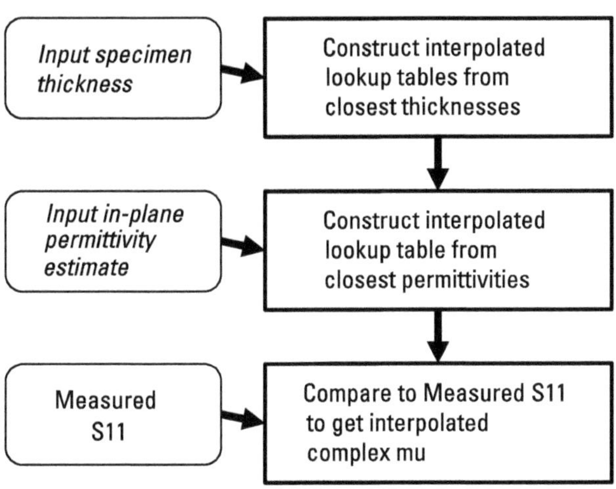

Figure 7.12 General flow of different interpolation stages to invert complex permeability from the precalculated CEM simulations.

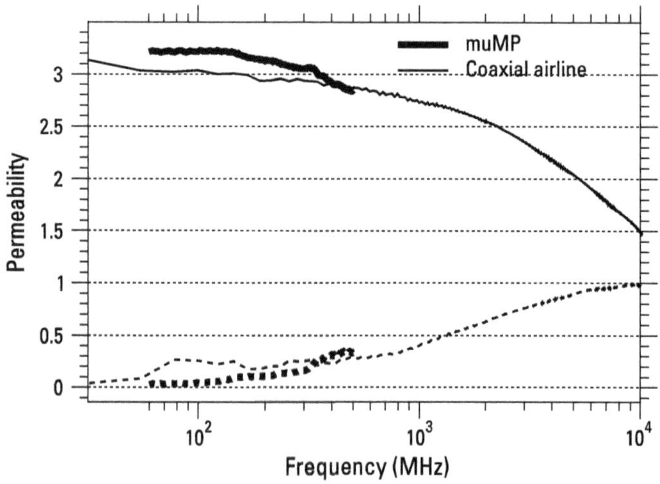

Figure 7.13 Inverted real (solid lines) and imaginary (dashed) magnetic permeability of a commercial carbonyl iron-based absorber measured with a mu measurement probe and a coaxial airline.

be caused by both measurement uncertainties of the two methods and by the fact that the two methods are measuring different areas of the specimen.

7.5 CEM Inversion Example: Slotted Rectangular Coaxial Line

In manufacturing applications, electromagnetic measurements can be used to verify materials as they are being made, providing QA or verifying specification compliance. Too often, such QA requires witness coupons that may or may not be fully representative of the manufactured materials or components. A more desirable situation is one where measurements are made directly on the manufactured materials or parts. Conventional microwave-measurement techniques such as waveguide or coaxial airline are destructive. Fortunately, with the additional fixture design possibilities enabled by CEM inversion, it is possible to design nondestructive, in-line methods that monitor material properties as the materials or components are being manufactured. Verifying material performance early in a manufacturing process reduces costs associated with rework. In a similar vein, identifying potential manufacturing problems before they create defective materials also reduces costs and improves efficiency.

An example of a material-measurement fixture designed specifically for monitoring intrinsic properties of materials as they are manufactured is a slotted rectangular coaxial line, or R-coax. Figure 7.14 shows a conceptual drawing of a slotted R-coax. This is a type of coaxial airline, but with a rectangular cross-section rather than the circular cross-section devices discussed in Chapter 5. Like a circular coaxial transmission line, there is a center and outer conductor that supports propagating waves at all frequencies. The fundamental mode has transverse electric and magnetic fields, and the upper frequency range of this device is a function of when the first higher-order mode can be excited. Exciting a higher-order mode is dependent on the dimensions of the waveguide and complicates data analyses. Small imperfections in system construction can excite these higher-order modes, so even with CEM-inversion methods, it is still better to keep to the frequency range where only the fundamental mode propagates. Like conventional circular coaxial lines, the system is eventually connected to a vector network analyzer, which usually has a 50-Ω impedance. Therefore, the transmission line dimensions are typically designed to also have an impedance of 50Ω to be well-matched to the input impedance of the network analyzer.

Unlike a conventional coaxial transmission line, Figure 7.14 has a slot cut through it. While the slot is visible going through the outer conductor, it also goes through the center conductor and out the other side so that a sheet specimen can be pulled through. In this configuration, only a portion of the transmission line cross-section is filled with material so that energy travels both through the material and through air. This same slot feature could also be put into a circular coaxial line; however, the rectangular cross-section provides some field concentration at the sample location so that overall sensitivity to thin materials is increased. The advantage of this geometry is that

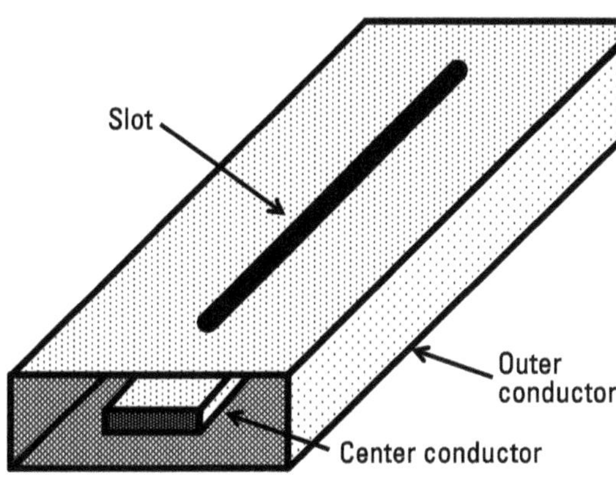

Figure 7.14 Sketch of rectangular coaxial airline fixture with a slot going through it.

materials manufactured in a roll-to-roll configuration can be continuously drawn through the slot and measured.

An inversion database is created by modeling materials inserted through the slot as a function of their dimensions and their intrinsic properties. In the case of an FDTD-simulation method, a simple dielectric model of a single relative real permittivity, ε', and a simple conductivity, σ, are assumed. These are then mapped to real and imaginary permittivity by

$$\varepsilon = \varepsilon' - i\frac{\sigma}{\varepsilon_0 \omega} \tag{7.2}$$

where ω is the angular frequency, and ε_0 is the absolute permittivity of free space. As in any of the CEM methods discussed so far, computational electromagnetic simulations are run for a wide range of different permittivity and conductivity combinations to span the expected range of specimen properties. Additionally, the CEM simulations must also be conducted for the size(s) of specimens expected to be fed through the slot and can even include the geometry of the transitions on either side of the slot that connect the R-coax to 50-Ω RF cables.

Running these different CEM simulations then creates a database of S-parameter values that correspond to the different combinations of intrinsic properties and specimen size. While technically it is possible to use either reflection or transmission S-parameters, dielectric materials are most easily characterized with the transmission S-parameter or S_{21}. This is because a specimen position within a slot may vary, causing additional phase variation in the measured S_{11} signal that only has to do with specimen position and not its material properties.

Alternatively, sometimes thin-resistive sheet materials are used in various antenna or absorber applications. A slotted rectangular coaxial line can also be used to determine the sheet impedance of a material, which is a combination of real and imaginary impedance. In a CEM code this complex impedance can be represented as a parallel combination of resistance, R, and capacitance, C,

$$Z = \frac{1}{\frac{1}{R} + i\omega C} \tag{7.3}$$

Therefore, a CEM-inversion database is constructed by a series of simulations of different combinations of R and C.

An example of inverted complex impedance from an R-coax system is shown in Figure 7.15. This data is from a commercial window tint material

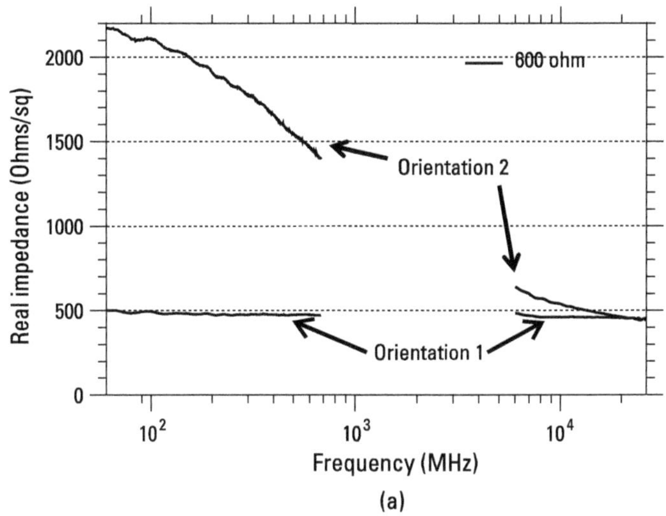

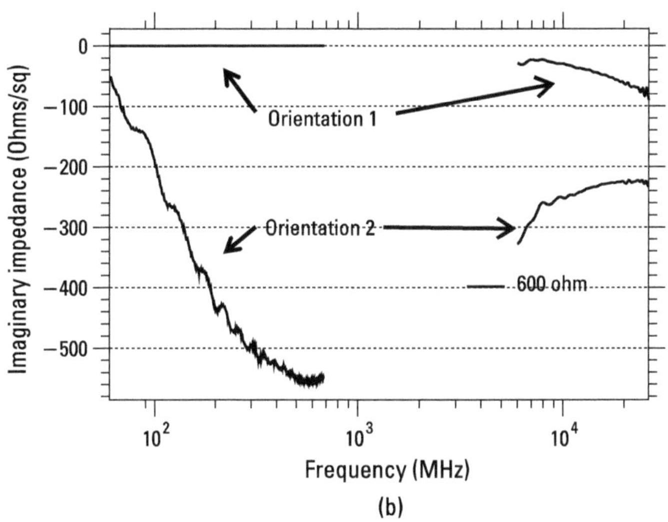

Figure 7.15 (a) CEM-inverted real and (b) imaginary sheet impedance of an anisotropic window tint compared to high-frequency free space results.

that consists of a thin metal layer deposited on to a polymer substrate. The low-frequency data is from an R-coax measurement, and the high-frequency curves are from measurement of the same material in a free-space focused-beam system. There are two sets of curves because the specimen was measured in two different orientations—with the E-field polarization parallel to the down-web direction and parallel to the cross-web direction. In other words, the material is anisotropic and exhibits significantly different impedance depending on which direction the E-field is relative to the specimen. In the R-coax system, the E-field is oriented so that it runs between the center conductor and the outer conductor. A specimen inserted through the slot therefore experiences an E-field only in one direction. For this example, the sheet specimen was cut into a square so that both the cross-web and down-web orientations could be characterized. The same specimen measured in a high-frequency free-space setup shows results that are consistent between the two frequency bands. In a manufacturing situation where a continuous roll of material is being measured, only the down-web field orientation is possible with this fixture.

Sometimes resistive materials are deposited on a thick substrate, and there is a need to treat the resistive layer separately from the substrate. A CEM method such as this can handle this kind of complication by creating an inversion database that also includes a known substrate of a known thickness. The inversion models this two-layer system in the database so the resistive layer can be extracted as an independent material separate from the substrate. An R-coax can be used to characterize the properties of a bare dielectric substrate as well, and the rectangular configuration gives it good sensitivity even to very thin substrates. For example, thin polyethylene terephthalate substrates have been characterized in the 0.5–3-GHz range with reasonable accuracy even when as thin as 25 microns [11]; 25 microns at 500 MHz is less than one twenty-thousandth of a wavelength in thickness. The R-coax can also be used to determine the dielectric properties of magneto-dielectric materials. In this case, the magnetic field lines travel around the center conductor so they are orthogonal to the plane of the material specimen. With this orientation, there is not much interaction between the magnetic field and the material permeability. So, a database that assumes a nonmagnetic specimen still gives reasonable results even when it is measuring a magnetic material. As with the mu measurement probe example in Section 7.4.2, additional CEM runs can be made to account for estimated permeability so that if it is high, it can be included as a user input to improve accuracy of the permittivity inversion.

As of this writing, these CEM-inversion methods are relatively new, and only a few commercially available material measurement devices exist that use them. However, they represent a new paradigm in the field of RF materials characterization. The examples in this chapter show that devices utilizing CEM inversion provide measurement capabilities that cannot be achieved with the more mature transmission-line or free-space techniques. For that reason, their use will likely grow, both in laboratory situations, but more importantly in field or manufacturing applications.

References

[1] Harrington, R. F., *Field Computation by Moment Methods*, Piscataway, NJ: Wiley-IEEE Press, reprint, May 1993.

[2] Yee, K. S., "Numerical Solution of Initial Boundary Value Problems Involving Maxwell's Equations in Isotropic Media," *IEEE Trans on Antennas and Propagation*, Vol. AP-14, No. 3, May 1966, pp. 302–307.

[3] Chew, W. C. et al., *Fast and Efficient Algorithms in Computational Electromagnetics*, Norwood, MA: Artech House, Inc., 2001.

[4] Sumithra, P., and D. Thiripurasundari, "A Review on Computational Electromagnetics Methods," *Advanced Electromagnetics*, Vol. 6, No. 1, March 2017, pp. 42–55.

[5] "CST Studio Suite Electromagnetic Field Simulation Software," Waltham MA: Dassault Systems Brochure, 2021; "QuickWave Complete EM Simulation," Warsaw, Poland: QWED Brochure, 2020; "Ansys HFSS," Cannonsburg, PA: Ansys Brochure, 2011; "EM Simulation Numerical Methods for Comprehensive Design," State College PA: Remcon Brochure, 2022.

[6] Tavlov, A., et al., *Computational Electrodynamics, The Finite-Difference Time-Domain Method* (Third Edition), Norwood, MA: Artech House, 2005.

[7] Deshpande, M. D., et al., "A New Approach to Estimate Complex Permittivity of Dielectric Materials at Microwave Frequencies Using Waveguide Measurements," *IEEE Transactions on Microwave Theory and Techniques*, Vol. 45, No. 3, March 1997, pp. 359–366.

[8] Hyde, IV, M. W., et al., "Nondestructive Electromagnetic Material Characterization Using a Dual Waveguide Probe: A Full Wave Solution," *Radio Science*, Vol. 44, No. 3, June 2009, pp. 1–13.

[9] Amert, A. K., and K. W. Whites, "Characterization of Small Specimens Using an Integrated Computational/Measurement Technique," *IEEE Antennas and Propagation Society International Symposium (AP-S/USNC-URSI)*, Orlando, FL, July 2–13, 2013, pp. 706–707.

[10] Schultz J. W., and J. G. Maloney, "A New Method for VHF/UHF Characterization of Anisotropic Dielectric Materials," *Antenna Measurement Techniques Association (AMTA) Symposium Proceedings*, Long Beach, CA, Oct. 11–16, 2015.

[11] Geryak, R. D., et al., "New Method for Determining Permittivity of Thin Polymer Sheets," *Antenna Measurement Techniques Association (AMTA) Symposium Proceedings*, Daytona FL, Oct. 24–9, 2021.

[12] Stuchly, M. A., and S. S. Stuchly, "Coaxial Line Reflection Methods for Measuring Dielectric Properties of Biological Substances and at Radio and Microwave Frequencies—A Review," *IEEE Trans. Instrumentation and Measurement*, Vol. IM-29. No. 3, Sept. 1980, pp. 176–183.

[13] Pelster, R., "A Novel Analytic Method for the Broadband Determination of Electromagnetic Impedances and Material Parameters," *IEEE Trans Microwave Theory and Techniques*, Vol. 43, No. 7, 1995, pp. 1494–1501.

[14] Schultz, J. W., "Anomalous dispersion in the Dielectric Spectra of Conductive Materials," *IEEE Transactions on Instrumentation and Measurement*, Vol. 47, No. 3, June 1998, pp. 766–768.

[15] Choi, C. Y., "Capacitive Plate Dielectrometer Method and System for Measuring Dielectric Properties," U.S. Patent 20080111559A1, May 2008.

[16] "1-Port Series," Indianapolis, IN: Copper Mountain Technologies Data Sheet, 2022.

[17] "Network Analyzer Products Catalog," Santa Rosa CA: Keysight, 2021; "Measurement Excellence that Drives Innovation," Munich, Germany: Rhode & Schwartz Network Analyzer Portfolio, 2019; "Vector Network Analysis Product Portfolio," Morgan Hill, CA: Anritsu, 2022.

[18] Baker-Jarvis, J., et al., "Dielectric and Conductor-Loss Characterization and Measurements on Electronic Packaging Materials," NIST Technical Note 1520, July 2001.

[19] Wasylyshyn, D. A., "Effects of Moisture on the Dielectric Properties of Polyoxymethylene (POM)," *IEEE Trans. Dielectrics & Elec. Insulation*, Vol. 12, No. 1, Feb. 2005, pp. 183–193.

[20] Schultz, J. W., et al., "A Comparison of Material Measurement Accuracy of RF Spot Probes to a Lens-Based Focused Beam System," *Antenna Measurement Techniques Association (AMTA) Symposium Proceedings*, Tucson AZ, Oct. 12–17, 2014.

8

Impedance Analysis and Related Methods

8.1 Impedance Analysis

Impedance analysis methods are based on the idea that a material specimen in a fixture can be configured to be accurately modeled with circuit elements. A circuit model includes some combination of resistors, capacitors, and inductors. Chapter 2–7 discussed measurement methods that use VNAs to determine scattering parameters, and the fixtures are usually described in terms of transmission-line models. In contrast, the methods in this chapter use measurement instrumentation that follows the paradigm of measuring complex impedance. Impedance analysis equipment often operates at lower frequencies—less than 1 GHz—and frequently even at kilohertz frequencies and below. At these frequencies, many materials have frequency-dependent behaviors that provide insights into the microscopic phenomena behind their macroscopic behaviors. The methods outlined in this chapter, which include both dielectric and magnetic techniques, are summarized in Table 8.1.

8.2 Dielectric Spectroscopy

The field of study that focuses on frequency-dependent dielectric properties over broad bandwidths is known as dielectric spectroscopy. Some of the

Table 8.1
Summary of Impedance Analysis and Related Methods

Method	Description
Parallel-plate capacitor	Specimen sandwiched between two parallel conductive electrodes.
	Determines dielectric permittivity of solids.
Concentric-cylinder capacitor	Specimen between two concentric cylindrical electrodes.
	Determines dielectric permittivity of liquids.
Interdigitated capacitor	Specimen on top of coplanar interdigitated electrodes.
	Determines dielectric permittivity or impedance during processes such as cure and film formation; can be nondestructive or embedded sensor.
High-frequency capacitor	Extension of parallel-plate capacitor method to higher frequencies by better accounting for parasitic impedances.
	Determines dielectric permittivity of solids.
Toroidal-inductance permeameter	Measures inductance of a toroidal-shaped specimen.
	Determines magnetic permeability of solids.
Thin-film permeameter	Measured inductance by inserting specimen in a current loop.
	Determines magnetic permeability of solids (thin films).

earliest dielectric measurements were made by Drude [1], who among other things, studied dielectric dispersion behavior of materials at optical frequencies. One of the first and most well-known works to relate experimental dielectric response to molecular phenomena was published by Debye [2], who won the 1936 Nobel Prize in chemistry for his work. Debye's text, along with other works [3–7], dealt primarily with the dielectric behavior of small molecules. Dielectric spectroscopy of polymers and other complex materials were studied more extensively in the 1950s and 1960s [8–10].

Dielectric spectroscopy is now a well-established method for studying not only electronic properties, but also for providing fundamental understanding of molecular dynamics processes in ceramics, polymers, composites, and other complex material systems [11, 12]. In this way, dielectric spectroscopy is analogous to another thermal analysis method called dynamic mechanical analysis (DMA). While DMA uses mechanical oscillation to probe frequency or temperature dependencies, dielectric spectroscopy uses an electrical oscillation. From this analogy, dielectric spectroscopy is sometimes also called dielectric analysis (DEA).

One advantage of dielectric spectroscopy is the extreme breadth of the frequency range (<1–~10^{12} Hz) over which it can be used. Although different techniques and fixtures are required to span this entire range, these many decades of timescale enable the probing of a wide variety of molecular processes. Dielectric spectroscopy can also follow molecular dynamics from low-viscosity liquids to rubbery solids to hard glassy solids. It can provide information on miscibility of mixtures, onset of flow, ionic transport, molecular relaxations, thermal transitions, and chemical reaction rates. In some cases, dielectric measurements have better sensitivity to changes in material properties than other thermal-analysis techniques. Sections 8.2–8.3 give a broad overview of dielectric spectroscopy, and more detailed information can be found in [1–14] or in the growing base of literature.

In dielectric spectroscopy, a sample is placed in contact with two or more electrodes with an applied, time-varying voltage. In most cases, these voltages and the resulting currents are measured to determine electrical impedance, and the impedance then relates to sample permittivity after accounting for the electrode geometry. With the extreme frequency range over which dielectric measurements can be made, there is no single instrument that covers the entire range. Measurements can be made either in frequency domain or in time domain. In frequency domain, sinusoidal voltages perturb the material, and its frequency-dependent responses are measured as complex impedance values. While higher-frequency measurements use network analyzers, lower-frequency impedance analysis instrumentation can be subcategorized into two primary methods: frequency-response analyzers and impedance bridges. Frequency-response analyzers use two phase-sensitive voltmeters to measure the voltage and current response of the specimen. Impedance bridges use an adjustable compensation impedance to balance the unknown specimen in a bridge circuit to determine its impedance. Additional information on these techniques can be found in reviews by Kremer and Arndt [13] and Pochan et al. [14]. In time-domain spectroscopy, a voltage step is applied to a sample, and its response is measured as a function of time. Conversion to frequency domain data is accomplished via a Fourier transform. Time-domain measurement in dielectric spectroscopy is analogous to stress-relaxation measurements in mechanical analysis. While the time-domain technique can extend to microwave frequencies, it is most accurate at very low frequencies (subhertz to tens of hertz).

8.2.1 Dielectric Parameters

Dielectric spectroscopy involves a sample in contact with a pair of electrodes. A voltage applied across the electrodes creates an E-field within the specimen, so it polarizes in response to that voltage. The dielectric apparatus then directly or indirectly measures the current across the electrodes. When the applied voltage is sinusoidal, the response current is also sinusoidal, as shown for a hypothetical material in Figure 8.1. The measured current is the same frequency or period as the applied voltage but is shifted along the timescale. After conversion into frequency domain, this time delay is equivalent to a phase shift. The quantities that relate the current response to dielectric material quantities are the phase shift, δ, and relative amplitude. The real part of the time-waveforms for the voltage, $V(t)$, and current, $I(t)$, can be expressed as

$$\mathrm{V}(t) = V_0 \cos(\omega t) \tag{8.1}$$

$$\mathrm{I}(t) = I_0 \cos(\omega t + \delta) \tag{8.2}$$

where V_0 and I_0 are the amplitudes of $V(t)$ and $I(t)$, and ω is the angular frequency of the sinusoidal waveform in radians per second.

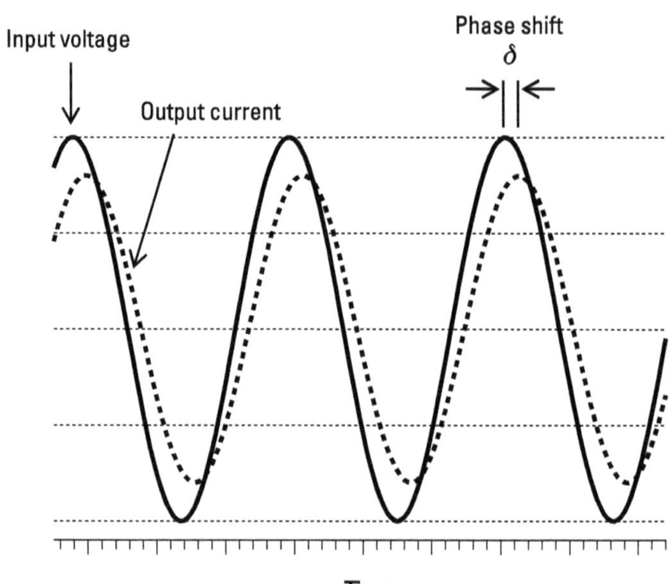

Figure 8.1 Voltage/current response of a material measured with impedance analysis.

For a non-steady-state system such as sinusoidal oscillation, voltage and current are complex quantities. For later analyses, voltage and current are better expressed in terms of their ratio, also known as impedance. In other words, the sample and electrodes can be represented as a set of individual or lumped-circuit elements with a characteristic impedance, Z. Because the response is measured at a nonzero frequency, the impedance is complex valued,

$$Z = \frac{V}{I} = Z' + iZ'' \tag{8.3}$$

where the prime indicates the real part, and the double-prime indicates the imaginary component. Based on the electrode geometry, the measured impedance can then be converted to real and imaginary dielectric permittivity.

An impedance-measurement device such as a bridge circuit must balance the specimen with known circuit elements. So materials within the electrodes are usually modeled as electrically equivalent to a resistor and capacitor connected in parallel, as shown in Figure 8.2, which provides a model that accounts for the complex behavior of many materials. Other more complicated circuits have also been used to model dielectric behavior [15]; however, the parallel circuit shown in Figure 8.2 is appropriate for a great many materials. While less common, it is also possible to use a capacitor and resistor in series to model materials, and an equivalence between the series and parallel circuit models can be determined [10].

With elementary circuit analysis, the capacitance, C, and resistance, R, of the parallel model in Figure 8.2 are related to the complex impedance by

$$\frac{1}{Z} = \frac{1}{R} - i\omega C \tag{8.4}$$

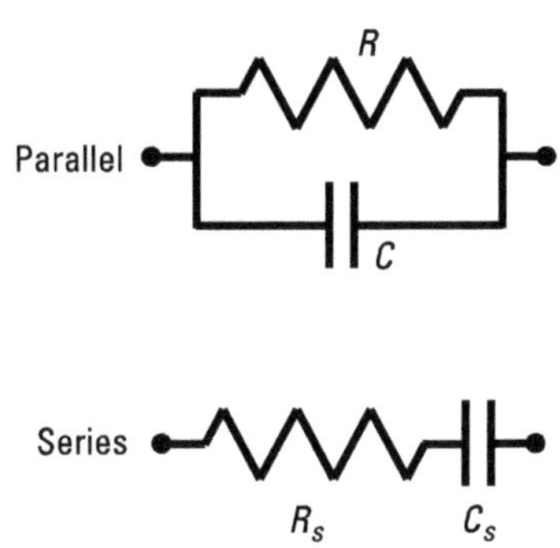

Figure 8.2 Parallel- and series-circuit models that can be used to characterize many material specimens in impedance analysis.

Equation (8.4) can be rearranged and combined with (8.3) to get

$$Z' = \frac{R}{1+\omega^2 R^2 C^2} \tag{8.5}$$

$$Z'' = \frac{-\omega R^2 C}{1+\omega^2 R^2 C^2} \tag{8.6}$$

Combining (8.5) and (8.6) yields the following result

$$\left(Z' - \frac{R}{2}\right)^2 + (Z'')^2 = \frac{R^2}{4} \tag{8.7}$$

which is the equation for a circle of radius, $R/2$, centered at $Z' = R/2$ and $Z'' = 0$. Assuming that the specimen under test is capacitive in nature, then C is positive and Z'' is negative. Figure 8.3 plots this equation in the complex impedance plane. This is also sometimes called a Nyquist plot, and common convention calls for plotting the negative of the imaginary impedance versus real impedance. The semicircle curve assumes that R and C are constant as a function of frequency. However, when this simple parallel model is applied to real data, R and C values are usually not exactly constant, resulting in semicircles that are skewed with centers somewhat below the Z' axis. As will be shown later (e.g., in Section 8.23), complex impedance plots such as Figure 8.3 are helpful for obtaining insight into the processes that contribute to a sample's dielectric behavior.

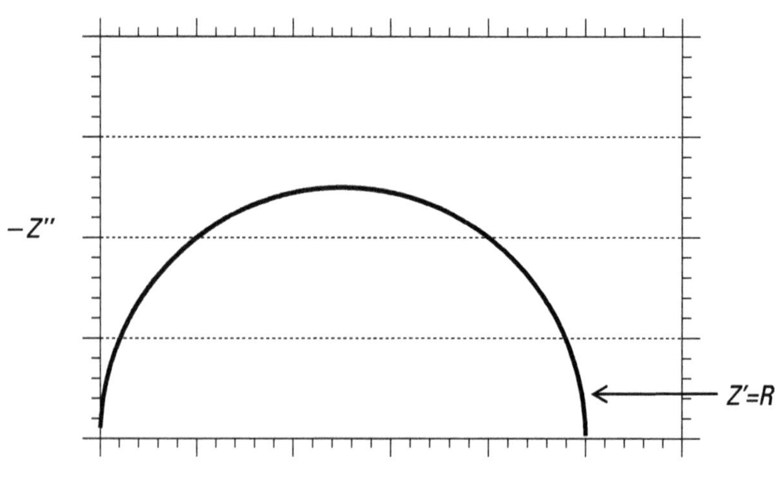

Figure 8.3 Parallel RC model plotted on the complex impedance plane.

8.2.2 Electrode Fixtures

The most common impedance-measurement geometry is two parallel planar electrodes that sandwich a thin, flat material specimen. This parallel-plate configuration is shown in Figure 8.4. The parallel plate geometry is popular because of the simplicity with which dielectric permittivity can be calculated. Using the area of the electrodes, A, and the spacing between them, d, the permittivity can be derived in the electrostatic limit from Gauss' law, and is a simple function of the measured capacitance, C,

$$\varepsilon_r = \frac{Cd}{\varepsilon_0 A} \tag{8.8}$$

Equation (8.8) is ideal and ignores fringing fields at the edges of the parallel plates; moreover, (8.8) is extended to frequency-dependent permittivity by allowing the capacitance and, therefore, the relative permittivity to be complex. It can also be generalized by translating the complex capacitance to impedance by $Z = -i/\omega C$, which leads to

$$\varepsilon_r = \frac{(Z'' - iZ')d}{\omega |Z|^2 \varepsilon_0 A} \tag{8.9}$$

where ω is the angular frequency and $Z = Z' + iZ''$.

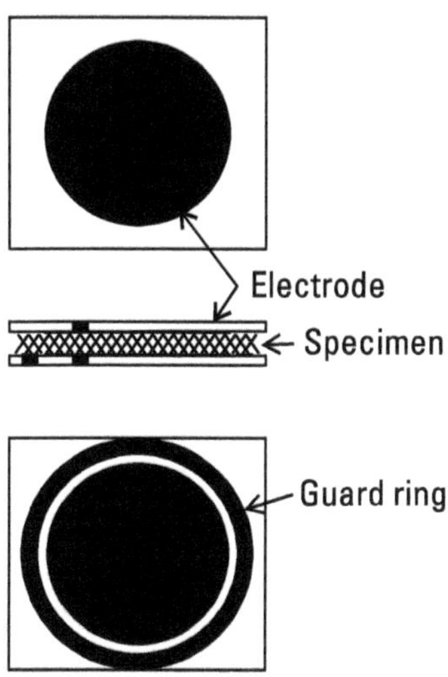

Figure 8.4 Parallel-plate electrode geometry, which is typical of impedance-analysis measurements for dielectric spectroscopy.

The finite size of the electrodes can cause an adverse effect on the measured impedance due to fringe fields along the electrode perimeter. Designing the electrode separation or specimen thickness, *d*, to be much smaller than the horizontal dimensions can minimize this. In addition, incorporating a guard ring into one of the electrodes, as shown in Figure 8.4, can reduce fringe-field effects. The most effective guard electrode has a width at least twice the thickness of the specimen, and the unguarded electrode must extend to the outer edge of the guard [9]. Even with a guarded electrode, best accuracy is achieved when calibrating the electrode area by first measuring the impedance or capacitance in air (i.e., without the sample). However, this is not always practical.

When calculating permittivity from unguarded parallel-plate electrode measurements, analytical corrections can be applied to the data. Many of these corrections are described in detail in the measurement standard of ASTM D150 [16]. For good accuracy, the samples should be flat and in good contact with the electrodes. Air gaps between the electrodes and sample cause the effective capacitance to decrease resulting in a reduction of the calculated permittivity. Electrodes can be constructed of metal and placed in contact with the sample, or they may be evaporated, sputtered, or painted directly onto the sample. Guidance for electrode construction and application is found in a number of ASTM standards [16–18].

Another electrode geometry that is more appropriate for liquids or solutions is the concentric-cylinder configuration, pictured in Figure 8.5. Calibration for these types of fixtures measures the empty cell or the cell filled with a nonpolar liquid with a known dielectric constant. The cylindrical geometry leads to the following expression relating permittivity to capacitance

$$\varepsilon_r = \frac{(Z'' - iZ')\ln(b/a)}{\omega |Z|^2 \varepsilon_0 2\pi l} \tag{8.10}$$

where *a*, *b*, and *l* are dimensions defined in Figure 8.5. Some liquids have high ion mobility, so it is desirable to use a four-probe arrangement in these cases where the electrodes are arranged with two outer electrodes inducing the sinusoidal voltage and two inner electrodes measuring the response. This four-probe arrangement reduces anomalous polarization artifacts, called electrode blocking. In both the parallel plate and concentric cylinder geometries, the loss tangent, which is defined as the imaginary permittivity divided by the real, can also be directly related to the real and imaginary impedance by

$$\tan\delta = \frac{\varepsilon''}{\varepsilon'} = \frac{Z'}{Z''} \tag{8.11}$$

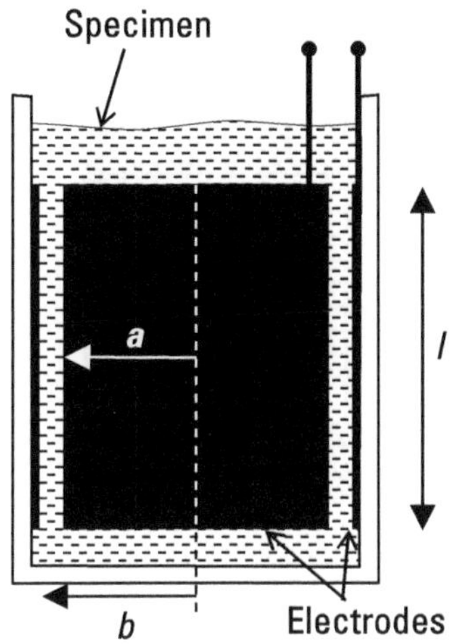

Figure 8.5 Concentric-cylinder electrode geometry, typical of impedance analysis measurements for liquids.

which is derived by simple rearrangement of either (8.9) or (8.10).

A third electrode geometry, useful for measuring dynamic changes in material systems such as polymer cure monitoring, is the interdigitated-comb electrode. This type of electrode was first introduced by Armstrong [19] and refined by later researchers [20–22]. A simple schematic of such an electrode is shown in Figure 8.6. Unlike the parallel-plate electrodes, the geometry of comb electrodes is complicated, and large fringe fields exist in the substrate of the electrodes. Hence sophisticated conformal mapping or numerical methods are required to quantitatively calculate dielectric permittivity [23, 24]. Their primary advantage is that they can be mounted noninvasively to the surface of a structure or can be easily incorporated into laminated structures such as fiber-reinforced composites or adhesive joints for cure monitoring [25, 26]. They have also found application in chemical sensing [27] and in biological systems [28, 29]. In some of these applications, valuable information such as molecular relaxation or changes in composition can be monitored with just the raw impedance data, so that quantitative calculation of permittivity is not necessary.

8.2.3 Error Sources

Experimental measurements are sometimes subject to instrument artifacts that can be difficult to distinguish from actual material behavior. This section

Figure 8.6 Schematic representation of an interdigitated electrode for dielectric spectroscopy.

briefly describes some of the assumptions and instrument effects that can occur when making dielectric measurements. To begin, a fundamental assumption that is usually made is that the measurement is linear. In other words, the measured impedance is independent of applied voltage, which is usually true for most conditions. However, it is possible for dielectrics to experience breakdown at high enough voltages (> 10^6 V/cm) and when electrode spacing is small, the applied voltage does not have to be that large to get to this condition. The measured behavior can also become nonlinear when there are electrochemical reactions resulting from the applied voltage, which is also restricted to higher voltages.

A second assumption typically made when interpreting dielectric data is that the measured property is time-invariant. However, when the measurement follows changes in sample properties as a function of time or temperature, time invariance is not strictly followed. In this case, the question of time invariance becomes one of determining whether the sample properties are approximately constant within the time it takes to measure the impedance at a given frequency. Instruments may average over several cycles at each frequency step, so the time for a given measurement is the number of averaging cycles times the inverse of the frequency. For high frequencies, this measurement time is still less than a fraction of a second. However, for frequencies near or below 1 Hz, this time can become significant.

Another important effect that may occur in conductive materials is the blocking of charge carriers by the electrodes. For example, ionic conductors with a short distance between the electrodes at sufficiently low frequencies may experience a pileup of negative ions at the positive electrode or positive ions at the negative electrode during each cycle of the periodic voltage. Blocked electrode effects can obscure the bulk properties of the sample. This blocking effect is especially prevalent in samples where ion mobility is high, such as with low–molecular weight materials or at high temperatures [30, 31]. It can also happen in materials that have some electron conduction and when there is a resistive contact barrier between the specimen and the electrodes. As charges accumulate at the electrodes, the sample becomes polarized, and a large false contribution to the dielectric constant is measured.

Besides the sample itself, two major experimental factors may be varied to control the electrode polarization. The first factor is the timescale—as the period of the oscillating voltage is increased, the charges have more time to accumulate at the electrodes. Thus, the blocking effect is minimized by measuring at higher frequencies (shorter time scales). The specific frequency at which blocking becomes important depends on the concentration of charge carriers and the material viscosity or charge mobility. The second factor that influences electrode polarization is the sample geometry. As the electrode separation decreases, the amount of charge carriers that pile up in a given cycle increases. So, increasing the spacing between electrodes is a way to minimize blocking effects. On the other hand, increasing the electrode spacing can have other adverse effects, such as decreased sensitivity and increased fringe fields. Thus, optimization of the electrode spacing is a compromise that must also account for the specimen properties and frequency range of interest.

Since blocking effects cause erroneous results, detecting when blocking occurs is an important part of data interpretation. To understand the effect of blocking, the parallel resistor and capacitor model of Figure 8.2 can be modified. Though there exist more complicated models of blocking effects [15], the simplest model of blocking is made by adding a second capacitance in series with the original circuit. This model is pictured in Figure 8.7, and the resulting real and imaginary impedance is given by

$$Z' = \frac{R_{\text{bulk}}}{1+\omega^2 R_{\text{bulk}}^2 C_{\text{bulk}}^2} \text{ and } Z'' = \frac{-\omega R_{\text{bulk}}^2 C_{\text{bulk}}}{1+\omega^2 R_{\text{bulk}}^2 C_{\text{bulk}}^2} + \frac{-1}{\omega C_{\text{block}}} \tag{8.12}$$

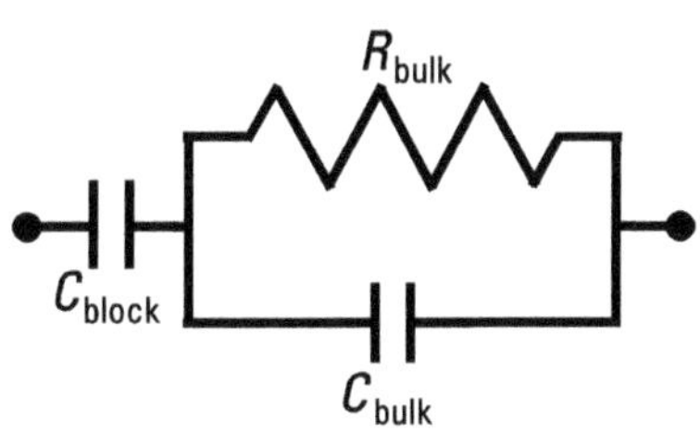

Figure 8.7 Circuit model of a material specimen that includes electrode blocking.

where the subscript bulk indicates material dependent properties, and the subscript block indicates the anomalous blocking effects.

Figure 8.8 shows a complex-impedance plot of (8.12), which can be compared to the nonblocking model in Figure 8.3. For the model that includes blocking, there is still a semicircle that corresponds to the impedance of the material under test. However, there is also a vertical line or tail on the right side of the plot, caused by electrode blocking effects. This tail is from the lowest-frequency data where the blocked charge carriers have time to accumulate. For comparison, Figure 8.9 shows actual data from a polymer latex emulsion while it is drying. This data was measured while the latex still had sufficient water in the film to allow high ion mobility. The tail is consistent with the blocking capacitor in the circuit model and demonstrates the usefulness of the model in diagnosing electrode blocking effects.

The blocking effect is dependent on the separation between electrodes, so it does not provide useful information about the specimen's intrinsic properties. Because it relates to how fast the ions accumulate at the electrodes, it is also time- (or frequency-) dependent. In Figures 8.8 and 8.9, each data point corresponds to a different measurement frequency, with the electrode-blocking effects at the lower frequencies where the ions have more time to accumulate. At higher frequencies, electrode blocking no longer occurs, and the data is a valid measure of the material's intrinsic characteristics.

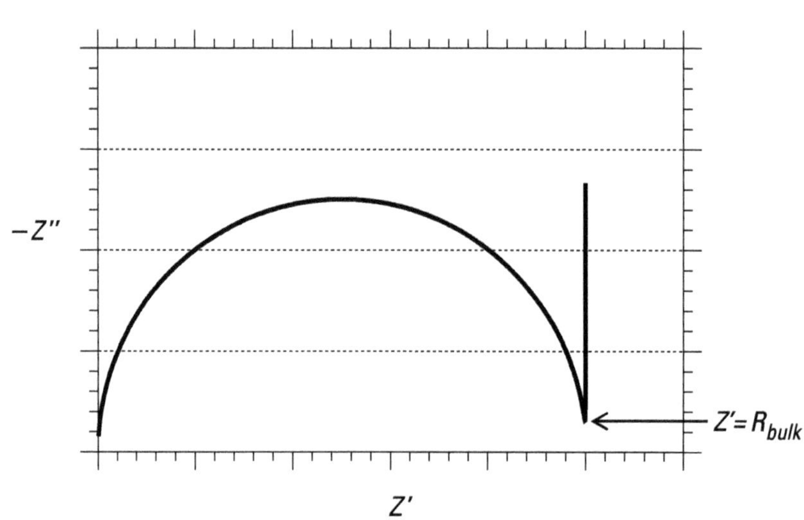

Figure 8.8 Complex impedance plot of the circuit model for a simple material with electrode blocking.

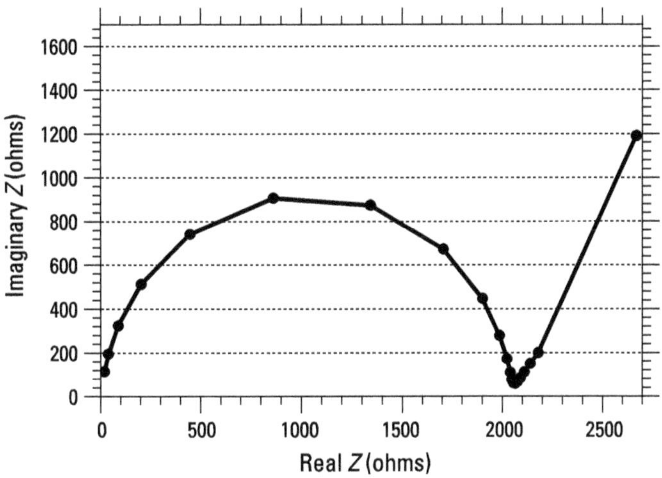

Figure 8.9 Measured complex impedance of a latex sample during drying, showing behavior qualitatively like the circuit model of Figure 8.8.

Another source of error that adversely affects measurement accuracy in more lossy or conductive specimens is parasitic impedance in the electrodes and their connection to the analyzer. Parasitic impedance of a device is undesirable behavior that can be described by adding additional circuit elements (resistive, capacitive, and/or inductive) to the circuit model of the fixture. For example, at sufficiently high frequencies, even carefully designed electrodes will have unwanted impedances from the wires connecting the sample electrodes to the measurement instrument. Wires will have some inductance depending on their radius and length. Junctions between the wires and the electrodes or to the impedance measurement instrumentation can have contact resistance. Figure 8.10 shows the complex impedance plot of a notional specimen under ideal conditions and when there is a 1-Ω contact resistance and a 10-nH inductance in series with the specimen. As this model shows, there can be both an offset in the Z' axis as well as a high-frequency tail when these parasitic effects are included. Such anomalous behaviors have been observed for conductive specimens in the literature [32, 33].

Figure 8.10 shows the importance of characterizing potential parasitic impedances in a fixture since they result in measurement error. The inductive effect is frequency-dependent and primarily affects the high-frequency behavior. In a complex impedance plot such as Figure 8.10, the data toward the left is higher in frequency, and the tail that crosses the zero resulting in

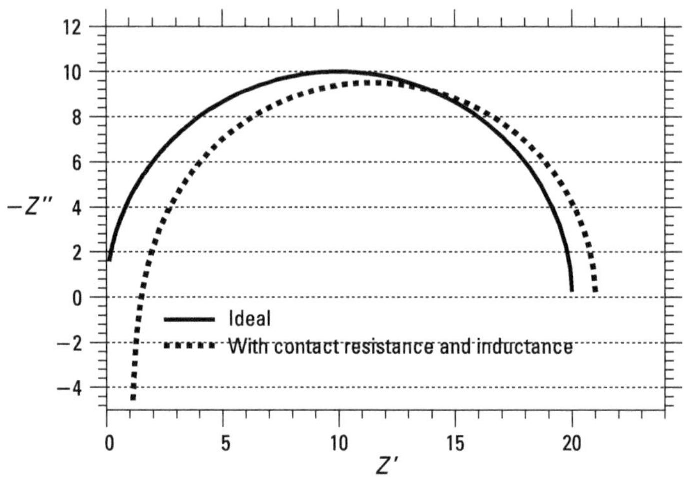

Figure 8.10 Complex-impedance plot of a circuit model describing an ideal sample and one where there is parasitic resistance and impedance associated with the electrode fixture.

net-positive imaginary impedance is from this parasitic inductance. On the other hand, contact resistance is a broadband effect impacting both low- and high-frequency impedance and is responsible for the rightward shift of the semicircle. Sometimes there can be parasitic capacitance as well, such as fringe fields around the edge of a capacitive fixture or blocking effects from charge-carrier accumulation at the material/electrode interface. In some cases, these effects dominate at low frequencies, and in others they are primarily at high frequencies. While complex impedance plots such as Figure 8.10 are helpful in diagnosing parasitic effects, additional analyses or modeling of the fixture may also be warranted to fully quantify these phenomena and their effect on the measurement accuracy.

8.3 Dielectric Spectroscopy Applications

Historically, one of the primary functions of dielectric spectroscopy is to characterize relaxation phenomena in materials. Relaxations are manifested by changes in properties, such as glassy to rubber hardness in polymers, so they are important for understanding the usefulness of a material for a given application. The frequency dependence of relaxation was described with general models such as the Debye equation (1.1) earlier in Chapter 1. Polymers and other dielectrics are more realistically described by variants of the Debye, also described in Chapter 1 (Section 1.3), and often a distribution of relaxation parameters is the most quantitative way to describe a transition.

While the real and imaginary permittivity are often plotted as a function of frequency, it can also be helpful to plot real and imaginary permittivity in a complex plane, much like the aforementioned complex-impedance or Nyquist plots. This idea was proposed by Cole and Cole [34] so these plots are sometimes referred to as Cole-Cole plots. Like the complex-impedance plots, they represent the data by plotting the imaginary permittivity, ε'' against the real permittivity, ε'. To illustrate this, we can plot the Cole-Cole generalization of the Debye along with an additional conductivity term,

$$\varepsilon = \varepsilon_U + \frac{\varepsilon_R - \varepsilon_U}{1 + (i\omega\tau)^{1-\alpha}} - i\frac{\sigma}{\omega\varepsilon_0} \tag{8.13}$$

where ε_U is the unrelaxed permittivity at high frequency, ε_R is the relaxed permittivity at low frequency, τ is the relaxation time constant, σ is the conductivity, and α is an empirical fitting parameter. Figure 8.11 shows a complex permittivity or Cole-Cole plot representation of (8.13) for different values of α. When $\alpha = 0$, (8.13) is the same as the Debye model and shows a perfect semicircle. Other values of α allow for a compressed semicircle with its center below the ε' axis. Since (8.13) also has a conductivity term, the Cole-Cole plot in Figure 8.11 also shows a rapid rise or tail in the imaginary permittivity on the right side due to this conduction. In cases where there are

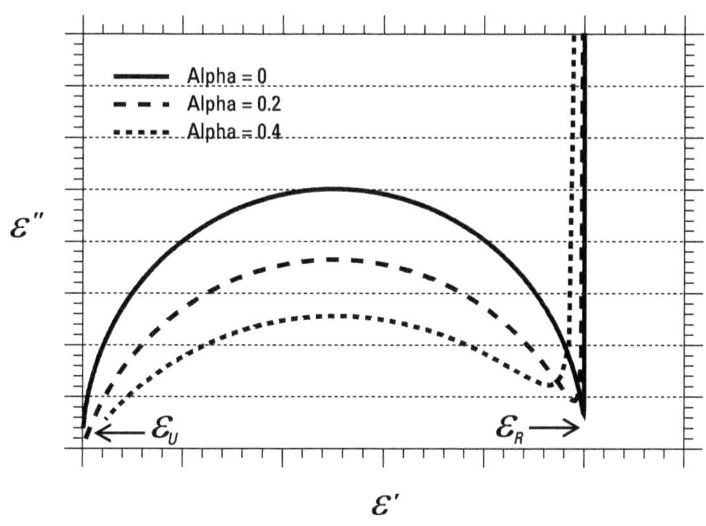

Figure 8.11 Cole-Cole plot of (8.13) for different values of α.

multiple relaxation phenomena, they will appear as separate semicircles on the Cole-Cole plot along the ε' axis. Thus, plotting data in this format provides a convenient way to differentiate the various relaxation and conduction phenomena for a given material.

Besides Cole-Cole, various researchers have suggested additional generalizations to the relaxation equations, and the Havriliak-Negami equation (1.5) also described in Chapter 1 (Section 1.3) is an example of this. It provides two additional parameters for better fitting measured dielectric data. Nonlinear numeric-regression methods must be applied to fit measured data to these functions, and the choice of the best empirical model can depend on the nature of the dielectric relaxation being studied. It is also important to scrutinize convergence criteria and confidence intervals to gain an understanding of the sensitivity of the various fitted parameters. Additionally, numeric experiments such as constructing additional input data sets with noise to reflect measurement uncertainties provides insight into the relative importance of the various fitting parameters.

8.3.1 Polymer Physics

An important aspect to polymer properties is their relaxation behavior. For example, amorphous polymers have a primary transition called the glass transition, sometimes labeled as the α-transition. Other relaxations can often be detected at lower temperatures or higher frequencies, and these may be labeled β, γ, and δ, in order of progression. While there continues to be debate as to the exact physical details of the glass transition, it is generally attributed to cooperative motion of polymer molecules. Relaxations at lower temperatures or higher frequencies are usually attributed to more localized motion of molecular segments. Semicrystalline polymers also have some amorphous phase, so there are glass and subglass relaxations in these materials analogous to the fully amorphous polymers. There are also additional high-temperature processes related to the crystalline phase. The positions of these transition temperatures are important for optimizing processing conditions in manufacturing. Other more subtle properties are also determined by relaxation behavior. For example, if a polymer is to be used for absorbing vibration, then it must have a large amount of damping at the temperature and frequency of use. Conversely, if a polymer is to be used in applications such as transmission of electromagnetic energy, then the damping or loss should be minimized.

Electrical-relaxation behavior measured by dielectric spectroscopy is analogous to the mechanical-relaxation behavior measured by DMA. For most transitions, both techniques show a step in the real part of the data and a peak in the imaginary part. However, while the real part of the dielectric permittivity goes from a low value to a high value, the real part of the dynamic mechanical modulus goes in the opposite direction [35]. So, when dielectric data is compared to DMA, it is common to convert the DMA data into mechanical compliance. Compliance as a function of frequency can also be fit to relaxation equations similar to those described in Section 1.3. Alternatively, the permittivity can be converted into dielectric modulus,

$$M = \frac{1}{\varepsilon} \tag{8.14}$$

An example comparing dielectric permittivity and modulus is shown in Figure 8.12. This data is for a thermoplastic polymer, polyethylene terephthalate, that was measured sandwiched between parallel-plate electrodes. The x-axis is temperature, but the specimen was actually heated from low to high temperature at a rate of 4 degrees Celsius per minute. Thus, it started as a glassy polymer and then went through the glass transition to the rubbery

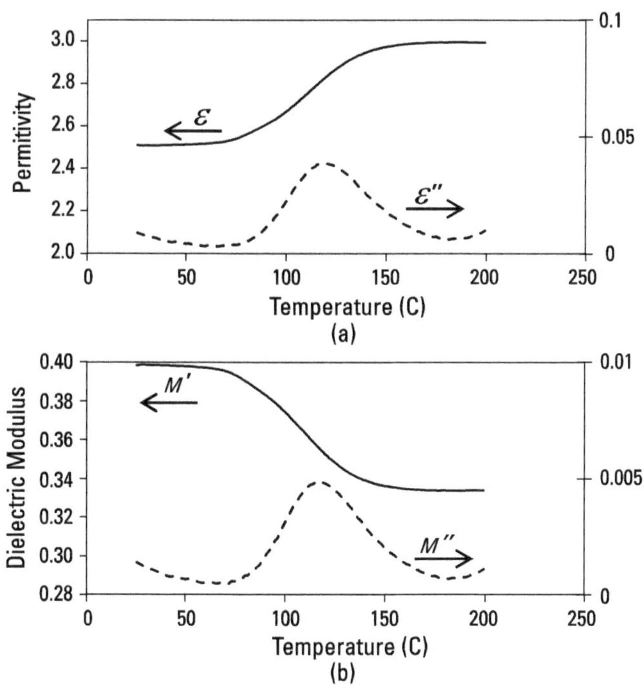

Figure 8.12 Comparison of complex permittivity and dielectric modulus of polyethylene terephthalate.

state at higher temperature. The plotted data represents the permittivity and modulus at 100 KHz, with the real permittivity and modulus on the left axis showing a step change, and the imaginary components or loss corresponding to the right axis. Near the glass transition, the material shows increased energy absorption. The dielectric modulus being higher at lower temperature and decreased at higher temperature is qualitatively analogous to mechanical modulus; however, it is not quantitatively equivalent. It is possible to have relaxations that are dielectrically strong but mechanically weak or vice versa. In other words, the mechanisms associated with relaxation may reflect differently for mechanical compliance versus dielectric permittivity. Thus, dielectric and mechanical spectroscopy are complementary techniques for measuring molecular relaxations in polymers.

As Figure 8.12 shows, polymer relaxations are studied not just in terms of frequency, but also in terms of their temperature dependence. Thus, dielectric spectroscopy often includes a way to heat or cool a specimen. In terms of temperature, the glass-transition behavior like that in Figure 8.12 can be described by the well-known Williams-Landel-Ferry (WLF) equation [10, 36],

$$\log\frac{f(T)}{f\left(T_g\right)}=\frac{-C_1\left(T-T_g\right)}{C_2+\left(T-T_g\right)} \tag{8.15}$$

where $f(T)$ is the relaxation rate at temperature, T, and C_1 and C_2 are fitted constants. Equation (8.15) has been used not only for dielectric spectroscopy, but also for various dynamic mechanical and rheological data. The glass transition can also be described with the Vogel-Fulcher equation [37],

$$\log f(T)=A-\frac{B}{T-T_0} \tag{8.16}$$

where T_0 is the Vogel temperature (usually 30–70C below T_g), and A and B are fitted constants. The form of (8.16) can be obtained by algebraic rearrangement of the WLF equation. For subglass relaxations, the relaxation rate is often better approximated with an Arrhenius relationship,

$$f(T)=Ce^{-\frac{E_a}{kT}} \tag{8.17}$$

where C is a fitted parameter, and E_a is the apparent activation energy of the transition.

Nonconductive materials can be described in terms of intrinsic dipoles, where frequency or temperature dependence of dielectric permittivity is a function of the dipoles' response to the incident field. Dipole movement may be from relative displacement of electrons along a molecular structure, or it can be from mechanical rotation or vibration of the molecular structure itself. It can also be caused by a variation in the dipole density, whether from a rearrangement of the dipoles between different semicrystalline or amorphous phases or from a change in the free volume within an amorphous polymer, which changes the dipole density and is therefore related to the physical density of the material [38]. Preferential alignment of dipoles within a polymer or even other organic materials such as liquid crystals can exist from strain, shear, or other orienting effects. The dielectric response may then be influenced by a change in the order or disorder of the aligned dipoles [39]. These mechanisms of order or disorder and molecular motion or vibration and density can also be framed in terms of thermodynamics. For example, the frequency dependence and/or relaxation time can indicate whether a dipole motion is due to highly local motion (faster) or cooperative motion (slower), which is related to a greater activation entropy [38].

When measured over a sufficiently broad frequency or temperature range, polymers may show multiple transitions. While the relaxation spectra may be thought of as a fingerprint of a particular polymer, it is dependent on more than just the chemical structure. Factors such as morphology and thermal history can have a significant effect on the relaxation spectra, so they should be considered or controlled when making measurements. For example, relaxation-peak positions and intensities in thermal measurements are dependent on the temperature scan rate of the experiment, as well as the thermal history of the polymer sample before the experiment was initiated.

Multiple relaxations may also occur when a polymer is a heterogeneous mixture. Polymer mixtures or blends have become prevalent in engineering applications because of their useful properties such as increased toughness. Most blends are multiphase, meaning that under a microscope, regions that are rich in one polymer are visibly separate from regions that are rich in a different polymer. It is therefore possible for a blend to have multiple glass transitions, each corresponding to a separate phase. There may be limited miscibility, however, resulting in a corresponding shift in the position of each glass transition. On the other hand, when a blend is completely miscible, there will only be a single phase and therefore only a single glass transition. The degree of mixing in a miscible blend can be estimated by applying the well-known Fox equation [40],

$$\frac{1}{T_g} = \frac{w_1}{T_{g1}} + \frac{w_2}{T_{g2}} \tag{8.18}$$

where w_1 and w_2 are the respective weight fractions of the two components, T_{g1} and T_{g2} are their respective glass transition temperatures (i.e., in pure form), and T_g is the shifted glass-transition temperature.

8.3.2 Cure and Process Monitoring

Polymers and polymer composites have become increasingly important in modern engineering applications. As a result, there has been an increasing need for monitoring the cure of polymers in various manufacturing processes. The dielectric technique has been refined to be more than just a laboratory measurement and is suitable for use in manufacturing processes. With suitable electrodes it can be used to monitor thermoset polymerization in situ [41]. Thus, it is applicable for process-control monitoring in industrial settings such as batch reactors, presses, autoclaves, ovens, and molding operations [42–45]. Dielectric spectroscopy has also been used to monitor UV cure of photopolymers [46].

Dielectric monitoring of cure indirectly measures the change in ion and dipole mobility in a resin as it polymerizes. Ion and dipole mobility are directly related to the viscosity of the resin, which is a function of molecular weight or degree of cure. Dielectric spectroscopy has two distinct advantages over direct mechanical measurement of the viscosity. First, a mechanical viscosity measurement is usually limited to either low viscosities (i.e., viscometer or rheometer) or high viscosities (i.e., DMA), while a dielectric measurement can work over both regimes. Second, with the small thin interdigitated electrode sensors similar to Figure 8.6, dielectric analysis can monitor cure in situ, while mechanical techniques require a separate sample with a specific geometry.

An example measurement of epoxy cure is provided in Figure 8.13, which shows the real and imaginary permittivity of a resin that is heated from room temperature at a rate of 4 degrees C/min. The real part of the permittivity drops from low-temperature to high-temperature as the crosslinking reaction takes effect. The imaginary permittivity or loss starts out low, but then increases as temperature increases, indicating a decreased viscosity within the uncured resin. Eventually the imaginary permittivity peaks, indicating a corresponding peak in the specimen conductivity at the height of the crosslinking reaction a little below 100C. At higher temperature the reaction completes, and the specimen solidifies into a cured polymer network with little ion or dipole mobility.

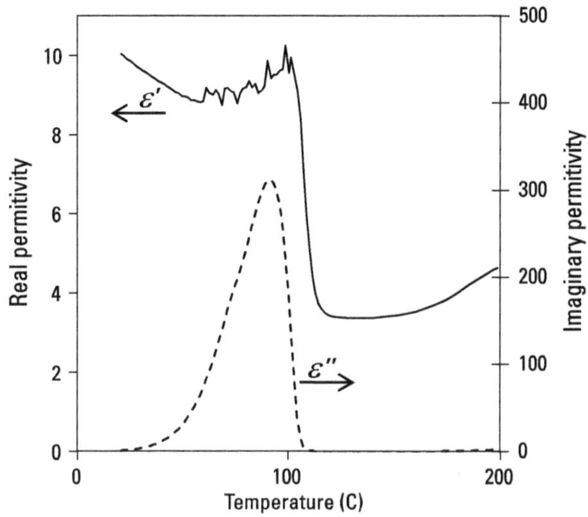

Figure 8.13 Dielectric spectroscopy of an epoxy resin curing as it is heated from room temperature to 200C.

Since dielectric monitoring of cure involves measurement of the ion and dipole mobility, a parameter of interest is the resistivity, which is directly related to the imaginary permittivity. Resistivity is technically a scalar quantity that is defined only for a constant current or zero frequency. However, an apparent resistivity, ρ, can be defined at arbitrary frequency by

$$\rho = \frac{1}{\sigma} = \frac{1}{\omega \varepsilon_0 \varepsilon_r''} \tag{8.19}$$

where σ is conductivity, ω is angular frequency, and ε_0 is the permittivity of free space.

With the dramatic changes that occur to dielectric properties during cure, dielectric spectroscopy is well-suited for determining chemical kinetic rate constants. For example, kinetics information can be obtained by assuming a correlation between extent of cure and various dielectric data, such as log of resistivity or relaxed permittivity. Vitrification can be determined by following the glass transition with frequency domain, relaxation, data. That said, these concepts should be used with caution, since empirical correlations can oversimplify the complexities of the curing reaction. In particular, cure process can include multiple steps occurring simultaneously, and a direct relationship between a dielectric parameter and cure kinetics does not always exist [31].

8.3.3 Film Formation and Environmental Effects

Another application for dielectric spectroscopy is in the dynamic characterization of film formation, such as with polymer-latex emulsions. With their ease of application and environmental friendliness, water-based latexes have become a popular alternative to solvent-based coatings. Characterization of these systems and their performance is an important need for industrial applications, and dielectric spectroscopy is a useful technique for measuring aspects of these material systems. For example, one way in which dielectric analysis has been proven effective is in the determination of latex particle size [47]. Dielectric analysis is also useful in the study of film formation or drying of polymer latices [48–50].

Figure 8.14 shows an example of the resistance measured from a latex coating versus drying time. This data was for a sample film cast on inter-digitated-comb electrodes. Since the film thickness shrinks as it dries, it is impractical to calculate a resistivity or complex permittivity in this type of measurement. Additionally, while the resistance data could have been calculated at a fixed frequency, increased range and accuracy were obtained by fitting the complex-impedance plot of each frequency scan to extrapolate the data to zero frequency. The data in Figure 8.14 show regions of differing slopes indicating discrete stages of drying: initial evaporation, percolation, and diffusion. Thus, even though it is qualitative, measurement data such as this can still be related to the underlying physical phenomena responsible for film formation. Relative trends such as this are often all that are required to understand the material behavior.

Materials also can undergo chemical and physical changes because of exposure to the environment. For example, radiation or high voltage may cause crosslinking or chain scission in polymers [51]; humidity may cause

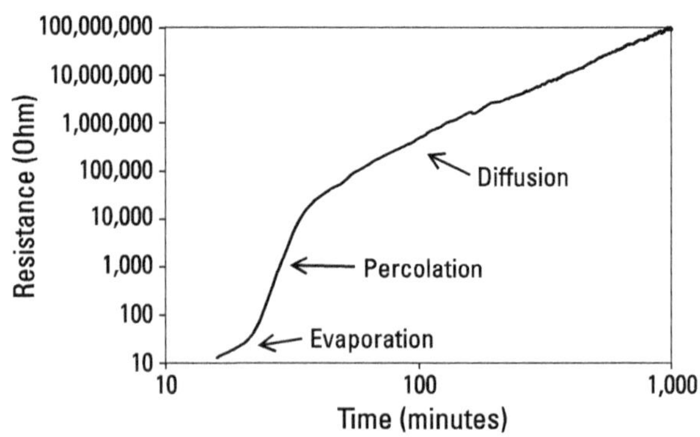

Figure 8.14 Resistance versus time for a polymer latex film measured as it dried.

swelling because of water sorption [52]; or materials may age because of additive migration. Dielectric spectroscopy can provide a useful way to monitor these changes. In these types of studies, samples may be measured at different times after exposure to different conditions, or a single polymer sample may be monitored continuously as it is undergoing change. When using interdigitated electrodes for in situ dielectric sensing, there is the additional advantage that measurements can be made without interrupting the experiment or process.

8.3.4 High-Frequency Dielectric Analyses

Even with techniques such as guarded electrodes and fringing-field corrections, the capacitance method is typically restricted to frequencies below 1–10 MHz. Above these frequencies, the parasitic impedances of the fixture begin to dominate the response. In some cases, the frequency range for this method has been extended to as high as 1 or 2 GHz by use of a coaxial line terminated with a carefully designed electrode that minimizes parasitic impedances. Such fixtures are available commercially and have been used by numerous researchers. However, accuracy at the upper frequency of the measurement range is eventually limited as the electrodes and specimen become a sizeable fraction of the operational wavelength and the simple capacitor approximation is no longer valid. Extending capacitance methods to high-frequency also requires a multistep calibration that compensates for both the RF analyzer and separately for the capacitance fixturing.

Alternatively, this idea of a capacitive termination of a coaxial line can be combined with the CEM-inversion techniques described in Chapter 7. For example, polymer and composite materials have been measured in a capacitive fixture combined with CEM-inversion to frequencies as high as 6 GHz [53]. A cross-section of a high-frequency fixture is shown in Figure 8.15. Like conventional dielectric analysis it has both a top and bottom electrode with a specimen sandwiched in between. Unlike conventional capacitive fixtures, however, the bottom electrode is at the end of a coaxial transmission line and is electromagnetically shielded. Data is collected from this fixture by making a reflection measurement when terminated by the capacitor at the end. Calibration is done by measuring a known material such as Teflon™. Then using the CEM-inversion method described in Chapter 7, the reflection amplitude and phase are compared to a table of computational simulation results for different combinations of thickness, and dielectric properties.

An example of the data that such a fixture can generate is shown in Figure 8.16. These two plots show real and imaginary permittivity measured of polyoxymethylene (POM), polyetherimide (PEI), and fiberglass, which

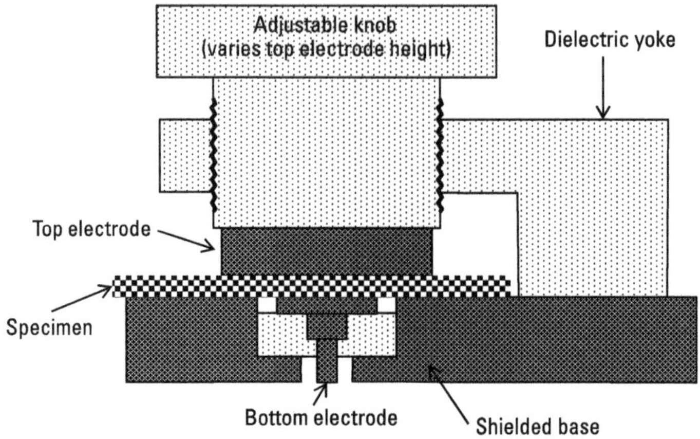

Figure 8.15 Cross-section drawing of a high-frequency capacitive fixture for measuring dielectric permittivity in the gigahertz range

is a nonwoven mix of glass fibers and epoxy. Consistent with the dispersion models described earlier, the more that the real part changes across the frequency range, the higher the imaginary part is. Most polymers are relatively low-loss at these frequencies; however, POM is known to have a more moderate dielectric loss. While this fixture does measure up to 6 GHz, the plotted data indicates that there is increased measurement uncertainty at frequencies above a few gigahertz, presumably because of additional parasitic impedances not accounted for by the computational model.

8.4 Permeameter Methods

Broadband-impedance analysis methods can also be applied to the measurement of magnetic permeability. Rather than a capacitive fixture, the measurement of magnetic properties requires an inductive one. Inductors consist of devices that direct current in a loop, and the inductance of such a device depends on the geometry of the current loop as well as on the magnetic permeability of what material is surrounded by the loop. Such a device is called a permeameter, and traditionally these methods were restricted to lower frequencies. However, more recent efforts have also extended these methods into the range of several gigahertz.

There are three prevalent permeameter design paradigms for characterizing microwave permeability as a function of frequency. One class of

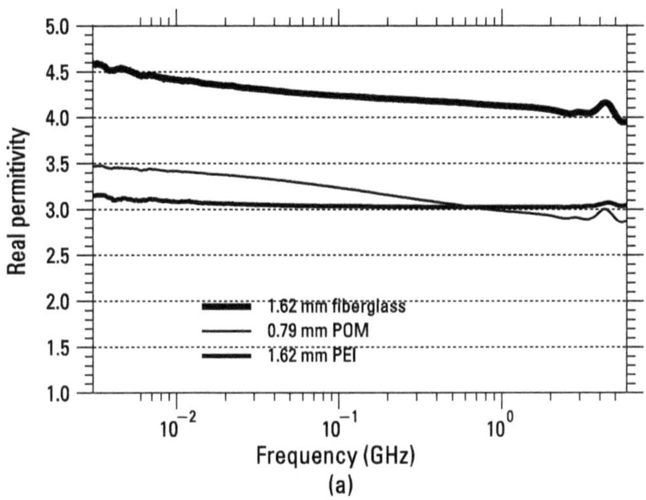

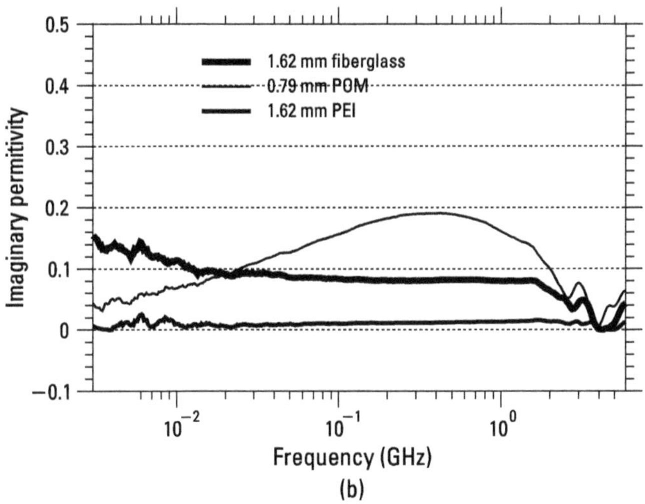

Figure 8.16 Real and imaginary permittivity of dielectric specimens measured with the high-frequency fixture of Figure 8.15.

permeameter is based on separate drive and pickup coils. The drive coil provides a time-dependent magnetic field in which a sample is placed. The pickup coil then senses the magnetic flux through the sample [54, 55]. A second permeameter methodology uses just a single coil and can be modeled with transmission line theory [56, 57]. Both techniques measure an apparent permeability and use a second measurement of a known sample to calibrate to actual permeability. Comparisons of these techniques [58] show that the two-coil technique performs better at lower frequencies while the single-coil/transmission line method is more accurate at higher frequencies.

A third alternative permeameter method is from a RF reflectometer constructed from a shorted coaxial airline where the specimen is a toroid and positioned adjacent to the short [59]. In this geometry, the current loop from the incident wave goes through the center of the toroid and around the outside. This method uses the known relationship between inductance, L, and permeability, μ_r/μ_0 of a toroid [60],

$$L = \frac{\mu_r t \ln(b/a)}{\mu_0 2\pi} \tag{8.20}$$

where t is the toroid thickness, and b and a are the outer and inner diameters respectively. The method measures the difference between the filled and unfilled fixture to determine a change of inductance, which is related to impedance. This then provides the necessary information for calculating μ_r.

While the toroid method works well for bulk magnetics, thin films are also a common configuration of these materials, and the previous two methods (drive/pickup coils and single loop) may be more appropriate. An example of the single-loop fixture for measuring films is sketched in Figure 8.17. This fixture works by making a short section of microstrip that ends in an electrical short. A film specimen is then placed between the ground plane and the upper conductor, and an analyzer measures the reflection with and without the specimen present. Unlike the toroid geometry, there is not a simple relationship between the magnetic permeability and impedance of the film. However, if the specimen is placed adjacent to the shorted end of the microstrip line, the following expression for the reflection scattering parameter can be written [61]

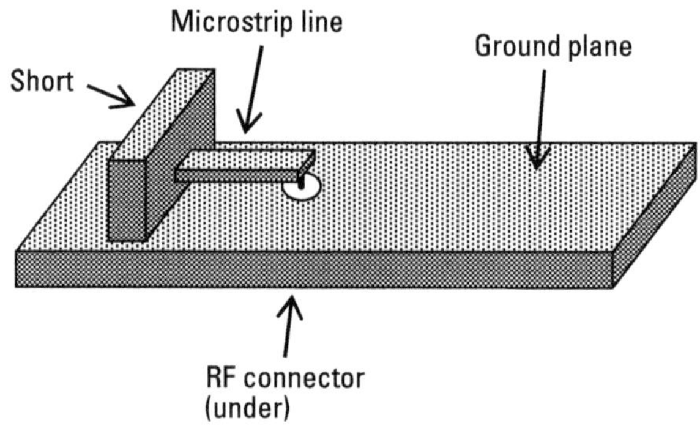

Figure 8.17 Notional sketch of a microstrip permeameter used to measure magnetic films.

$$S_{11} = \left(\frac{\tanh \gamma d_1 - \beta}{\tanh \gamma d_1 + \beta} \right) e^{-2\gamma_0 d_2} \text{ and } \beta = \frac{\gamma \mu_0}{\gamma_0 \mu^a} \quad (8.21)$$

where $\gamma = \gamma_0 \sqrt{\mu^a \varepsilon^a}$ is the usual propagation constant within the stripline transmission line and ε^a and μ^a are the relative, apparent permittivity and permeability, respectively. The apparent permittivity and permeability are functions of the actual permittivity and permeability, as well as the stripline geometry and specimen thickness. In addition, d_1 and d_2 are the widths of the two regions of the stripline shown in Figure 8.18.

When there is no specimen, the propagation constant of the empty permeameter can be derived from (8.21). With appropriate algebraic rearrangement, it leads to

$$\gamma_0 = \frac{-\ln\left(-S_{11}^{\text{empty}}\right)}{2\left(d_1 + d_2\right)} \quad (8.22)$$

In addition, a Taylor series expansion of the hyperbolic tangents around a small angle assumption can further simplify (8.21). Doing this and rearranging leads to

$$\mu^a = \frac{S_{11}^{\text{sample}} e^{2\gamma_0 d_2} + 1}{\gamma_0 d_1 \left(1 - S_{11}^{\text{sample}} e^{2\gamma_0 d_2}\right)} \quad (8.23)$$

Thus, measurements are made of both the empty permeameter and with the specimen inserted so that the above equations can calculate an apparent

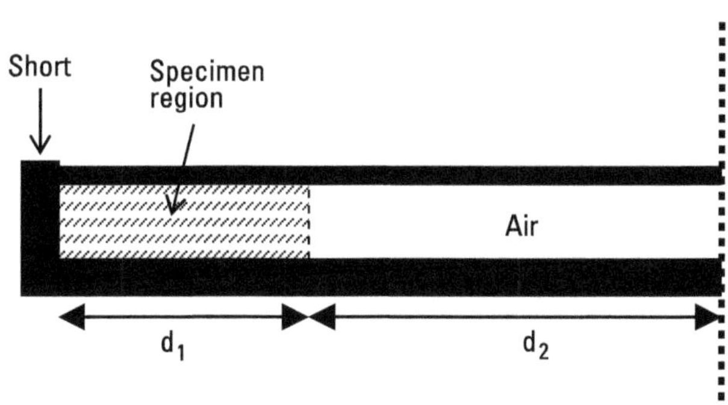

Figure 8.18 Transmission line representation of a microstrip permeameter with different regions.

permeability. It is also possible to insert a metal short in the middle of the stripline to act as the short circuit. In this case (8.22) must be adjusted to account for the alternate position of the short circuit. The disadvantage of this approach is that contact resistance between the shunt and the two conductors can be an error source. The advantage is that placing the reference plane closer to the front of the sample can potentially reduce some other uncertainties from the ideal propagation assumption.

The material specimen does not necessarily fill the space between the two conductors, so the apparent permeability will be lower than the actual permeability. Assuming a magnetic thin film supported on a dielectric substrate, the measurement cross-section will look something like Figure 8.19. In other words, the apparent permeability includes contributions not only from the magnetic specimen, but also the supporting substrate and surrounding air. Examples of measured apparent permeability for a commercially available magnetic foil are shown in Figure 8.20. This apparent permeability of the magnetic film shows a magnetic relaxation occurring within the measured frequencies (1 MHz–1 GHz). For comparison, Figure 8.20 also provides permeameter measurements of a nonmagnetic aluminum foil and paper, and they show apparent permeabilities of 1 with no significant imaginary part. All the data shows an anomalous upturn in the real apparent permeability at high frequencies. This permeameter has a microstrip line that is approximately 5-cm-long, and the highest measurable frequency is limited by that length. A higher-frequency permeameter can be constructed by reducing the length of the microstrip section so that it is electrically shorter.

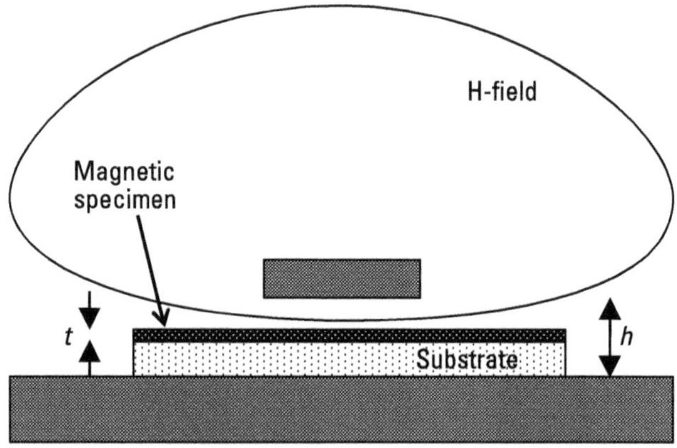

Figure 8.19 Cross-section of single loop permeameter measurement of a thin magnetic film.

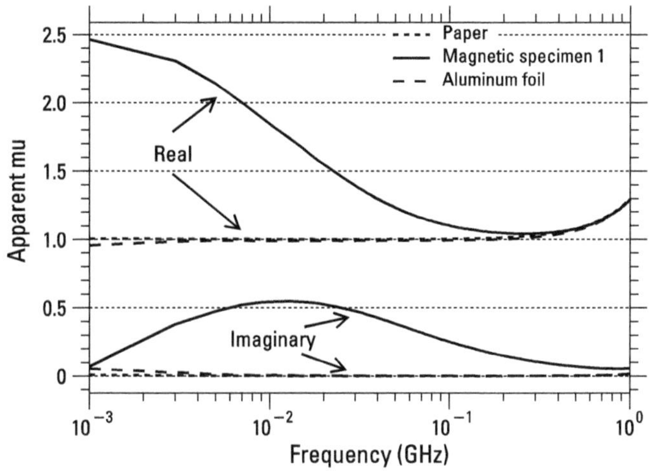

Figure 8.20 Measured apparent permeability of a commercially available magnetic metal film.

Translation of the apparent permeability into actual permeability then requires one more step to account for the physical dimensions of the specimen and the permeameter fixture. At the simplest level, a second measurement of a known material with approximately the same dimensions can be used to calibrate the apparent permeability [56]. This results in a proportionality constant, K, to relate the two permeabilities

$$\mu = K_1\left(\mu^a - 1\right) \tag{8.24}$$

A more sophisticated method expression can be used to better account for thickness differences between the known and unknown specimens [57],

$$\mu = \frac{1}{K}\frac{h}{t}\left(\mu^a - 1\right) \tag{8.25}$$

where h and t are defined in Figure 8.17. These expressions are only valid when the known specimen is somewhat similar to the unknown specimen under test. For example, they fail in the limit when $\mu^a \to 1$, in which case the actual permeability goes to 0 instead of 1. More importantly, these expressions assume a linear relationship between the apparent and actual permeability. A more accurate model can be obtained by using computational electromagnetic simulations to study a given permeameter fixture geometry. This can still lead to a more general analytical expression that relates the two parameters [62],

$$\mu = \frac{1}{K}\frac{h}{t}\left(\frac{\mu^a - 1}{\mu^a(C-1)+1}\right)+1 \tag{8.26}$$

where C is another fitted constant. Alternately, the full CEM-inversion methods described above for the capacitive fixture or in Chapter 7 can be applied to this type of measurement fixture by simulating multiple combinations of specimen thickness and complex permeability for a given permeameter geometry to create an inversion database.

References

[1] Drude, P., *The Theory of Optics*, translated by C. Riborg Man and R. A. Millikan, Mineola, NY: Dover, 1902.

[2] Debye, P., *Polar Molecules*, Mineola, NY: Dover, 1929.

[3] Smyth, C. P., *Dielectric Behavior and Structure*, New York, NY: McGraw Hill, 1955.

[4] Smith, J. W., *Electric Dipole Moments*, London, United Kingdom: Butterworths, 1955.

[5] Fröhlich, H., *Theory of Dielectrics*, Oxford, United Kingdom: Oxford University Press, 1958.

[6] Böttcher, C. J. F. et al, *Theory of Electric Polarization, Volume 1, Dielectrics in Static Fields*, Amsterdam, Netherlands: Elsevier, 1973

[7] Böttcher, C. J. F., and P. Bordewick, *Theory of Electric Polarization, Volume 2, Dielectrics in Time-Dependent Fields*, Amsterdam, Netherlands: Elsevier, 1978.

[8] Von Hippel, A., *Dielectrics and Waves*, Hoboken, NJ: John Wiley & Sons, 1959.

[9] Von Hippel, A., *Dielectric Materials and Applications*, Hoboken, NJ: John Wiley & Sons, 1954.

[10] McCrum, N. G., et al., *Anelastic and Dielectric Effects in Polymeric Solids*, Hoboken, NJ: John Wiley & Sons, 1967; republished by Mineola, NY: Dover, 1991.

[11] Hunter Woodward, W. H. (ed.), *Broadband Dielectric Spectroscopy: A Modern Analytical Technique*, ACS Symposium Series; Washington, D.C.: American Chemical Society, 2021.

[12] Kremer, F., and A. Schonhals (eds.), *Broadband Dielectric Spectroscopy*, Berlin, Germany: Springer-Verlag, 2003.

[13] Kremer, F., and M. Arndt, "Broadband Dielectric Measurement Techniques," in *Dielectric Spectroscopy of Polymeric Materials*, J. P. Runt and J. J. Fitzgerald (eds.), Washington, D.C.: American Chemical Society, 1997, pp. 67–79.

[14] Pochan, J. M., et al., *Determination of Electronic and Optical Properties* (Second Edition), B. W. Rossitter and R. C. Baetzold, (eds.), Physical Methods of Chemistry Series, VIII, Hoboken, NJ: Wiley, 1993.

[15] Hill, R. M., and C. Pickup, "Barrier Effects in Dispersive Media," *J. Mater. Sci.*, Vol. 20, 1985, pp. 4431–4444.

[16] "Standard Test Methods for AC Loss Characteristics and Permittivity (Dielectric Constant of Solid Electrical Insulation," ASTM D 150, 2018.

[17] "Standard Test Method for Permittivity (Dielectric Constant) and Dissipation Factor of Solid Dielectrics at Frequencies to 10 MHz and Temperatures to 500 C," ASTM D 2149, 1996.

[18] "Standard Test Methods for Relative Permittivity and Dissipation Factor of Expanded Cellular Polymers Used for Electrical Insulation," ASTM D 1673, 1994.

[19] Armstrong, R. J., "Method of Laminating Employing Measuring the Electrical Impedance of a Thermosetting Resin," U.S. Patent 3,600,247, 1971.

[20] Senturia, S. D., and S. L. Garverick, "Method and Apparatus for Microdielectrometry," U.S. Patent 4,423,371, 1983.

[21] Kranbuehl, D. E., "Method of Using a Dielectric Probe to Monitor the Characteristics of a Medium," U.S. Patent 4710550, 1987.

[22] Mamishev, A. V., et al., "Interdigital Sensors and Transducers," *Proceedings of the IEEE*, Vol. 92, No. 5, May 2004, pp. 808–845.

[23] Lee, H. L., "Optimization of a Resin Cure Sensor," MS thesis, Massachusetts Institute of Technology, 1982.

[24] Kim, J. W., "Development of Interdigitated Capacitor Sensors for Direct and Wireless Measurements of the Dielectric Properties of Liquids," PhD dissertation, University of Texas at Austin, December 2008.

[25] Kranbuehl, D. E., "Cure Monitoring," in *International Encyclopedia of Composites*, S. Lee (ed.), New York: VCH, 1990, pp. 531–543.

[26] Sorrentino, L., et al., "Local Monitoring of Polymerization Trend by an Interdigital Dielectric Sensor," *Int. J. Advanced Manufacturing Technology*, Vol. 79, 2017, pp. 1007–1016.

[27] Santos-Neto, I. S., "Interdigitated Electrode for Electrical Characterization of Commercial Pseudo-Binary Biodiesel-Diesel Blends," *Sensors* 21, No. 21, 2021, p. 7288.

[28] Bao, X., et al. "Broadband Dielectric Spectroscopy of Cell Cultures," *IEEE Trans. On Microwave Theory and Techniques*, Vol. 66, No. 12, December 2018, pp. 5750–759.

[29] Claudel, J., et al., "Interdigitated Sensor Optimization for Blood Sample Analysis," *Biosensors*, Vol. 10, No. 12, 2020, p. 208.

[30] Fontanella, J. J, et al., "Electrical Relaxation in Poly(Propylene Oxide) With and Without Alkali Metal Salts," *Solid State Ionics*, Vol. 50, 1992, pp. 259–271.

[31] Senturia, S. D., and N. F. Sheppard, Jr., "Dielectric Analysis of Thermoset Cure," *Advances in Polymer Science*, Vol. 80, Berlin, Germany: Springer-Verlag, 1986.

[32] Jonscher, A. K., "Analysis of the Alternating Current Properties of Ionic Conductors," *J. Mater. Sci.*, Vol. 13, 1978, pp. 553–562.

[33] Schultz, J. W., "Anomalous Dispersion in the Dielectric Spectra of Conductive Materials," *IEEE Trans. on Instrumentation and Measurement*, Vol. 47, 1999, pp. 766–768.

[34] Cole, K. S., R. H. Cole, "Dispersion and Absorption in Dielectrics," *J. Chem. Phys.*, Vol. 9, 1941, pp. 341–351.

[35] Chartoff, R. P. J. D. Menczel, and S. H. Dillman, "Dynamic Mechanical Analysis (DMA)," in *Thermal Analysis of Polymers: Fundamentals and Applications*, J. D. Menczel and R. B. Prime (eds.), Hoboken, NJ: John Wiley & Sons, 2008.

[36] Ferry, D., *Viscoelastic Properties of Polymers* (Third Edition), Hoboken, NJ: Wiley, 1980.

[37] Vogel, D. H., "Das Temperaturabhaengigkeitsgesetz der Viskositaet von Fluessigkeiten," Physikalische Zeitschrift, Vol. 22, 1921, p. 645; Fulcher, G. S., "Analysis of Recent Measurements of the Viscosity of Glasses," *J. American Ceramic Society*, Vol. 8, No. 6, 1925, pp. 339–355.

[38] Vassilikou-Dova, A., and I. M. Kalogeras, "Dielectric Analysis (DEA)," in *Thermal Analysis of Polymers: Fundamentals and Applications*, J.D. Menczel and R. B. Prime (eds), Hoboken, NJ: John Wiley & Sons, 2008.

[39] Chartoff, R. P., J. W. Schultz, and J. S. Ullett, "Methods and Apparatus for Producing Ordered Parts from Liquid Crystal Monomers," U.S. Patent US6423260B1, 2002.

[40] Fox, T. G., "Influence of Diluent and of Copolymer Composition on the Glass Temperature of a Polymer System," *Bull. Am. Phys. Soc.*, Vol. 1, 1956, p. 123.

[41] Smith, N. T., and D. D. Shepard, "Dielectric Cure Analysis: Theory and Industrial Applications," *Sensors*, Vol. 12, No. 10, 1995, pp. 42–48.

[42] Sorrentino, L., et al., "Local Monitoring of Polymerization Trend by and Interdigitated Dielectric Sensor," *Int. Journal of Advanced Manufacturing Technology*, Vol. 79, 2015, pp. 1007–1016.

[43] Kranbuehl, D. E., et al., "In Situ Sensor Monitoring and Intelligent Control of the Resin Transfer Molding Process," *Polymer Composites*, Vol. 15, 1994, pp. 299–305.

[44] Crowley, T. J., and K. Y. Choi, "In-Line Dielectric Monitoring of Monomer Conversion in a Batch Polymerization Reactor," *J. Appl. Polymer Sci.*, Vol. 55, 1995, pp. 1361–1365.

[45] Kranbuehl, D., J. Rogozinski, and A. Meyer, "Monitoring the Changing State of a Polymeric Coating Resin During Synthesis, Cure, and Use," *Proceedings of the XXIVth International Conference in Organic Coatings*, Athens, Greece, 1998, pp. 197–211.

[46] Schuele, D., R. Renner, and D. Coleman, "Monitoring of Photopolymerization through Dielectric Spectroscopy," *Mol. Crys. Liq. Crys.*, Vol. 299, 1997, pp. 343–352.

[47] Sauer, B. B., et al., "Polymer Latex Particle Size Measurement through High Speed Dielectric Spectroscopy," *J. Applied Polymer. Sci.*, Vol. 39, 1990, pp. 2419–441.

[48] Cansell, F., et al., "Study of Polymer Latex Coalescence by Dielectric Measurements in the Microwave Domain: Influence of Latex Characteristics," *J. Appl. Poly. Sci.*, Vol. 41, 1990, pp. 547–563.

[49] Kranbuehl, D., et al., "In Situ Sensing for Monitoring Molecular and Physical Property Changes During Film Formation," *Film Formation in Waterborne Coatings*, T. Provder (ed.), Am. Chem. Soc. Sym. Ser. Vol. 648, 1996, pp. 96–117.

[50] Schultz, J. W., and R. P. Chartoff, "Dielectric and Thermal Analysis of the Film Formation of a Polymer Latex," *J. Coatings Technology*, Vol. 68, 1996, pp. 97–106.

[51] Kwaaitaal, T., and W.M.M.M. van den Eijnden, "Dielectric Loss Measurement as a Tool to Determine Electrical Aging of Extruded Polymeric Insulated Power Cables," *IEEE Trans. Electrical. Insulation*, Vol. EI-22, 1987, pp. 101–105.

[52] Maffezzoli, A. M., et al., "Dielectric Characterization of Water Sorption in Epoxy Resin Matrices," *Polymer Eng. Sci.*, Vol. 33, No. 2, 1993, pp. 75–82.

[53] Schultz, J. W., "A New Dielectric Analyzer for Rapid Measurement of Microwave Substrates up to 6 GHz," *Antenna Measurement Techniques Association (AMTA) Symposium Proceedings*, Williamsburg VA, November 4–9, 2018.

[54] Kawazu, T., et al., "A New Microstrip Pickup Coil for Thin-Film Permeance Meters," *IEEE Trans. Magnetics*, Vol. 30, No. 6, 1994, pp. 4641–643.

[55] Yamaguchi, M., et al., Development of Multilayer Planar Flux Sensing Coil and Its Application to 1 MHz–3.5 GHz Thin Film Permeance Meter," *Sensors and Actuators*, Vol. 81, 2000, pp. 212–215.

[56] Pain, D., et al., "An improved Permeameter for Thin Film Measurements up to 6 GHz," *J. Appl. Phys.*, Vol. 85, No. 8, 1999, pp. 5151–5153.

[57] Bekker, V., et al., "A New Strip Line Broad-Band Measurement Evaluation for Determining the Complex Permeability of Thin Ferromagnetic Films," *J. Magnetism and Magnetic Mat.*, Vol. 270, 2004, pp. 327–332.

[58] Yamaguchi, M., et al., "Cross Measurements of Thin-Film Permeability up to the UHF Range," *J. Magnetism and Magnetic Mat.*, Vol. 242–245, 2002, pp. 970–972.

[59] Hoer, C. L., and A. L. Rasmussen, "Equations for the Radiofrequency Magnetic Permeameter," *J. of Research of the NBS—C. Engineering and Instrumentation*, Vol. 67C, No. 1, January–March 1963.

[60] Goldfarb, R. B., and H. E. Bussy, "Method for Measuring Complex Permeability at Radio Frequencies," *Review of Scientific Instruments*, Vol. 58, No. 4, April 1987, pp. 624–627.

[61] Baker-Jarvis, J., "Transmission/Reflection and Short-Circuit Line Permittivity Measurements," NIST Technical Note 1341, 1990.

[62] Schultz, J. W., "Computational Analysis of a Permeameter Material Measurement Fixture," *Antenna Measurement Techniques Association (AMTA) Symposium Proceedings*, Boston, MA, November 16–21, 2008.

About the Author

John W. Schultz is the chief scientist at Compass Technology Group, a small engineering company that specializes in electromagnetic material measurements and the development of measurement devices. He received a BA in physics (1987) from the University of Maryland, a MS in physics (1990) from the University of Texas at Austin, and a PhD in materials engineering (1997) from the University of Dayton. At the beginning of his career, he worked as an intel analyst both for several defense companies and then for the U.S. Air Force at the Air Force Information Warfare Center. From 1998 to 2013 he was part of the research faculty at the Georgia Tech Research Institute, where he attained the rank of tech fellow. Since 2013, he has led research and product development efforts at Compass Technology Group. He is lead author on dozens of journal and conference publications and hundreds of technical reports, and he has over half a dozen patents.

Index

Artech House Microwave Library

Behavioral Modeling and Linearization of RF Power Amplifiers, John Wood

Chipless RFID Reader Architecture, Nemai Chandra Karmakar, Prasanna Kalansuriya, Randika Koswatta, and Rubayet E-Azim

Chipless RFID Systems Using Advanced Artificial Intelligence, Larry M. Arjomandi and Nemai Chandra Karmakar

Control Components Using Si, GaAs, and GaN Technologies, Inder J. Bahl

Design of Linear RF Outphasing Power Amplifiers, Xuejun Zhang, Lawrence E. Larson, and Peter M. Asbeck

Design Methodology for RF CMOS Phase Locked Loops, Carlos Quemada, Guillermo Bistué, and Iñigo Adin

Design of CMOS Operational Amplifiers, Rasoul Dehghani

Design of RF and Microwave Amplifiers and Oscillators, Second Edition, Pieter L. D. Abrie

Digital Filter Design Solutions, Jolyon M. De Freitas

Discrete Oscillator Design Linear, Nonlinear, Transient, and Noise Domains, Randall W. Rhea

Distortion in RF Power Amplifiers, Joel Vuolevi and Timo Rahkonen

Distributed Power Amplifiers for RF and Microwave Communications, Narendra Kumar and Andrei Grebennikov

Electric Circuits: A Primer, J. C. Olivier

Electronics for Microwave Backhaul, Vittorio Camarchia, Roberto Quaglia, and Marco Pirola, editors

An Engineer's Guide to Automated Testing of High-Speed Interfaces, Second Edition, José Moreira and Hubert Werkmann

Envelope Tracking Power Amplifiers for Wireless Communications, Zhancang Wang

Essentials of RF and Microwave Grounding, Eric Holzman

Frequency Measurement Technology, Ignacio Llamas-Garro, Marcos Tavares de Melo, and Jung-Mu Kim

FAST: Fast Amplifier Synthesis Tool—Software and User's Guide, Dale D. Henkes

Feedforward Linear Power Amplifiers, Nick Pothecary

Filter Synthesis Using Genesys S/Filter, Randall W. Rhea

Foundations of Oscillator Circuit Design, Guillermo Gonzalez

Frequency Synthesizers: Concept to Product, Alexander Chenakin

Fundamentals of Nonlinear Behavioral Modeling for RF and Microwave Design, John Wood and David E. Root, editors

Generalized Filter Design by Computer Optimization, Djuradj Budimir

Handbook of Dielectric and Thermal Properties of Materials at Microwave Frequencies, Vyacheslav V. Komarov

Handbook of RF, Microwave, and Millimeter-Wave Components, Leonid A. Belov,Sergey M. Smolskiy, and Victor N. Kochemasov

High-Efficiency Load Modulation Power Amplifiers for Wireless Communications, Zhancang Wang

High-Linearity RF Amplifier Design, Peter B. Kenington

High-Speed Circuit Board Signal Integrity, Second Edition, Stephen C. Thierauf

Integrated Microwave Front-Ends with Avionics Applications, Leo G. Maloratsky

Intermodulation Distortion in Microwave and Wireless Circuits, José Carlos Pedro and Nuno Borges Carvalho

Introduction to Modeling HBTs, Matthias Rudolph

An Introduction to Packet Microwave Systems and Technologies, Paolo Volpato

Introduction to RF Design Using EM Simulators, Hiroaki Kogure, Yoshie Kogure, and James C. Rautio

Introduction to RF and Microwave Passive Components, Richard Wallace and Krister Andreasson

Klystrons, Traveling Wave Tubes, Magnetrons, Crossed-Field Amplifiers, and Gyrotrons, A. S. Gilmour, Jr.

Lumped Element Quadrature Hybrids, David Andrews

Lumped Elements for RF and Microwave Circuits, Second Edition, Inder J. Bahl

Microstrip Lines and Slotlines, Third Edition, Ramesh Garg, Inder Bahl, and Maurizio Bozzi

Microwave Component Mechanics, Harri Eskelinen and Pekka Eskelinen

Microwave Differential Circuit Design Using Mixed-Mode S-Parameters, William R. Eisenstadt, Robert Stengel, and Bruce M. Thompson

Microwave Engineers' Handbook, Two Volumes, Theodore Saad, editor

Microwave Filters, Impedance-Matching Networks, and Coupling Structures, George L. Matthaei, Leo Young, and E. M. T. Jones

Microwave Imaging Methods and Applications, Matteo Pastorino and Andrea Randazzo

Microwave Material Applications: Device Miniaturization and Integration, David B. Cruickshank

Microwave Materials and Fabrication Techniques, Second Edition, Thomas S. Laverghetta

Microwave Materials for Wireless Applications, David B. Cruickshank

Microwave Mixer Technology and Applications, Bert Henderson and Edmar Camargo

Microwave Mixers, Second Edition, Stephen A. Maas

Microwave Network Design Using the Scattering Matrix, Janusz A. Dobrowolski

Microwave Power Amplifier Design with MMIC Modules, Howard Hausman

Microwave Radio Transmission Design Guide, Second Edition, Trevor Manning

Microwave and RF Semiconductor Control Device Modeling, Robert H. Caverly

Microwave Transmission Line Circuits, William T. Joines, W. Devereux Palmer, and Jennifer T. Bernhard

Microwaves and Wireless Simplified, Third Edition, Thomas S. Laverghetta

Millimeter-Wave GaN Power Amplifier Design, Edmar Camargo

Modern Microwave Circuits, Noyan Kinayman and M. I. Aksun

Modern Microwave Measurements and Techniques, Second Edition, Thomas S. Laverghetta

Modern RF and Microwave Filter Design, Protap Pramanick and Prakash Bhartia

Neural Networks for RF and Microwave Design, Q. J. Zhang and K. C. Gupta

Noise in Linear and Nonlinear Circuits, Stephen A. Maas

Nonlinear Design: FETs and HEMTs, Peter H. Ladbrooke

Nonlinear Microwave and RF Circuits, Second Edition, Stephen A. Maas

On-Wafer Microwave Measurements and De-Embedding, Errikos Lourandakis

Parameter Extraction and Complex Nonlinear Transistor Models, Günter Kompa

Passive RF Component Technology: Materials, Techniques, and Applications, Guoan Wang and Bo Pan, editors

PCB Design Guide to Via and Trace Currents and Temperatures, Douglas Brooks with Johannes Adam

Practical Analog and Digital Filter Design, Les Thede

Practical Microstrip Design and Applications, Günter Kompa

Practical Microwave Circuits, Stephen Maas

Practical RF Circuit Design for Modern Wireless Systems, Volume I: Passive Circuits and Systems, Les Besser and Rowan Gilmore

Practical RF Circuit Design for Modern Wireless Systems, Volume II: Active Circuits and Systems, Rowan Gilmore and Les Besser

Principles of RF and Microwave Design, Matthew A. Morgan

Production Testing of RF and System-on-a-Chip Devices for Wireless Communications, Keith B. Schaub and Joe Kelly

Q Factor Measurements Using MATLAB, Darko Kajfez

Radio Frequency Integrated Circuit Design, Second Edition, John W. M. Rogers and Calvin Plett

Reflectionless Filters, Matthew A. Morgan

RF Bulk Acoustic Wave Filters for Communications, Ken-ya Hashimoto

RF Circuits and Applications for Practicing Engineers, Mouqun Dong

RF Design Guide: Systems, Circuits, and Equations, Peter Vizmuller

RF Linear Accelerators for Medical and Industrial Applications, Samy Hanna

RF Measurements of Die and Packages, Scott A. Wartenberg

The RF and Microwave Circuit Design Handbook, Stephen A. Maas

RF and Microwave Coupled-Line Circuits, Rajesh Mongia, Inder Bahl, and Prakash Bhartia

RF and Microwave Oscillator Design, Michal Odyniec, editor

RF Power Amplifiers for Wireless Communications, Second Edition, Steve C. Cripps

RF Systems, Components, and Circuits Handbook, Ferril A. Losee

Scattering Parameters in RF and Microwave Circuit Analysis and Design, Janusz A. Dobrowolski

The Six-Port Technique with Microwave and Wireless Applications, Fadhel M. Ghannouchi and Abbas Mohammadi

Solid-State Microwave High-Power Amplifiers, Franco Sechi and Marina Bujatti

Stability Analysis of Nonlinear Microwave Circuits, Almudena Suárez and Raymond Quéré

Substrate Noise Coupling in Analog/RF Circuits, Stephane Bronckers, Geert Van der Plas, Gerd Vandersteen, and Yves Rolain

System-in-Package RF Design and Applications, Michael P. Gaynor

Technologies for RF Systems, Terry Edwards

Terahertz Metrology, Mira Naftaly, editor

Understanding Quartz Crystals and Oscillators, Ramón M. Cerda

Vertical GaN and SiC Power Devices, Kazuhiro Mochizuki

The VNA Applications Handbook, Gregory Bonaguide and Neil Jarvis

Wideband Microwave Materials Characterization, John W. Schultz

Wired and Wireless Seamless Access Systems for Public Infrastructure, Tetsuya Kawanishi

For further information on these and other Artech House titles, including previously considered out-of-print books now available through our In-Print-Forever® (IPF®) program, contact:

Artech House
685 Canton Street
Norwood, MA 02062
Phone: 781-769-9750
Fax: 781-769-6334
e-mail: artech@artechhouse.com

Artech House
16 Sussex Street
London SW1V 4RW UK
Phone: +44 (0)20 7596 8750
Fax: +44 (0)20 7630 0166
e-mail: artech-uk@artechhouse.com

Find us on the World Wide Web at: www.artechhouse.com